Life and Mind: Philosophical Issues in Biology and Psychology
Kim Sterelny and Robert A. Wilson, Series Editors

Rock, Bone, and Ruin

Rock, Bone, and Ruin

An Optimist's Guide to the Historical Sciences

Adrian Currie

The MIT Press
Cambridge, Massachusetts
London, England

This book was set in ITC Stone Sans Std and ITC Stone Serif Std by Toppan Best-set Premedia Limited. Printed and bound in the United States of America.

Library of Congress Cataloging-in-Publication Data

Names: Currie, Adrian, author.
Title: Rock, bone, and ruin : an optimist's guide to the historical sciences / Adrian Currie.
Description: Cambridge, MA : The MIT Press, [2018] | Series: Life and mind: philosophical issues in biology and psychology | Includes bibliographical references and index.
Identifiers: LCCN 2017021974 | ISBN 9780262037266 (hardcover : alk. paper)
Subjects: LCSH: Paleontology--Philosophy.
Classification: LCC QE711.3 .C68 2017 | DDC 560--dc23 LC record available at https://lccn.loc.gov/2017021974

10 9 8 7 6 5 4 3 2 1

Contents

Acknowledgments

Philosophy is a deeply social activity (despite its reputation), and I have benefited from the intellectual beneficence of many philosophical communities, as well as a number of scientific ones. In the six years or so that I have been developing the ideas in this book, I have talked to many more people than I can name here (or have remembered to name here!)—so apologies if I have left you out. Any mistakes, of course, are my own. Indeed, given the numerous philosophical and scientific ideas I have attempted to synthesize, I expect there to be a fair few errors—I beg your patience in that regard.

Some of the research for this book was supported by grants from the European Commission and the Templeton World Charity Foundation.

Jo Bulbulia, Brett Calcott, David Dick, Zoe Drayson, Marc Ereshefsky, Ben Fraser, Liz Irvine, Colin Klein, Arnon Levy, Alison McConwell, Daniel Nolan, Derek Turner, Michael Weisberg and Alison Wylie have all read significant chunks during various stages of writing, and I am grateful to them for their feedback, charity, and hilarious asides. Thanks also to the four anonymous reviewers at MIT Press for useful, thoughtful, and kind comments. A special thanks goes to the students of Phil 557/667 2016 in Calgary, who read large sections of the book (very patiently!), and provided both useful feedback and a lot of fun.

Bits and pieces of this book have been presented in various contexts. These have included (many) meetings of the Australasian Association of Philosophy in both New Zealand and Australia, seminars at the departments of Philosophy at the Australia National University, Sydney University, IRH Bucharest, Otago University, Kansas State University, the University of Calgary, and the University of Washington, as well as at the

Philosophy of Science Association, the International Society for the Philosophy, History and Social Studies of Biology, Philosophy of Biology at Dolphin Beach, the Society for the Philosophy of Science in Practice, the Canadian Society for HPS, and meetings of the Sydney-ANU Philosophy of Biology working group.

I am also grateful to Shahar Avin, Sam Baron, Ben Borkovic, Pierrick Bourratt, David Braddon-Mitchell, Don Brinkman, Rachel Brown, Andrew Buskell, Justin Caouette, Carol Cleland, Steph Collins, Bailey Delehanty, Megan Delehanty, Alkistis Elliott-Graves, Patrick Forber, Paul Griffiths, Jon Herington, Stephan Kubicki, Adam Hochman, Andrew Inkpen, Ben Jeffares, Anton Killin, Sabina Leonelli, Kate McMahon, John Matthewson, Mark Migotti, Jay Odenbaugh, Maureen O'Malley, Emily Parke, Kathryn Reese-Taylor, Alex Sandgren, Leah Schwartz, Kim Shaw-Williams, Rosa Terlazzo, Jessica Theodore, Aaron Thomas-Balduc and Ken Waters for discussion of these ideas. Participation in reading groups and more informal discussions at the ANU, Sydney University, Calgary University, and Bucharest IRH, as well as CSER and Cambridge HPS, have been invaluable.

Many of the ideas in this book have wormed their way into pieces I have written for *Extinct*, the philosophy of paleontology blog (http://www.extinctblog.org). I'd like to thank the readers of the blog, and of course my fellow journeypeople: Derek Turner, Joyce Havstad, and Leonard Finkelman. An extra thank you to Leonard for preparing the figures in chapter 5.

I'd also like to thank the team at MIT Press, who have done an excellent job. In particular, Philip Laughlin who championed the book, and Gregory McNamee, whose superb copyediting smoothed out many of my idiosyncrasies without losing too much personality (and allowing a few contractions and exclamation points to remain!).

Two people deserve particular mention. First, I am greatly indebted to Kim Sterelny. He was my supervisor both through my master's and doctoral theses and has remained a steady influence—and friend—ever since. As a source of constructive and (frustratingly) incisive criticism he knows no equal, and his relaxed (yet blunt) guidance and cognitive largesse should, I think, take a fair bit of the blame for the best bits of this book.

Second, I've been extraordinarily lucky to have in Kirsten Walsh a partner-in-crime who is brilliant, critical, and curious. There was a stage during this book's editing when we met weekly to go over each chapter with a fine-toothed comb, and that process was one of the most intellectually rewarding experiences I have had. She has greatly influenced my ideas and is forever encouraging and kind.

Finally, I dedicate this book to my parents, who always give me enough rope but are there to help when I leave myself hanging.

1 The Tooth of the Platypus

Here is a puzzle. The "historical sciences"—geology, paleontology, archaeology, and so on—have made extraordinary progress in understanding both the events and entities of the deep past and the processes that produce them. However, philosophers and scientists often emphasize how impoverished historical evidence is. The deep past's signal is often ambiguous, incomplete: degraded, decayed, and deteriorated. Why such success, despite such terrible evidence?

The puzzle dissolves when we see that historical evidence is not impoverished; we have rich, varied, and powerful ways of reconstructing the past. To see this, we must dig into how historical scientists respond to the challenges they face in generating knowledge. Along the way, we will challenge preconceptions about what successful science is like.

I am going to start with an example of historical reconstruction that has two surprising features. First, the physical remains the reconstruction is based upon are prima facie inadequate. However, second, I will demonstrate that it is an example of how straightforward scientific reconstruction of the past can be, when the right find, techniques, and background theories are available. It will then act as a helpful contrast for the really hard cases that I will tackle later.

1.1 A Dress Rehearsal

In the Australian Aboriginal dreamtime, the duck-billed platypus did not have an auspicious beginning. She was the product of the duck *Tharalkoo* being ravished by the water rat *Bigoon*. If this story includes the generation of extinct platypus, then *Tharalkoo* and *Bigoon* must have been monstrous. *Tharalkoo*'s namesake, *Obdurodon tharalkooschild*, was a species of duck-billed platypus that haunted Australia's Miocene around 5 to 15 million years ago. Her most distinctive feature is size: Modern platypuses reach half a meter in length at most, while *O. tharalkooschild* managed a full meter. Journalists inevitably christened her "Platyzilla." But size is not *O. tharalkooschild*'s only interesting feature. The teeth that extant platypus are born with are quickly replaced with horny pads suitable for dispatching the tiny soft-shelled crustaceans they prey upon. But not *O. tharalkooschild*. She retained her teeth into adulthood, suggesting to paleontologists that she consumed "soft-bodied aquatic animals such as insect larvae or even small vertebrates such as frogs, small lungfish or even baby turtles" (Pian et al. 2013, 1257).

I will start by summarizing the central arguments of this book, and investigation of Platyzilla is a good opportunity for a dress rehearsal.

Pian et al. present a rich picture of *O. tharalkooschild*'s morphology, ancestry, and ecology. This is remarkable considering the physical remains we have of her. They consist of a molar found at the Riverslea world heritage site in Queensland. That's it: a single tooth. One thing about mammals,

Figure 1.1
Peter Schouten's reconstruction of *O. tharalkooschild*, with fossilized molar (inset).

however, is that their teeth are enormously informative. Mammalian dentition is highly specialized, and it sports features that track a multitude of morphological and ecological traits.

O. tharalkooschild's molar declares its owner to be an ornithorhynchid monotreme, one of the platypuses. To see why, let's take a glance at some mammalian tooth anatomy.[1] Mammalian molars have *cusps*, raised edges along a tooth's sides. In marsupial mammals, these consist of three cusps arranged in a rough pyramid. Two features of molars concern us here: the *trigonid* and the *talonid basin*. The *trigonid* is the major set of three cusps on each lower molar. On the lower side of the tooth, the *talonid* (or *talonid basin*) is a further set of three cusps (see figure 1.2). In chewing, the talonid basin connects with the relevant structure in the upper tooth (the *trigon*, the inverse of the trigonid in the upper molars). Platypus molars are distinctive because the crests forming the talonid basin are the same width and height as the crests formed by the trigonid. This feature is present in not only extant infant platypuses but also other extinct members of the lineage with adult teeth, such as *Obdurodon dickinsoni*. However, no other mammals exhibit this mimicry. The feature, then, is a clear signal of ornithorhynchid-hood.

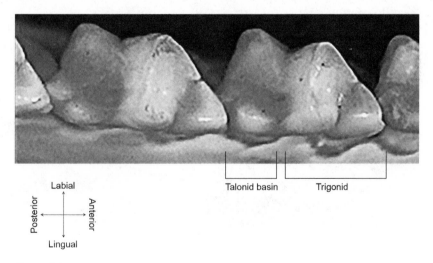

Figure 1.2
Trigonid and other lower-molar features. (Myers et al., 2017. The Animal Diversity Web. Accessed at http://animaldiversity.org.3/3/2017.)

But what makes *O. tharalkooschild* unique among platypuses? In other words, why does she represent a new species? In this case, size matters. Other extinct platypus, such as *O. dickinsoni* and *O. insignis*, boast lower molar crown lengths of 8.71 millimeters and 7.2mm respectively. *O. tharalkooschild*, by contrast, manages 11.7mm (see figure 1.3). Ratios between tooth and body size remain fairly stable across mammals, and so bigger teeth reliably indicate bigger organisms. Further, *O. tharalkooschild's* size does not fit within the expected range of variation within other platypus species. Even the biggest *O. insignis* or *O. dickinsoni* were much smaller than *O. tharalkooschild*. The best explanation of the size differences between specimens, then, is separate evolutionary history rather than within-species variation.[2] There are other features as well, but they need not concern us here.

O. tharalkooschild is reconstructed on the basis of comparisons between her and her relatives' dental morphology. Some of this morphology marks her as a platypus—and thus a bearer of that lineage's duck-billed, flat-tailed morphology.[3] Other features suggest she is a new kind of platypus, deserving her own species name. More speculatively, hypotheses about Platyzilla's diet are generated by considering the molar's functional morphology (in combination with comparisons to living platypuses). What could such teeth have been used for, given what we know about *O. tharalkooschild's*

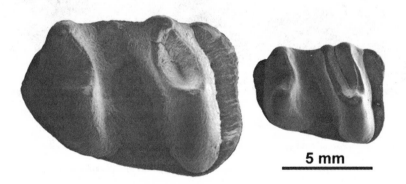

Figure 1.3
Comparison of *O. tharalkooschild* molar (left) with another ancient platypus, *Obdurodon insignis*. (Detail from figure 2, Pian, Archer & Hand 2013, 1257; © Taylor & Francis.)

environment? They appear well suited to consuming the vertebrate biota that were characteristic of the Riverslea region at that time, and this provides some warrant for the hypothesis that she ate small turtles and the like.

There are thus robust, unambiguous regularities connecting the features of mammalian teeth with their morphology, taxonomic grouping, and (to an extent) ecology. Paleobiologists exploit these to build sophisticated and plausible reconstructions of extinct lineages. Once we see this, it is no surprise that we know so much about *O. tharalkooschild*. Mammalian teeth are a rich source of information about the past. Mammalian paleobiologists are *lucky* to work with a feature that is both commonly fossilized and so informative. They are lucky on two counts. First, they have the requisite background theory and technology to exploit the teeth. Second, the fossils not only formed but also went on to survive the ravages of time, to be discovered by people savvy enough to see their value.

But what happens when historical scientists are not so lucky?

Everyone agrees that in some circumstances we know a lot about the past, but some are pessimistic about our capacity to succeed in less evidentially advantaged contexts. I argue that such pessimistic attitudes are mistaken. Let's start by looking at the sorts of claims that might underlie pessimism, by considering some further questions about *O. tharalkooschild*.

Unlucky historical scientists must deal with ambiguous, degraded signals from the past. Moreover, on the face of it, they cannot improve their situation unless new remains are found or new technologies or theories are produced. In short, if their evidence is restricted to "traces"—downstream consequences of past events—the progress of investigation depends on lucky finds. Derek Turner, for one, has adopted this kind of view: "When it comes to acquiring knowledge of the past, the only evidence we have to go on consists in observable records, remains, and traces" (Turner 2007, 158).

To see how limiting this can be, consider a different question concerning *O. tharalkooschild*. What explains her gigantism, evolutionarily speaking? Pian et al. point out that *O. tharalkooschild*'s gigantism seems to conform to a trend in Australian marsupials: a general size increase through the mid-late Cenozoic. At that time, there were significantly larger representatives of modern Australian marsupials, kangaroos, wombats, and their ilk, as well as enormous animals that left no modern representatives. *Thylacoleo carnifex,* for instance, was a large, stocky predator with an incredible crushing bite (see Wroe et al. 2005 and Currie 2015b), while the hippopotamus-sized *Diprodoton* was the largest marsupial to have ever lived (Long et al. 2003). Australia's marsupials increased in size up until the Quaternary extinction (occurring over the last hundred thousand years or so). Given *O. tharalkooschild*'s example, the same pattern may also have occurred in the monotremes. How should we explain these size increases?

As we will see in more detail in chapter 7, paleobiologists sometimes discuss size increase in terms of *Cope's rule*. According to the rule, the average size of individuals in a lineage tends to increase over evolutionary time (Alroy 1998; Hone & Benton 2005). Assuming the rule sometimes holds, what could explain it? On one kind of view, Cope's rule is a *driven* trend: Environmental factors conspire to cause size increases. For instance, perhaps lineages tend to emerge at relatively small sizes and shift toward an optimal size favored by natural selection. On another view, Cope's rule is a *passive* trend: The pattern is explained by internal population dynamics rather than environmental pressure. It could be, for instance, that lineages have lower size limits, and that new taxa typically evolve close to that limit. Assuming there are equal chances of size increases or decreases, we should expect populations on average to contain larger individuals as time passes, since increases will be retained but decreases will "bounce" along the lower bound. That is, although size increase is open to selection, the lower bound ensures that lower sizes are not. Under those conditions, chance alone would be sufficient to generate a pattern of increasing size (see McShea 1994 for discussion of passive and driven trends).

How could we tell whether an instance of Cope's rule is driven or passive? How might we explain *O. tharalkooschild*'s gigantism? What resources do we have to reconstruct her evolutionary past? It would seem that we need detailed, long-term evidence of historical processes to answer such questions—and discovering what would count as a trace of such processes is tricky indeed.[4] A single molar, certainly, seems both unsuitable and insufficient. Moreover, our access seems dependent on what traces we have. Elliot Sober's work on phylogenetic reconstruction (1988) echoes these worries about traces. The degradation of evidence leads to "a set of epistemological difficulties which are peculiar to the historical sciences" (1988, 2). This peculiarity is due to the differential survival of traces: "The knowability of the past depends on whether the physical processes linking past to present are information preserving or information destroying" (1988, 1).

When we encounter such "information destroying" processes, then, we should doubt our capacity to uncover the past. Dick Lewontin makes similar complaints about attempts to reconstruct the evolution of human cognition: "History, and evolution is a form of history, simply does not leave sufficient traces, especially when it is the forces that are at issue. Form and even behavior may leave fossil remains, but forces like Natural Selection do not" (Lewontin 1998, 132). Moreover, what of Pian et al.'s speculation about Platyzilla's diet? Short of getting lucky, say finding a fossilized *O. tharalkooschild* stomach with its victims intact, it isn't clear what could raise their hypothesis from mere conjecture. Many historical targets, then, seem unrecoverable: The past's signals degrade, distort, and decay, leaving scientists with a biased, patchy record. Moreover, it is a record that we cannot complete or significantly improve, because we cannot generate new evidence as experimentalists do.

Experiments grant scientists control of their epistemic fates. Instead of relying on what nature provides, they can generate their own evidence (Currie & Levy, under review). Experimental manipulation allows repeated tests, thus controlling for fluky results. It also grants control over variables, producing targeted evidence that discriminates between the causal roles of a system's components. It is commonly thought that historical investigation is not amenable to experiment: We cannot manufacture evidence about the deep past. As Turner says, "Although they can develop new technologies for identifying and studying potential smoking guns … historical scientists can never manufacture a smoking gun" (Turner 2007, 158). After all, the events targeted by historical scientists occurred in the deep past, and the processes they are concerned with happen over long time periods. Such targets can neither be isolated nor controlled. Jared Diamond and James Robinson put the point starkly: "the cruel reality is that manipulative experiments are impossible in many fields widely admitted to be sciences. That impossibility holds for any science concerned with the past, such as evolutionary biology, paleontology, epidemiology, historical geology, and astronomy; one cannot manipulate the past" (2010, 1). The apparent upshot is that historical scientists cannot make their own luck: They are at the mercy of the processes by which time covers her tracks.

It seems, then, that we can identify three grounds for pessimism about "unlucky" historical investigations, that is, investigations targeting past events that have left meager traces (I will be precise about what I mean by "pessimism" and "optimism" in the next subsection).

1.   Our available evidence about the past is limited to traces.

Prima facie, the difference between inferring *O. tharalkooschild*'s size and explaining its gigantism in evolutionary terms is that the material remains—the single molar—license claims about how big it was, but not about why. If historical evidence depends upon the material remains of past events and entities alone, then in unlucky scenarios we are unlikely to resolve our ignorance.

2.   Much information from the past has degraded or disappeared.

This is an empirical bet: Lewontin (surely) does not think that "forces" like natural selection *never* leave traces—we are not utterly ignorant of evolutionary history—he is best read as claiming that the remains of intangible, large-scale processes like evolution are unlikely to survive into the present. If the bet is justified, then in unlucky scenarios much of the past will remain a mystery.

3.   Historical scientists cannot manufacture evidence.

Unlike scientists who can experimentally control their targets, historical scientists cannot improve their epistemic situation by generating new evidence (yes, 3 is a corollary of 1). They are, then, helpless in the face of time's trace-destroying processes.

If 1–3 hold, then when historical scientists are unlucky—when the past has not graced us with clear and unambiguous signals—we should be pessimistic about our capacity to know much about it. We should not expect such investigations to progress, but rather to founder on a lack of evidence.

I argue that the propositions underlying pessimism are false. Historical scientists can rely on evidence other than traces, they can manufacture evidence, and empirical bets such as Lewontin's are not licensed. This, I argue, grounds optimism about historical science. Even in "unlucky" circumstances, where scientists target entities in the deep past that are not amenable to experimental investigation and have left only incomplete and degraded material remains, we should not expect investigation to halt, falter, run in circles. Rather, we should expect investigation to progress as new knowledge is uncovered. In short, I am an optimist about the historical sciences, and I hope to convince you of the same.

The first and third claims are false because historical scientists can and do appeal to *surrogative*, non-trace evidence. I discuss these in two contexts. First, *analogous* evidence. Here, instead of appealing to contemporary events that are causally connected to our past target, we appeal to instances of the same type as our target. Pian et al. suggest that *O. tharalkooschild's* large molars were used in the consumption of small vertebrates. They cite the molar's functional viability for that purpose: A critter with those kinds of molars would be well equipped for consuming the soft-shelled turtles that shared Platyzilla's environment. This evidence could be supplemented by examining other animals with similar dental arrangements; are there patterns connecting those features with certain diets? In doing so, they would conceive of *O. tharalkooschild's* teeth as tokens of a type: Critters with *tharalkooschild*-type dental morphology pursue prey of a certain nature. These appeals to convergence constitute evidence for historical hypotheses that outrun available trace evidence.[5] The role and nature of analogies will be discussed in chapters 7 and 8.

Second, historical scientists manufacture evidence by conducting controlled studies of *surrogates*—most obviously, modeling. One way in which Pian et al. could test their claim that *O. tharalkooschild* preyed upon vertebrates would be to draw upon their knowledge of platypus anatomy to construct a biomechanical model of the platypus's jaw. They could use this to calculate the strength of her bite and then test whether it could produce the necessary pressure to crush small turtles. Further manipulations of such a model could produce more information: It might, for instance, provide a basis for hypotheses about *O. tharalkooschild's* killing style.[6] I argue that such studies provide evidence about the past. In chapters 9 and 10, I will explore the use of models and the construction of "surrogate experiments" in historical science.

Moreover, focusing on traces, as many accounts of method in historical science do, misses the important role that scientists' general picture of the past plays in supporting historical investigation. The support behind Pian et al.'s reconstruction comes in part from how they fit the puzzle together; they consider *O. tharalkooschild* as a whole, living organism embedded in an environment and part of an ecological system. As I argue in chapter 6, historical scientists not only exploit lines of evidence linking the present to the past but also link together past events, entities, and processes.

Proposition 1—that historical evidence is restricted to traces—fails, because of the important role that various surrogates can play in supporting historical hypotheses and the exploitation of relationships between past events. Proposition 3, that historical scientists cannot manufacture evidence, is false because some surrogates (simulations most obviously) play similar roles to controlled experiments. This puts historical scientists in control of their epistemic fate: they are able to actively improve their evidential situation.

The second motivation for pessimism is an empirical bet about the availability and richness of traces. In circumstances where events in the past are unlikely to leave systematic, accessible traces—and those traces are likely to be homogeneous, restricting us to similar types of evidence—pessimism is more tempting. However, I argue that, even under such conditions, making bets against historical science is a mistake.

One reason for doubt is that because we think of historical reconstruction in simplistic ways, we are likely to underestimate the epistemic worth of the information scientists have. This is due to relying on overly simplified accounts of method. As we shall see in chapter 6, many philosophers and scientists argue that past hypotheses gain support from unifying groups of traces; historical scientists provide "common-cause" explanations (Cleland 2002, 2012; Tucker 2004, 2011). On this view, the primary reason to believe hypotheses about *O. tharalkooschild* is that her existence best explains both features of the fossilized molar and comparisons between it and other specimens. Undoubtedly, this plays an important role in historical science, but we shouldn't take common cause explanation to be *the* method of historical enquiry.

First, historical scientists care about the *varieties* of evidence they unify: Consilience, or the drawing together of multiple streams of evidence, matters (Forber & Griffith 2011; Wylie 2010, 2011). One source of evidence about *O. tharalkooschild*'s diet comes from its molars. Certain kinds of tooth suit certain kinds of prey. Thus, under certain conditions, I can infer from dental morphology to dietary habits. Another potential source is size: It could be that larger platypuses tend to take larger prey (for instance, imagine that bigger modern platypus hunt larger crustaceans than their diminutive conspecifics). If that were true, then two information streams would converge on the result that *O. tharalkooschild* preyed on vertebrates. As I will discuss in chapter 6, the convergence of independent lines of evidence can make for well-supported hypotheses. Second, as I have said, historical scientists care a lot about the picture of the past they present: How hypotheses "hang together" also plays its part. What we think we know about *O. tharalkooschild* fits into a larger picture of platypus evolution. The relationships between these various aspects—the dependencies between, say, *O. tharalkooschild*'s morphology and that of other extinct platypus such as *O. dickinsoni*—sometimes play a critical role.

I will argue that pessimism is due in part to an impoverished picture of historical methodology. Historical scientists are not only, or even primarily, in the business of unifying traces: They also weave together evidence streams and consider their emerging picture of the past holistically. To understand our epistemic access to the past, we must recognize this rich and opportunistic methodology.

Another reason to doubt pessimistic empirical bets about historical science is due to facts about *us*: We are not in a good position to judge whether a particular investigation will succeed or not. One reason for this, articulated in chapter 4, is the continual updating and improvement of background theory connecting traces to the past. The development of new technology and theory opens up new, unexpected avenues of investigation. By using CT scanners and other technology, mammalian paleontologists are able to examine teeth at extremely fine grains. Different feeding patterns over short time scales cause differences in micro-wear, and this enables scientists to uncover information about the day-to-day diet of extinct animals. Prior to this capacity, knowing what *O. tharalkooschild* had for lunch seemed a pipe dream. Moreover, in chapter 10 I will argue that historical investigation is scaffolded. That is, evidential relevance is often revealed in stages; relatively well-confirmed (or sometimes ultimately false!) hypotheses form "scaffolds" from which new evidence may be located. As a result, which lines of evidence might ultimately matter for an investigation are not discernible until various other empirical and theoretical aspects are on the table (Currie 2015b). For instance, the bite force required to dispatch soft-shelled turtles is only relevant to *O. tharalkooschild* reconstruction once it has been reasonably established that (1) she retains her molars into adulthood and (2) her and the turtles' geographic ranges overlapped. It is only from the perspective of such thresholds that we can discern which evidence matters; such hypotheses form the background conditions required for evidential relevance.

Our ignorance about evidential relevance and new theoretical and technological developments undermine pessimistic bets about historical science—and encourage optimistic bets. I will argue that we are biased toward pessimistic judgments about historical evidence. As total evidence should either increase or stay the same (as opposed to decrease) as new technologies and background theories come online, and as investigative scaffolds are reached, we will tend to underestimate the total evidence available.

In short, I shall provide arguments that undermine pessimism about historical science and support a more optimistic take on our capacity to uncover the past. Along the way, we shall deepen our understanding of the methodology and epistemology of these sciences. My argument against pessimism is also an argument that philosophical accounts of the aims, methods, and epistemic resources of the historical sciences have been impoverished; my positive account is a richer picture of those methods and resources. The historical sciences have been extremely fruitful, and I will both explain why and provide counsel on how to extend our reach even further.

But what do I mean by optimism and pessimism, how exactly do I argue for them, and why should we care? Let's make this explicit.

1.2 The Argument of the Book

I am going to argue against a position about our capacity to reconstruct the past—pessimism—and for a different position—optimism. In this section, I will explain what those positions are, discuss the value of establishing optimism, then turn to the arguments themselves.

In one important sense, this is not a book about the historical sciences. Rather, it is a book about how science works under non-ideal circumstances. Philosophers of science have traditionally focused on best-case scenarios: the paradigmatic examples of good science that are illustrative of the general models of reasoning they are interested in. Galileo's telescope, Newton's universal gravitation, Darwin's *Origins*, Franklin, Watson and Crick's DNA molecule, and so on. But science is often carried out in less than ideal epistemic circumstances—indeed, even those paradigm cases turn out to be much less than ideal once run over with a skeptical historian's eye. The historical sciences, as we will see, are full of examples of scientific success in non-ideal circumstances. It is intended, in a sense, as a corrective to our focus on science's Big Theories and Big Episodes. And as we will see, looking at scientific practice from the perspective of these sciences emphasizes science's disunified, opportunistic side.

Let's start by distinguishing between pessimism and optimism in terms of *attitudes* and in terms of *predictions*. Taken as an attitude, optimists highlight the positive aspects of a situation, pessimists the negative. The optimist who thinks the glass is half-full, and the pessimist who sees it as half-empty, do not have a primarily *empirical* dispute. They agree how much water there is, but they adopt different non-epistemic attitudes toward the contents of the glass. By contrast, pessimists and optimists in the predictive sense disagree about the probable outcome of some project. The pessimist expects failure, the optimist success. I might be pessimistic (in the predictive sense) about my capacity to eat an entire po' boy from Oakland's Souley Vegan, but I can nonetheless be optimistic (in the attitude sense) about the experience.

I am interested in *predictive* or *epistemic* pessimism and optimism, when applied to the historical sciences in unlucky circumstances. The pessimist predicts that our attempts to reconstruct the past will often fail; the optimist predicts that we will often succeed. In chapter 11, I will more carefully distinguish between different kinds of optimistic claims, but for now it will suffice to note that when I refer to optimism or pessimism I mean a stance about the future success of historical science, not a general attitude about that success. Note that optimism and pessimism, as I have discussed them, appear to require a view of what "success" is. As I will explain more directly below (and in chapters 12 and 13), scientific success is diverse and often indirect. I am not interested in providing an analysis or account of success here, but take it that Pian et al.'s study is successful insofar as it generates some epistemic goods: knowledge of a new critter and its properties, as well as a possible new inroad to macroevolutionary pattern in Australia and beyond.

It's worth contrasting pessimism and optimism with a distinction more familiar to philosophers: scientific realism and antirealism. There are three major differences between the former distinction and the latter. Scientific realism is a rich, complex topic, and the conceptions I will identify will

have exceptions. Regardless, the basic points are that the realism debate
centers on a fairly narrow range of epistemic goods (i.e., whether scientific
theories are true), is often interested in our attitudes toward our best current
science, and positions are traditionally supported by appeal to inductions
across scientific history. The positions I am interested in—pessimism and
optimism—depart from the realism/antirealism debate insofar as they deny
each of these points in the specific ways I will outline. I will treat the debate
more carefully (and in a more up-to-date fashion) in chapter 13—the point
of my brief foray here is to get a contrast with pessimism and optimism.

First—and most important—realism and antirealism concern themselves
with a narrower range of questions than optimism and pessimism. Specifically, they are concerned with the truth, or approximate truth, or otherwise of scientific claims. Realists and antirealists disagree over whether
we should *believe* scientific claims, whether scientific theories are true, and
whether the entities posited by those theories exist. The pessimist and the
optimist, by contrast, are worried about a wider range of epistemic goods—
knowledge is not restricted to "truth" or some analogue. Scientists are interested in generating good explanations, adequate representations, precise
predictions, new technologies, successful techniques, telling interventions,
effective cures, and so forth. Getting stuff right (or "true" or "veridical")
plays a role, but we should not take it as the central epistemic good. Indeed,
sometimes the precision that truth requires must be sacrificed to achieve
other epistemic goals—that is, there are plausibly trade-offs between truth
in terms of precision and other things we want from science. Some popular
examples are between generality and precision (Weisberg 2006; Mitchell
2002), fine-grained and coarse-grained description (Jackson & Pettit 1992),
understanding and precision (Potochnik 2015), and so forth. In analyzing *O. tharalkooschild* as an instance of Cope's rule, absolute precision—
veridicality—is not required in understanding her size or how she fits into
the relevant patterns. In some contexts, such detail could in fact detract
from our capacity to understand (or manipulate, or generate explanations
from, etc.) Cope's rule and Platyzilla's relationship to it.

Second, realism and antirealism are primarily concerned with attitudes
toward our best current science, while optimism and pessimism are about
the future course of science. The realist argues that the best science as it
currently stands is most likely true (at least regarding the things that the
realist in question takes science to say true things about), or that we should

believe in the entities posited in our best current science. The optimist cares more about the capacity of science to progress—working out which scientific investigations will deliver epistemic dividends. As such, the two positions care about different things. If our current theories prove to fail, this is often bad news for the realist—but it does not straightforwardly undermine optimism.

Third, traditionally, at least, realism and antirealism are supported by inductive appeals to scientific history. The realist might appeal to the enormous success of science—perhaps pointing to the capacity of scientific theories to unify and explain a wide range of phenomena, or to underwrite our technological capacities. As realists such as Smart (1963) or Putnam (1982, 1975) may have put it, it would be miraculous if these achievements were not due to science getting stuff right, that is, its being true. The antirealist might appeal to another aspect of scientific history—the repeated failures of scientific theories to survive in the long run (Laudan 1981). Even the most successful theories of the past, our antirealist might point out, have been superseded, and so why should we think current science is any different?[7] I will argue against appeals to science's history in support of realism or antirealism in the postscript.

By contrast to scientific realism and antirealism, optimism and pessimism are established by analysis of (1) an *epistemic situation* and (2) the available *epistemic resources*. We can understand an *epistemic situation* as the challenges scientists face when generating epistemic goods in a particular context.[8] In the lab, for instance, these challenges involve controlling for experimental artifacts, and ensuring that the introduced controls and isolation do not undermine the study's external validity. For historical science, as we have already seen, this can involve dealing with incomplete, degraded trace evidence. *Epistemic resources* are the knowledge, capacities, sources of evidence, and techniques scientists have at their command. As mentioned earlier, many philosophers and scientists think that the capacity to experiment is an epistemic resource that historical scientists lack. To establish optimism about some scientific enterprise, we argue that the available epistemic resources are sufficient to generate the relevant epistemic goods in that epistemic situation; to establish pessimism, we argue that such resources are insufficient. I will discuss "sufficiency" later.

In my discussion of *Obdurodon tharalkooschild*, I articulated a common view about the epistemic situation of historical scientists, and their epistemic resources. (1) Historical scientists face incomplete and messy evidence—because of the degradation of traces—and (2) they have impoverished methods of responding to that mess—because their resources are limited to traces. This picture underlies pessimism about the future success of historical science—there is much we are unlikely to know.

So, as a realist, I might argue that science generates truths and that we should believe in both the theories and entities of our best current science. As an optimist, I argue that scientists, within a particular epistemic situation, will succeed in generating a range of epistemic goods. Realists traditionally establish their positions on the basis of past scientific success, whereas the optimist argues on the basis of the epistemic resources available for that scientific task given the relevant epistemic situation. These views, as well as their antirealist and pessimist analogues, are logically independent. I can discuss one without directly attending to the other.

So much for what optimists and pessimists claim. Why should we care about them? For one thing, scientists draw methodological conclusions from pessimistic attitudes. These attitudes are often expressed tacitly and often expressed outside of print. For instance, scientists often worry about work being mere "storytelling," in part because of the kinds of worries I will discuss. The examples I pick out are, I think, representative instances of this general kind of attitude.

Henry Gee's view on cladistics is an example of pessimism that is taken to have methodological upshots (2000). He argues that paleontological narratives attempting to connect the ancestry of fossil specimens to each other and with living lineages are mere speculation—fossils are only "points in time," and we lack license to fill the gaps between them. Because of this, he defends something similar to pattern cladism. On this view, the ways that biologists organize taxa should not be considered anything like hypotheses about the tree of life or about the processes that produce taxa. Rather, they are merely ways of capturing *patterns*, of organizing the biological world. They are empirically and metaphysically innocent. He discourages formulation of the kinds of narratives we will see in chapter 2, on the basis that they "can never be tested by experiment, and so they are unscientific. ... No science can ever be historical" (2000, 5–8). The clean mathematics of cladistics, without the speculative gloss of phylogenetics, is more scientific by Gee's lights, and thus (so he claims) paleontologists ought to restrict their methods accordingly.

This is a common theme: Historical scientists cannot know much, history's information is destroyed, and therefore they should adopt a conservative method, lest they slip into free-form, unencumbered, and unscientific storytelling. Here is Lewontin on human cognitive evolution again:

Despite the existence of a vast and highly developed mathematical theory of evolutionary processes in general, despite the abundance of knowledge about living and fossil primates, despite the intimate knowledge we have of our own species' physiology, morphology, psychology, and social organization, we know essentially nothing about the evolution of our cognitive capacities, and there is a strong possibility that we will never know much about it. (1998, 108–109)

So, Lewontin implies, we should spend our epistemic resources on questions that we at least have a chance of answering! This connection between pessimism and methodological conservatism is most striking in archaeology. Alison Wylie (1985, 2001, 2011; Chapman & Wylie 2016, chapter 2) has nicely documented and criticized this tendency. By her lights, the history of archaeology can be understood in terms of a tension between two extremes arising from pessimism. On the one hand, if we cannot know much, then archaeologists should be epistemic cowards, restricting themselves to only the most careful claims and methods. Often, the domain of archaeology proper is restricted to the mere cataloguing of artifacts. On the other hand, given that so little can be known, why not embrace a kind of epistemic subjectivism—if archaeology is only stories, why not at least make up good ones? In regard to the former strategy, Wylie says,

If they are committed to epistemically responsible (scientific) practice, they must confine themselves to the pursuit of narrowly descriptive goals. A variant on this theme is the recurrent claim that archaeologists *qua* archaeologists should be primarily concerned with documenting the contents of the archaeological record as completely and systematically as they can, deferring any more expansive goals to later stages of enquiry. (Wylie 2011, 377)

Christopher Hawkes's classic discussion of what he calls "text-free history," the study of past human activity without written records, is an example. Hawkes ranks the accessibility of aspects of humanity's past in terms of how *human* the fact in question is: "the more specifically human are men's activities, the harder they are to infer by this sort of archaeology ... the more human, the less intelligible" (1954, 162). I take it that by "human," Hawkes means specific, contingent, idiosyncratic activities that outrun our basic ecological needs. For example, he considers reconstructing the techniques and sustenance activities of prehistoric humans to be an arduous but achievable epistemic goal, with a relatively straightforward inferential structure. The institutional or, worse still, spiritual life of our ancestors is much more difficult to infer.

And the critical factor, standing between fair intelligibility and stark unintelligibility, is surely ecology, the study of the physical environment. So long as you can depend on that, as you can for the material aspects of man's life, his technology and his economic existence, your exercise of this sort of archaeology is rewarding. ... But now transcend that, and your returns diminish sharply. ... I do not say that you are left in the end with nothing that you can apprehend. But I do say that there cannot be much, and that we should ask ourselves just what there can be. (1954, 162–163)

Although Hawkes was writing in the 1950s, according to Wylie, such attitudes are a common thread running through archaeological musings on method; see, for instance, Holly Hayter's (1994) discussion of ethnographic analogy, and my response (Currie 2016a).

Given their often tacit or unpublished status, it is difficult to ascertain exactly how representative the pessimistic discussions above are. Naturally, there are scientists working today, and in the past, who adopt significantly more optimistic attitudes toward our capacity to learn about the past, but nonetheless pessimism is a strong current running through historical science.

Optimism matters, then. First, it matters for how historical scientists go about their business. As we will see, the epistemic claims that underlie optimism, for instance, the need for many strands of evidence, and our inability to predict what information ultimately will be evidentially relevant, motivates an emphasis on speculation. I turn Lewontin, Hawkes, Gee, and others' conservative message on its head. In chapter 11, I argue that historical investigation proceeds most fruitfully when scientists let their imaginations loose, since hypotheses must be put forward in order for us to figure out how to make empirical progress. Historical science is rich with examples of surprising discoveries made on the basis of outlandish hunches, guesses and hypotheses—indeed, getting things wrong can often be a crucial scaffold for getting things right. As we have seen, more conservative approaches to historical investigation are frequently underwritten by pessimistic views on our capacity to uncover the past.

Second, as we will see in chapter 12, establishing optimism has consequences for our reasons to do historical science and how we should encourage it. Two factors relevant for motivating a scientific investigation are (1) the chance of success and (2) the usefulness of that success. If we doubt that much progress will be made in a domain, then our epistemic resources might best be spent elsewhere. If the knowledge gained (for all of its intrinsic value) doesn't matter for other things we care about, then there might also be instrumental reasons to prefer other areas of inquiry. My arguments for optimism should lead us to expect historical science to be successful, and this should motivate us to actually do (and fund) historical investigation. Moreover, my discussion of the epistemic capacities of historical science has upshots for its instrumental value. For instance, the defense of analogy in chapter 8 can be extended to include the use of historical science to inform our understanding of challenges we face in the here and now: most obviously extinction and climate change. Moreover, the optimistic picture of what motivates success in historical science challenges some aspects of how science is structured and funded. Because progress in historical reconstruction is often indirect and unintended, funding models that emphasize specific epistemic returns are of limited utility. We should also be wary of centralizing our resources. In the circumstances I describe, we want diverse, often idiosyncratic research programs, and so a more distributed pattern will be more effective. Although the majority of the book takes an abstract angle, then, there will be more practical conclusions.

Third, establishing optimism (or pessimism for that matter) requires understanding both the epistemic situation and the epistemic resources available to historical scientists, and this is valuable in and of itself. For one thing, accounts of such sciences will aid in assessing comparisons between sciences. For instance, arguments that experimental science is somehow "better" than historical science, or that they are equal, or whatever, can be made only in light of the epistemic situation such scientists face. For another, accounts of how science works can shed light on how science ought to proceed, and on philosophical questions about the nature of science. In the postscript, I will illustrate this by arguing that my defense of optimism includes claims that undermine history-based arguments for both realism or antirealism. Such a use of science's history clashes with the context sensitivity of method and the indirectness of progress that reflection on the historical sciences demonstrates. That is, to use science's history to establish realism or antirealism we need to (1) make inductions across science—and if scientific method and success is heterogeneous this is very hard to do—and (2) be able to identify episodes of success and failure—a tricky task given that in scientific progress success and failure are often intertwined.

With the distinction between optimism and pessimism in hand, and with discussion motivated, let's look at the book's argument. In effect, I provide a positive and a negative argument. The negative argument proceeds by undermining pessimism about historical science in unlucky circumstances (that is, when trace evidence is poor), the positive argument motivates optimism on the basis that pessimism fails in unlucky circumstances. They could be put semiformally as follows:

1. According to pessimists, some domain D will likely be epistemically inaccessible because of our epistemic situation and resources pertaining to D;
2. However, pessimists have incorrectly characterized our epistemic situation and resources pertaining to D;
3. Therefore, pessimists have given us no reason to think domain D is inaccessible.

Philosophers and scientists have argued that historical scientists are restricted to trace evidence, and that trace evidence is often poor. On this basis one could argue for pessimism about those sciences. However, historical scientists are *not* restricted to trace evidence, and we will often be wrong about the paucity of evidence from traces, and so this argument for pessimism fails.

What about optimism, the view that many domains will be accessible, that in fact we will learn much about the past? Here is my argument:

1. Unlucky circumstances provide the best case for pessimism about historical science;
2. Pessimism is not motivated in unlucky circumstances;
3. If pessimism is not motivated even in the best case, then we should be optimists;
4. Therefore, we should be optimists about historical science.

If pessimism cannot be established *even in the best case*, this presumably motivates optimism. As we will see in chapter 11, there is much more to be said about varieties of optimism, and when they apply, but that discussion is best left for later, once we have a fuller picture of the epistemic situation and resources of historical science.

I want to close this subsection with a brief defense of the *normativity* of my arguments. I do not take myself to be merely describing scientific practice: this is not a cumbersome piece of scientific journalism! It's worth explaining why it isn't.

There are different ways of providing normative oomph to arguments like mine. One way would be to provide some general, abstract standard for optimism. There might be some threshold of success for scientific optimism, say, and this could be underwritten by an abstract characterization of the nature of evidence, or epistemic warrant. Another approach would be to contrast historical science with other cases that are better understood, or more familiar, which we can take as a yardstick. I will do neither of these things.

First, I do not think there is a generally applicable epistemic standard to be had; indeed, a theme of this book is that epistemic warrant is local and that there is very little we can say that is both general, abstract, and explanatory. Although we can identify particular epistemic goods, and discuss how they can be more-or-less reliably generated in particular epistemic situations, the thought that there is a general schema across goods and situations strikes me as unlikely. Relying on such a schema would be anathema to my view. Of course, such schema can be instrumentally useful: I can use Bayesian machinery without committing to its capturing the fundamental nature of evidential reasoning (indeed, my occasional appeals to the law of likelihood should be taken in this spirit). But given the antigeneralist message of the book, relying too heavily on a general, abstract account of evidence, sufficiency, or success would be odd.

Second, much work on historical reconstruction—particularly Cleland and Turner's—is based on contrasting its method and epistemic fortunes with that of the experimental sciences. Indeed this can be very useful: Experimental science is surely highly successful and well understood (or at least we philosophers have spent more time grappling with experiment than we have historical reconstruction!). I have nothing against such contrasts, but herein I do not want to follow that approach. This is because I want to confront historical science on its own terms. This affords us an interesting perspective on the nature of science: its progress, aims, and so forth. Moreover, contrasting historical science from other kinds of investigation from the outset might overly influence our understanding of historical reconstruction of science. I want to take it on its own terms.

But, if I refuse to provide some general standard or to contrast historical with other kinds of science, whence comes the normativity? I take myself to be explaining why science is practiced as it is, and why it works (when it does). Further, I take these considerations to motivate optimism. So what is my method, and why isn't it merely descriptive?

As discussed earlier, I proceed by picking out epistemic situations. An epistemic situation might be thought of as the set of challenges and advantages we have in generating knowledge in a given context. The "unlucky circumstances" I identify in chapter 5 are an example of an epistemic situation. The notions of epistemic situation and resources allow us to make contrastive claims *within* historical science. I opened with the case of *Obdurodon tharalkooschild* and argued that various features possessed by mammalian teeth afford remarkable access to the past: In such circumstances we are lucky. There are also trickier cases—unlucky circumstances—trying to reconstruct the evolutionary circumstances that led to *O. tharalkooschild*'s size, for instance. By showing that even here optimism is a reasonable position, I take myself to ground a more general optimism.

Further, an important motivation for caring about optimism in historical science is how it feeds into the actual practice of historical reconstruction. As Turner (2015) has discussed, an important issue facing historical scientists is working out which research avenues will be fruitful—they must make *predictions* about future success. In the Tyrrell Museum's unprepared fossil room, for instance, are hundreds of unprepared fossils, some dating back to the mid-1970s (see figure 1.4).

In unprepared storage, which research to pursue is not an abstract or highfalutin' philosopher's question. There are only a limited number of fossil preparators, and the activity is onerous (see Wylie 2015). Paleontologists must decide which fossils are likely to produce the most fruitful research. To do this, all that is required is a *comparative* metric. No general standard is required. The historical sciences themselves potentially benefit greatly by comparing different epistemic circumstances, and thus getting a handle on which features might matter for success. This task doesn't require comparisons with experimental science or a general standard of evidence. The capacity to make internal comparisons, which my approach underwrites, is what we want in this context.

Figure 1.4
Don Brinkman shows Leonard Finkelman and Kirsten Walsh through unprepared fossil storage at the Tyrrell Museum. (Photo by author.)

Historical scientists want to know how to make decisions about which research avenues will be fruitful, which are likely to lead to the successful generation of epistemic goods. To do so, we must be able to identify different methods for generating goods in different epistemic contexts, and compare them. Here is a simple example. When we're interested in understanding macroevolutionary patterns, such as Cope's rule, we need large, easily integratable data sets. And that needs a lot of specimens, spread over long time periods, without too much incompleteness. So, to generate the relative epistemic good, we want to identify those sections of life that are likely to meet that criteria. Typically, paleontologists break for successful, aquatic, hard-bodied invertebrates in these circumstances. As well they should. Critters like ammonites were populous, successful, and are more likely to fossilize than most other lineages. They are thus a good place to look for that kind of data—better than, say, soft-bodied, terrestrial, or less successful critters. But this decision is context-sensitive. Good knowledge generation depends on what your question is, what resources you have, and what evidence you are likely to discover.

Finally, on my view evidential warrant is local—it depends on a range of sometimes highly circumstantial factors including, but not limited to, the consequences of getting things wrong, the techniques, technologies and theoretical objects available to make epistemic contact with the world, the goals of research, and so forth. To generate normative claims (such as "we should be optimistic about our capacity to know lots in this domain") we need to know a lot about this rich contextual information. I will be filling in this detail, and generating normative claims about scientific practice and our philosophical attitudes toward them on that basis. This means that there will be no simple distinction between purely descriptive and norma- tive factors in this book. So be it: Science is a messy, opportunistic, social, and very human enterprise attempting to understand an often complex recalcitrant, stubborn world. Trying to understand such an endeavor with a simple account of evidential warrant would be folly.

1.3   The Book's Structure

My arguments draw on two major case studies, which are introduced in the next chapter. The first is paleobiological. The sauropods, stereotypical long-necked, long-tailed, enormous dinosaurs, are puzzling. Their gigan- tism easily outruns models of maximal terrestrial animal size. The emerg- ing explanation of sauropod size highlights a series of complex, contingent events, as well as environmental shifts, which conspired to build giants. The second case study is geological. I examine attempts to explain signs of glaciation in Neoproterozoic tropics, in particular the "Snowball Earth" hypothesis. The glaciation is accounted for as the result of a worldwide freeze, set off by the Neoproterozoic's distinctive geography.

Once the case studies are under our belt, I get to work. Chapters 3, 4, and 5 build toward an analysis of "unlucky circumstances"—the differ- ence between investigations of *O. tharalkooschild*'s size and investigation of its evolutionary history. I do this by constructing what I call the *ripple model of evidence*. Chapter 3 develops an account of "traces" and their evi- dential warrant. Chapter 4 draws on two apparently opposed views about historical hypotheses: Carol Cleland's view that they are overdetermined, and Derek Turner's that they are underdetermined. By critically examining and synthesizing these, I develop the model that is presented in chapter 5. That same chapter also develops the arguments for pessimism, my target for much of what follows.

Chapters 6 through 10 undermine the arguments for pessimism. Chapter 6 discusses the methodology historical scientists employ: How do they generate knowledge, and why does it work? I survey and criticize existing accounts and build on these by highlighting the role of "coherency testing". I argue that there is no methodological "essence" of historical science, rather an opportunistic, pluralistic strategy is adopted. Historical scientists are not obligate specialists when it comes to method; they are omnivores. Chapters 7 and 8 emphasize and defend the role of analogy in historical science. There, I consider the possible *parochialism* of historical investigation. If historical scientists are primarily interested in particular past events, and if those events exhibit "historicity"—that is, their behavior and occurrence turns on highly specific, path-dependent factors—does it follow that their sources of evidence are highly limited? I argue it does not. First, historicity does not restrict scientists to parochial investigation. Second, it does not restrict them to trace evidence. I argue that "analogous" evidence, which is relevant to our target since both are the same type of event, is a further source.

Chapter 9 undermines the claim that historical scientists cannot "manufacture" evidence, leading to a general discussion of the relationship between epistemic practices such as experiments and simulations in chapter 10. I argue that simulations can generate new evidence before introducing a framework for understanding this argument and showing how simulations and experiments can function in similar ways. Although I do not think we should take experiments and simulations to be the same kind of activity, I argue that under certain circumstances simulations play explicitly experiment-like roles—and do so successfully.

In chapter 11, I draw together my arguments against pessimism and discuss how, and to what extent, they promote optimism. I also draw methodological lessons for the practice of historical science in "unlucky" circumstances: Success requires speculation. That is, we should not respond to a lack of luck by adopting a restrictive, conservative empiricism but rather by embracing a creative, open-ended approach founded on what I shall call "empirically grounded" speculation.

In chapter 12, I discuss more practical upshots of my discussion. I first take optimism and the nature of historical science to provide reasons to actually *do* historical science. That is, the instrumental value of historical science is strengthened by my discussion. Optimism involves the belief that historical science will pay epistemic dividends. Moreover, my account motivates an interest in the indirect benefits of historical science. Even when the direct aims of an investigation fails, this can have indirect benefits by generating unexpected epistemic goods that can help us to understand the past, present, and future. I also provide some lessons about how we ought to fund historical investigation in light of my discussion. In brief, we should shift from centralized or concentrated studies with precise goals to targeting less specific goals and more distributed resources.

In the postscript, I will look at some Big Views about the nature of science. I will argue that traditional accounts of scientific progress get things wrong, and that positions on scientific realism or antirealism that rely on generalizing across science become extremely problematic once we recognize science's disparate, varied nature. The philosophy of science would have looked very different if it took paleontology or archaeology as its representative case.

Many of my arguments will be made in reference to the case studies from chapter 2, and this requires a few caveats. First, these are examples of *live science*, that is, they are subject to ongoing work. Thus aspects of the investigations, and their output, are likely to be tinkered with, transformed, or even outright discarded down the track. Happily, my arguments do not turn on the results of these investigations. Rather, I take them to be representative of the method I am concerned with. They are primarily *illustrative*.[9] Second, given that illustrative role, I might be accused of cherry picking: They could fail to be representative. I have selected these case studies because they are relatively unexamined by philosophers, and because they are difficult cases. Philosophers of science, especially when historical science is concerned, have tended to focus on a small cluster of case studies. Adding to this stock is important. Moreover, examining live scientific debate as well as more settled cases is essential if we are to understand science as it is practiced. Additionally, recall that I am interested in "unlucky" circumstances that lack robust signals from the past. Many important areas of historical science, such as the impressive synthesis of cladistics, molecular genetics, and computational methods that constitute modern systematics, use large, well-established datasets and well-confirmed theories. Such an example would not suit my project. In contrast, both sauropod gigantism and Snowball Earth are cases where our epistemic reach is challenged and stretched. Further, I will not restrict myself to those two cases. We shall meet the evolutionary development of the long-extinct ancestors of sea urchins, mass extinctions, the use of lasers to study Mayan ruins, Martian exogeology, marsupial-like saber-toothed "lions," ritual sacrifice in Mesoamerica, the use of *E. coli* to test macroevolutionary theory, the connection between shamanistic practices and art, enormous Mesozoic fleas, and the Cambrian explosion. Finally, philosophers never truly fly solo: our work is a social activity. My conclusions and arguments ought to encourage others to examine further cases. If these reveal idiosyncrasies in my view, so be it.

Another caveat concerns my use of the term "historical science." Just what *is* a historical science, and which sciences fall under that category? I will not concern myself with this question, and for good reason. I am not playing the demarcation game: I do not claim that there is a class of "historical" sciences that use a distinctive method, or have a distinctive aim, or whatever (compare with Cleland 2002). Rather, I examine a group of scientific investigations that face similar epistemic situations: That is, they must overcome a similar set of challenges in order to generate knowledge. The claims I make will apply across science—but do so patchily (see Currie & Turner 2016). They should apply wherever similar epistemic situations are faced. For instance, there might be reason to think that investigations of extremely distant targets, such as planets or the biota of the deep seas, have many of the same epistemic features as investigations targeting *temporally* distant targets. I suspect many of my arguments will therefore apply in these domains as well.

Some philosophers and historians of science will note the ironically "ahistorical" nature of my approach. Much of the issues I discuss have echoes—some rather concrete—from the past. Whewell emphasized consilience in the nineteenth century, and at around the same time Darwin and Lyell certainly had things to say about historical reconstruction, and indeed philosophers around the middle of the twentieth century worried about the nature of historical science (see Currie and Turner's discussion). Why haven't I tied the discussions herein to those historical figures? My reasoning is this: If you're going to do history, you should do it properly.

Figures like Whewell (and Darwin for that matter) had subtle, specific views about the nature of science and historical reconstruction that deserve to be understood in their own right. A book that both tackles the methodological and epistemic issues herein and took those figures seriously, as opposed to using them as foils or caricatures, would quickly collapse under its own weight. I think a project that looked at the kinds of views I discuss from a historical perspective would be important, useful, and enlightening. But it would be a different project from this one.

Having said this, it is worth briefly touching upon the aforementioned debate in the twentieth century. That discussion began with Carl Hempel's (1942) fitting historical method into his account of explanation by dismissing historical explanation—which he took to be "genic," that is, not law-based—as mere explanation "sketches." On his view, there is only one kind of scientific explanation[10]—that which utilizes lawlike premises—and so much the worse for historical reconstruction if it didn't fit the mold. Philosophers such as William Dray (1957) and W. B. Gallie (1959) argued that historical explanation had a distinct form. Here, there are two kinds of legitimate scientific explanation. The basic issues, which are carried forth in contemporary work on historical reconstruction (particularly Derek Turner and Carol Cleland's work), concern the comparative legitimacy of the historical sciences and the related issue of what science *is*. In establishing optimism about historical reconstruction, I take myself to be defending the legitimacy of historical science (although I have strong reservations about epistemic claims at such a coarse grain: establishing legitimacy requires our examining the local detail) and to be providing a sketch about what science is like: its success (at least in unlucky circumstances!) is due to opportunism, speculation, and what I call methodological omnivory. At base, then, I side with Dray and Gallie.

It is worth reiterating that although the direct purpose of this book is to argue for optimism about historical science, it would be a mistake to construe it as narrowly concerned with those sciences. It is, rather, an examination of the methods, strategies and success of science in non-ideal, "unlucky" circumstances. Moreover, a philosophical account of historical science is illuminative of wider issues in the philosophy of science. The epistemic upshots of historicity and contingency, the sanctioning of simulations, the relationship between overdetermination and underdetermination, the relationship between experiment and modeling, the nature of background theories, scientific realism and progress, how we ought to best promote successful science, and the purpose of idealization are all discussed, and built upon, within.

If you will permit me a shallow analogy, just as the tooth of the platypus is a rich vein of information for those interested in uncovering the deep past, the work of historical scientists provide diverse, fascinating, and important raw material for philosophical analysis. Explaining how such sciences succeed (when they do) and understanding what our attitude should be toward the knowledge they produce are impossible without a close investigation of their work. Without further ado, let's get to the case studies.

2   Snowballs and Sauropods

I examine, analyze, and (where appropriate) vindicate the practice of historical science from a philosopher's perspective. As such, my primary interest lies in how scientists generate knowledge. We know a lot about the deep past: how and why? I will answer by characterizing both a set of epistemic situations historical scientists find themselves in (see especially chapter 5) and the resources they can draw on in those contexts (chapters 6–10). On such an approach, we should expect a fair amount (actually, a lot!) of abstraction. Many scientific details will be omitted in order to foreground the philosophical meat. However, it is important to get the level of abstraction right. I want to understand the *material basis* of historical inference; that is, I seek historical science's justification in worldly facts and their discovery, as opposed to the formal structure of the reasoning involved (see Norton 2003 work for a defense of a "material" theory of inductive inference). My defense of optimism and account of the method of historical science, then, will not seek vindication in analytic or a priori analysis of the inferences involved (although there will be some of that), but in an examination of the practice of historical science and the facts that underlie and vindicate it.

To understand science in these terms, we had better have some common ground, and that is my aim in this chapter. I am going to take you through two case studies: paleobiological reconstruction of sauropod dinosaurs and the paleoclimatological "Snowball Earth" scenarios. This base of joint knowledge will allow me to refer to these debates in a more or less offhand way throughout the rest of the book. So pay attention—there will be tests later ...

Both cases, sauropod gigantism and Snowball Earth, differ from *Obdurodon tharalkooschild* insofar as they are more difficult: They occupy a comparatively tricky epistemic situation. As we will see, when scientists lack clear routes from material remains to the past, they draw on a wide variety of techniques, sources of evidence, and theory.

Before I begin, a note on time scales. Both cases occurred in the deep past—measured in hundreds of million years before the present (see figure 2.1). Geologists typically delineate such scales in terms of distinctive layers of rock, or "strata." and the types of fossils found therein. The sauropods lived during the Mesozoic period, the so-called Age of the Dinosaurs, from 252 to 66 million years ago (ma). The period's earliest bound is set by one of the most dramatic events in life's history: The Permian-Triassic extinction event. This involved the loss of over 90 percent of biodiversity in the world's oceans. It ends with another extinction event, one that wiped out all dinosaurs except the birds. The stage for Snowball Earth, the Neoproterozoic, is earlier still. It covers the period from 1,000 to 541 ma. This period saw the emergence of the earliest fossils, and it closed just

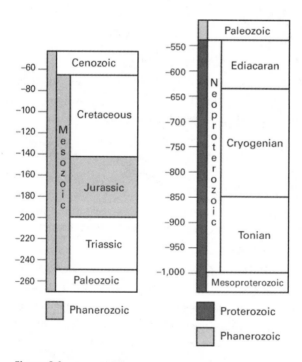

Figure 2.1
The Mesozoic period (on the left) and the Neoproterozoic (on the right). Scale in millions of years (ma). (Wikimedia Commons.)

before the Cambrian "explosion": the radiation of an extraordinary diversity of multicellular life (how much of an explosion depends on who you talk to).

Were this a book intended primarily for science education, I might call on metaphor to aid in our comprehension of such scales. Perhaps we could imagine the time scale of the universe as a string or a calendar or a day (Bill Bryson spends a particularly long time on this sort of thing). I am not writing that kind of book, so I will simply say that the geological and evolutionary time scales we are interested in are mind-boggling (although the cosmologist might still scoff).

2.1   Sauropod Gigantism

Sauropods are the ultimate out-group among terrestrial vertebrates simply because of their size. Their vast dimensions and sheer existence in the history of life oblige us to evaluate any potential limits of body size in terrestrial vertebrates. Whereas many other fossil forms can fit comparatively easily within existing frameworks, sauropods appear so far out of the range that they are a definite challenge. (Clauss 2011, 3)

Clade Dinosauria was remarkable on many counts. Dominance: From their Triassic arrival through to the disaster at the end of the Cretaceous, dinosaurs dominated ecosystems for 135 million years (and their pilgrims into the present, birds, haven't done too shabbily either). Diversity: Not counting birds (the most diverse nonaquatic vertebrates), more than 1,000 species and 500 genera of dinosaurs have been identified. They occupied deserts, rainforests, and took to the skies. However, given their reputation, dinosaurs are somewhat disappointing in terms of size. If we randomly sampled dinosaurs, most would be broadly dog-sized. That's quite large from the perspective of vertebrates generally, but nothing to write home about. When we compare dinosaurs to mammals occupying similar niches, we find similar sizes.

There are two major (and familiar) exceptions to this rule. Some theropods, such as the iconic *T. rex,* were enormous predators, outrunning mammalian counterparts by an order of magnitude. But the true giants were the sauropods: barrel-chested, tree-trunk-limbed *Diplodocus, Brachiosaurus, Apatosaurus,* and so on; often visible in the background of depictions of

Jurassic and Cretaceous landscapes, levering their long necks to reach the highest foliage. Sauropods were not only large, but successful:

because they are the largest herbivores ever, as well as the terrestrial vertebrates that dominated the megaherbivore niche of more land masses from the end of the Triassic until the end of the Cretaceous for an incredible 135 million years … [sauropods could] be regarded as the most successful vertebrate herbivores ever known. (Hummel & Clauss 2011, 11)

Sauropods were diverse, experimenting with a range of variation during their reign. Some sported enormous, bony spikes along their backs. Variation in tooth morphology points to their occupation of a variety of vegetarian niches (see Tutken 2011; Hummel & Clauss 2011). However, these were relatively minor variations on a theme: The basic sauropod morphology was remarkably stable. No sauropods chewed; their teeth were used for tearing or clipping vegetation. Their characteristic body plan was retained throughout the Mesozoic: quadrupedal, with columnar legs; a tiny head balanced on the end of a long neck, counterbalanced by a tail. These features were part of the sauropod lineage long before they became giants (see figure 2.2).

Figure 2.2 represents various stages of sauropod evolution, from a small biped in the early Triassic (A), to a full-blown giant in the late Jurassic (D). From the mid-late Triassic on (B), the standard sauropod phenotype is in place.

So, just how big were the biggest sauropods? This is a tricky question, and examining it is a nice way to introduce some scientific reasoning. Paleontologists typically understand "bigness" in terms of length and weight. Length is the easier of the two: Just measure your fossil. However, such measurements depend on having both a complete enough fossil, and the wherewithal to arrange it properly. And many of the biggest sauropods have left only the scantiest remains. The genus *Argentinosaurus*, for example, is a contender for the heavyweight title, but is known only from a few vertebra, scraps of ribcage, and the fibula (lower leg bone). These scraps are enormous. The fibula, for instance, is 1.55 meters long. To infer overall length from fibula length, paleontologists need to know, proportionately, how those variables relate to one another. Constructing such models requires that we compare *Argentinosaurus* remains with more complete sauropod finds, and the comparisons we choose make a difference to our results. An early estimate from Paul (1994), based on comparison with *A. altus*,

Figure 2.2
Basal Sauropoda. A: Basal Saurachian, late Triassic; B: Basal Sauropodomorph from late Triassic; C: Basal Eusauropod from mid-Jurassic; D: Brachiosaurus from late Jurassic. (From Rauhut et al. 2011, 120; © Indiana University Press.)

estimated *Argentinosaurus* length at 35 meters (and 80–100 tons!). However, in comparison with *Saltasaurus*, Carpenter (2006) estimated a top length of 30 meters. Regardless, *Argentinosaurus* was really, really big. Since the impressive discovery of *Dreadnoughtus*, smaller than *Argentinosaurus* but significantly more complete, a better comparison may be available (see chapter 4). Still, even at the most conservative estimates, sauropods rival baleen whales in length, certainly managing well over 25 meters.

How about weight? This is trickier than length, since an animal's size doesn't tell us directly about how much it weighs. This is because animals differ in density: A comparatively sized bird will be much lighter than a mammal, because birds have physiological and anatomical adaptations to flight that make for a light frame. We need to know, then, an animal's size, and the relationship between its size and its density. A common technique for estimating weight involves scanning full-sized models—themselves reconstructions—and representing them digitally. Scientists take cameras into museums and painstakingly capture the features of full-sized reconstructions so they can be digitally rendered and manipulated. The animal's body parts are represented by a series of three-dimensional geometric shapes with associated volumes. This volume, plus the estimated average density, provides an estimate of body weight (Soinski, Suthau & Gunga [2011] apply this technique to sauropods). Such an approach can be tested by applying it to extant animals; Gunga's team, for instance, got respectable results with a stuffed rhinoceros and the skeleton of an Indian elephant. The use of modeling to generate evidence about the past will be a common theme throughout, particularly chapters 9 and 10, as will the use of extant animals as analogues, particularly chapters 7 and 8.

Estimates of sauropod weight are plagued by clashes between analogues. Obvious analogues for sauropods are large grazing mammals, elephants, rhinoceros, and so forth. However, physiological reconstruction of sauropods tend to use birds as models, for as we shall see there is reason to think they boasted various birdlike weight-reducing adaptations such as "pneumatized" skeletons (that is, having hollow bones allowing the passage of air). If so, this partially undermines the relevance of large herbivorous mammals, since they are much heavier for their size than birds. And so, assuming we have good morphological reconstructions, although estimating sauropod volume is relatively straightforward, working out density is significantly more daunting. This is reflected in the wide variety of estimations. For instance, using this technique, Gunga et al. (2008) revised a weight estimate for *Brachiosaurus* from 74.4 metric tons to 38 metric tons. In chapter 8 I will discuss how scientists navigate between various analogues to provide well-supported reconstructions of the past. Just working out how "big" the biggest sauropods were illustrates both the difficulties involved in reconstructing the past and the creative ways in which scientists respond.

We can conservatively pin the largest sauropods at somewhere just short of 30 meters in length and 50 metric tons (Ganse et al. 2011). This is *big*. Although aquatic animals equal these sizes and overshoot them in weight, such animals are cheating. The sea contains significantly more, and more easily extractable, biomass (krill, for example), buoyancy removes the need for bracing and makes mobility less energy expensive, and water cools large bodies. These factors massively advantage sea monsters over land monsters

(McNab 2009). On land, the biggest mammals have managed is Paraceratherium, a rhinoceros-like beast. At her largest, she was a "mere" 8 meters in length and 18 tons in weight (see Fortelious & Kappelman 1993).

So, sauropods are uniquely gigantic land animals. Getting a handle on the how and the why of their size is challenging, and has bred a rich research program investigating various aspects of sauropod behavior, environment, physiology, and evolutionary history. We could understand this program as targeting two questions.

First, the *phenotype question*: What were sauropods like, and how were they viable? "Viability" explanations account for organisms in terms of what they need to survive and reproduce (see Wouters 1995, Currie 2015b for discussion). Roughly, we must consider sauropods as living, breathing critters: How did they move, support their weight, grow, and so forth? In answering the phenotype question, paleontologists both reconstruct sauropod traits, and explain how these traits contributed to the animals' functioning. Second, the *evolution* question: How did sauropods evolve? Here, we are concerned with "etiological" explanations over evolutionary time. Why did sauropods, and no other terrestrial lineage, evolve such extreme gigantism? We want to know what was different about the sauropod lineage that led them down their unique evolutionary path.

These two questions, and investigation of them, are intimately related (I discuss this point in fuller detail in Currie 2015b). As we shall see, for instance, Senter's contention that the sauropod neck evolved via sexual selection (the evolution question) relies in part on the real costs that living sauropods may have paid for them (the phenotype question).

I will approach this complicated and rich literature by dividing it into parts. We'll start with "extrinsic" drivers of evolutionary change before shifting to more internal, physiological aspects. In chapter 13 I will argue that if we were to start our philosophy of science from historical science, this would lead us to develop an indirect, piecemeal view of scientific progress. In service of this, as we go, notice the extraordinary array of different epistemic resources, reasoning styles and techniques paleontologists employ in uncovering the story of the sauropods.

2.1.1 Extrinsic Evolutionary Drivers

"Extrinsic" causes can drive size increase over evolutionary time. That is, events, pressures, and processes outside of the target lineage can make a difference to patterns of survival and success, and thus shape lineages. According to theories of biogeography, geographical changes can drive morphological evolution by affecting ecological conditions, resource availability, and migration patterns. Changes in atmospheric composition could have ontogenetic effects or evolutionarily favor larger or smaller sizes. Competition from other lineages, particularly arms races, can drive evolutionary change. Explanations of evolutionary change citing extrinsic causes are widespread but, as we shall see, have minimal relevance for sauropods. Let's look at the role of extrinsic factors in sauropod evolution.

Some early explanations of sauropod gigantism singled out atmospheric change, specifically increases in atmospheric O_2. Global oxygen levels fluctuate over geological time (between 13 and 35 percent), and spikes appear to track significant events in life's history. Some peaks in O_2 levels are plausibly the result of biological activity. For instance, a major increase in the Precambrian coincides with the rise of oxygen-producing microorganisms, and another with the rise of vascular land plants 300 million years ago. Other peaks may see the casual arrow reversed: Oxygen increases sometimes drove biological events rather than being their product. O_2 spikes correlate with the Cambrian explosion, both arthropod and vertebrate land invasions, and increases in animal size: arthropods in the Carboniferous, proto-reptiles in the Permian, and mammals in the Tertiary.

Berner et al. (2007) discuss experimental evidence that increased oxygen has ontogenetic effects in egg-laying reptiles and argue that this could account for Permian proto-reptile gigantism:

The gigantism has classically been attributed to an increase in diffusive capacity caused by an increase in atmosphere O_2 concentration. This may explain the effect seen in egg-laying vertebrates, because diffusion across the eggshell will be increased and have an effect on hatchling and therefore adult body size. (558)

They report a study that tested body size in alligators, given exposure to different oxygen levels during ontogeny. Roughly, alligator eggs are isolated from one another and placed in environments of differing oxygen content. The hatchlings are measured and compared. The study found a positive correlation between increased oxygen and body size. Such ontogenetic effects, if there is selection for larger morphs in the population, could scaffold evolutionary size increases. Although this extrinsic cause could matter for some size increases, such an explanation is not available for sauropods:

Atmospheric O_2 does not seem to have fluctuated significantly during the Mesozoic (Sander & Clauss 2008).

In fact, Sander and Clauss (2008) argue against any explanation of sauropod gigantism that relies on extrinsic causes of sauropod gigantism. Changes in temperature or geography do not track changes in sauropod size. Moreover, most dinosaurs were not especially large by mammalian standards. If an external force is to blame, why did other lineages remain diminutive? This is too quick: The relationship between phenotype and climate is varied and complex, and so a lack of simple correlations does not rule out connections. It does, however, demonstrate an epistemic problem. If the climate's effect on sauropod size is subtle, the low fidelity of our access to information about climate change and morphological evolution makes assessing the viability of such hypotheses difficult.

One big difference between reconstructing these aspects of sauropod evolution, and *O. tharalkooschild* features such as its platypus-hood, is the necessity of making inferences relying on such tenuous connections, such as that between climate and body size, especially when starting with partial knowledge of the targets. Sander and Clauss are right to reject the O_2 hypothesis as read, however: If increases in O_2 were a central cause of sauropod evolution, we ought to see at least some coupling between size increase and atmospheric fluctuation. Moreover, we would expect increases to be across the board rather than restricted to sauropods (and some theropods).

As it stands, the role of climate and geography in driving sauropod gigantism is at best subtle and difficult to track. In contrast, the role of Jurassic predation is considered an important extrinsic driver. As I mentioned earlier, sauropods are not the only mysterious giants of the Mesozoic. The theropods massively outdo mammalian predators in size (according to Burness et al. 2001, they were on average twelve times heavier than their Cenozoic counterparts). It is thought that predation from theropods drove two important aspects of sauropod size.

First, it is taken for granted that sauropod gigantism is, at least in part, an adaptation for predator avoidance. The late Jurassic saw herbivorous dinosaurs specialize dramatically. For instance, heavily armored ceratopsids (the iconic triceratops and her ilk) and ankylosauria emerged during this time, presumably in response to arms races with increasingly sophisticated predators. Instead of evolving protective armor, sauropods sought refuge in size; and as they got bigger, so would their predators, and so forth. Second, it has been suggested (Griebeller & Werner 2011; Sander & Clauss 2008) that sauropod growth *speed* is also an adaptation to predation. Sauropod growth was astonishingly fast. Hatchlings began life at around 5 kilograms, and bone histology suggests growth rates of around two tons per year. Griebeller and Werner characterize the sauropod life cycle as follows:

the survivorship curve of sauropods was most likely characterized by a phase in which predation dominated survival, as in modern birds ... because sauropods were most probably precocial and the high growth rates in body size suggest a significant predation risk for juveniles. This was followed by a phase in which sauropods were more or less safe from predation due to their large body size, as is the case in turtles ... crocodiles ... and large mammals (2011, 269)

Sauropods, then, could be characterized as having a vulnerable juvenile stage, followed by an invulnerable (to predation, at any rate) adulthood. If your safety depends on bulk, it pays to get bulky fast. Sauropods, then, evolved high-speed growth. In an interesting corollary, Sander and Clauss suggest that theropod gigantism was itself enabled by the bounty of sauropod infants (this can't be a full explanation, of course: one needn't be gigantic to eat 5kg hatchlings).

Examining possible extrinsic drivers of sauropod gigantism reveals some characteristic features of historical science. Scientists draw on surrogative evidence, in particular evidence from evolutionary convergences and other analogues, to support and construct hypotheses. We saw this in Griebeller and Werner's appeal to the survivorship curve of modern birds, for example. Sauropods are not considered in isolation. Rather, they are treated as an instance of a general phenomenon: evolutionary gigantism. If increases in atmospheric oxygen regularly track and sometimes explain other cases of gigantism, then it could be bought to bear for sauropods. In doing so, scientists take sauropod size increase as an instance of a process type (though see Currie 2014a).

Let's turn to features of the sauropods themselves.

2.1.2   Cranes, Sex, Vacuum Cleaners
Besides size, the most distinctive sauropod feature is surely their long necks. There are three (non–mutually exclusive) views on their evolutionary function. By the first, sauropod necks were "cranes," allowing for differentiated feeding heights. Second, they were "vacuum cleaners," functioning

to increase ground grazing cover. Third, the necks were "sexy," evolving in response to sexual selection.

By the crane theory, increased neck length functioned to differentiate sauropod grazing. Individuals with longer necks could access increased amounts of more diverse resources, and so outcompete their conspecifics. The most pressing objections to the crane theory concern physiological viability: There is reason to think that sauropods couldn't raise their necks. Here, our explanation of a trait's evolutionary purpose is constrained by physiology. Martin, Martin-Rolland, and Frey (1998) produced morphological reconstructions of sauropod necks from fossil evidence. They argued that sauropod necks were more structurally akin to horizontal beams than vertical masts (or cranes). If the necks are analogous to beams, they would have required significant bracing, massively restricting movement. From living biological analogues they also reconstructed musculature, concluding that the necks were rigid. From an engineering standpoint, they claimed, sauropod neck movement was restricted to below the shoulder. Seymour (2009) has concerns about blood pressure. Drawing on modern examples, he calculates that "A sauropod with its head in the trees would ... have to increase its metabolic rate to 175 per cent, and, at this level, expend 49 per cent of its total energy requirements just to circulate the blood" (318). Based on this model, it is highly unlikely that the metabolic cost of neck raising was feasible.[1] However, as Christian (2010) points out, contemporary long-necked animals like giraffes have adaptations mitigating circulation issues. He also argues that, assuming resources are scattered, the benefit of high grazing could outweigh increased costs (although I wonder why an animal as bulky and expensive as a sauropod would evolve under such conditions). A further anatomical concern is raised by Senter (2007): "Keystone-shaped cervical centra ('ventral bodies' in mammalian nomenclature) at the bases of their necks allow giraffes, camelids and birds to hold their necks vertically, but sauropod cervical centra lack such shapes, even among sauropods that are typically portrayed with vertical necks" (46).

Later mechanical reconstructions (Christian & Dzemski 2011) suggest there is no single story about neck-raising across the clade. *Brachiosaurus*, they argue, habitually held their necks above the shoulders (although not vertically as in some reconstructions). There is an emerging consensus, then, that some sauropods (diplodocus, for instance) kept a low profile while others did not. Christian and Dzemski go on to argue (reasonably) that neck-posture diversity represents diverse feeding strategies.

What was the point of horizontal necks, then? Ruxton and Wilkinson (2011) present a novel hypothesis. The inspiration comes from an analogy with cylinder vacuum cleaners popular in the middle of the twentieth century:

Because the machinery required to create the suction was large and heavy, the main body of the vacuum cleaner was positioned by the user in a central location within the room, and the user then moved a light head-part at the end of a long tube across the surrounding carpet. ... By analogy with cylinder vacuum cleaners, the long neck of sauropods might have been adapted to allow less movement of the exceptionally heavy body of these animals. (779)

Perhaps sauropods evolved long necks to increase ground browsing cover while minimizing locomotion. Ruxton and Wilkinson support their proposal geometrically. They calculate the benefits of increased browsing range against locomotion costs, finding that extensions of neck length bring substantial energy savings, but with diminishing returns. Reasonably, an upper limit might be reached due to constraints from "other factors (perhaps mechanical aspects of the support and control of a long structure—or breathing considerations)" (780). They test their model's predictions against the fossil record, finding that it roughly matches the spread of actual sauropod neck lengths. Here again, we see scientists testing hypotheses against traces, but also drawing on (sometimes surprising!) analogues and utilizing abstract, highly idealized models to probe the past. Hunting for analogues also aids in exploring alternatives—previously unconsidered cases open up new avenues of hypothesis construction and testing.

There is reason to think that giraffe neck length evolved via sexual selection, and Senter (2007) argues this could be true of sauropods as well. In a sexually selected character we should expect:

1. The character to be emphasized in one sex;
2. The character's involvement in dominance or courtship displays;
3. For there to be no immediate survival benefit;
4. For there to be some survival cost;
5. For the character to display positive allometry during ontogeny;[2]
6. A lack of correlation between phylogenetic increases in character size and general body size.

If a character is sexually selected for, this is typically due to one sex preferring a trait possessed by the other in breeding contexts (hence, 1 and 2). Hypotheses of sexual selection need to be distinguished from others. For instance, the trait might be subject to run-of-the-mill natural selection (hence 3 and 4). Further, ontogenetic and phylogenetic effects must be controlled for: The trait's change could be due to its being coupled with other traits that are fitness enhancing (hence 5 and 6).

Take a paradigmatically sexually selected character, such as the peacock's tail.[3] Comparing a drab peahen with a startling peacock surely shows that the character is primarily expressed in males (1), and they are certainly used in dominance displays (2). It is prima facie plausible that the tail does not aid the peacock's survival (3), in fact it is likely that it undermines their viability (4). I cannot find any data on positive allometry in peafowl, but it seems likely (5) and it is plausible that peacock tail length evolved independently of peacock overall size (6): Peafowl are not particularly sexually dimorphic once tails are taken out of the equation.

How do sauropod necks fare on these criteria? Can we unify the sauropod's neck and the peafowl's tail as instances of traits that evolved via mate preference? The fossil record is too fragmented for (1) and (2): identifying sex and behavior from fossils is notoriously tricky. However, analogies may be drawn from giraffe, which do display sexual dimorphism in neck length and make use of them in partner choice. The third and fourth indicators turn on the character's effect on an organism's viability. Senter argues that sauropod necks could come at significant survival costs. Some sauropod heads were at equal height to theropod mouths: "Longer necks at that convenient height would have provided longer targets, making it easier for a carnivore to find a place to bite then would have been the case with shorter-necked prey" (47). This is only a potential cost. Some long-necked lineages, giraffes and swans included, use their necks as weapons or in threat displays. Kim Sterelny has suggested to me that sauropods could, swanlike, charge down their enemies with lowered head and "suitably scaled-up hissing." Biting sauropod necks, then, might be a more serious proposition than Senter allows. The fifth indicator is difficult to access, although Senter claims that those cases of ontogenetic differentiation that are found in sauropod remains support allometry. Finally, Senter tackles the sixth indicator via a statistical analysis of limb versus neck length across the clade and finds correlations supporting phylogenetic allometry.

And so there is a weak case for sexual selection's role in the evolution of the distinctive sauropod neck. The evidence is indirect: Fossils alone provide little inroads to behavior and the benefits or otherwise of increased neck length, and only limited access to allometry. Senter relies upon reconstructions and leans heavily on analogous data from extant animals. Substantial claims about these information sources are required, although such hypotheses are certainly not empirically intractable.

Both the physiology and evolutionary function of the sauropod neck are contentious. I think it reasonable to accept that, at least to an extent, sauropod necks enabled more efficient grazing via differentiating browsing height and/or by increasing browsing range. As we shall see, one answer to how sauropods acquired sufficient resources is that their necks minimized the energy expenditure spent in gathering food.

In explanations and reconstructions of sauropod necks, historical scientists draw on contemporary analogues and link sauropod gigantism with other cases from biology and engineering. Comparisons and studies across the sauropod clade are also exploited. Paleontologists, then, do not rely only on the relationship between traces and the past, but on how their picture of the past hangs together. I will draw on this in chapter 6 when arguing against "trace-centrist" views of historical scientific method.

2.1.3   Dentition and Digestion

Further exploring sauropod physiology requires estimating the nutritional content of Mesozoic flora and understanding the nature of sauropod metabolic and digestive systems.

Sauropods had diverse dentition. The Diplodocoidea and titanosaurs sported a few, pencil-like teeth clustered at the front of their mouths, while basal sauropods had spoon-shaped teeth. This and other dental variation suggests niche-differentiation, as should be expected from a successful and diverse lineage. Sauropod lifeways were varied. One commonality, and a divergence from mammals, is that sauropods did not chew their food. This is significant. Nonmastication has an upshot for sauropod neck length. Mammalian reliance on chewing sets a lower limit on minimal head size versus body size, and thus neck length, as a heavy head would render a long neck unsupportable. Free from the constraints of specialized masticatory dentition, sauropods were able to evolve tiny heads, and thus long necks.

Moreover, without pausing to chew, sauropods increased their ingestion volume.

Sauropods relied upon symbiotic microbes in the gut for digestion. Sander, Christian, et al. (2011) summarize the evidence as follows:

(*a*) phylogenetic bracketing that indicates that symbiotic fermentation bacteria were the same as in modern herbivorous birds and mammals, (*b*) all large recent herbivores employ fermentative digestion and (*c*) the fact that sauropods would have needed to consume impossibly large amounts of plant matter without it ... (129)

In other words, gut fermentation was almost certainly present in the common ancestor of sauropods, birds, and mammals; they infer from contemporary guts to sauropod guts; and rely on physiological plausibility: No other options are on the table. However, whatever else we might know about sauropod digestion is highly speculative (Hummel & Clauss 2011), and extrapolated from much smaller contemporary lineages.[4]

This lack of mastication provides an unfavorable contrast between sauropods and mammals. Chewing allows mammals to "outsource" the digestion process, breaking up food before it reaches the gut. In extreme cases (ungulate ruminants) this process is extended by cud-chewing. Some authors (Sanders et al. 2001) have suggested there is evidence for an avian-like gastric mill in sauropods, but it is generally agreed that they did not swallow gizzard stones (see Wings & Sander 2007; this is endorsed in Hummel & Clauss 2011).

The relevance of this assumption [a lack of mastication and gastric mill] cannot be overestimated. The food that terrestrial herbivores ingest is basically reduced in size in one way: mechanical breakdown. Having passed the site of particle reduction—either the oral cavity with its dental apparatus, or the gastric mill with its gastroliths—there is generally little further breakdown of ingesta particles in terrestrial herbivores. ... In particular, long ingesta retention times in the gut, and therefore a long exposure to microbial fermentation, might well have compensated for the lack of particle breakdown. (Hummel & Clauss 2011, 13–14)

In other words, what birds achieve by swallowing stones, and cows achieve by chewing, sauropods achieved with size and patience. The implication is that sauropod gigantism is an adaptation for digestion; gigantism's function was to accommodate vast fermentation tanks. The output of methane and sulfurous gas (and associated smell) must have been prodigious! Indeed, as we will see in chapter 10, Wilkinson et al. (2012) have attempted to explain high Mesozoic temperatures via a sauropod-fueled greenhouse effect.

Midgley et al. (2002) argue that sauropod gigantism is an adaptation to the Jurassic's low-nutritional flora. Because nutritional content is low, high quantities of food must be ingested. A solution is to increase digestive capacity (and thus overall size). Assuming there is a positive relation between the costs of size increase and the payoff of increased digestive capacity, low nutrient content could drive selection for gigantism. Hummel, Gee, et al. (2008) argue in opposition that the gymnosperm- and fern-dominated flora of the Jurassic were not low in nutritional value (see also Gee 2011). They measure the nutrient value of plants by comparing the potential metabolic energy of the substance with the cost of digestion. Coal, for instance, would provide bountiful energy—if only it were digestible. Meat contains significantly less energy than coal, but at least some stomachs can break it down with relative ease. They determined the metabolic energy of plants using in vitro fermentation. A stomach is simulated in the laboratory with a fermentation vat, and the heat energy plant matter expels during "digestion" (fermentation) is taken as a proxy for metabolic energy. Representative Jurassic flora, particularly gymnosperms, some conifers, and some ferns performed as well as grass—the main nutrient source for many mammalian megaherbivores of the Cenozoic. Thus, the low nutrient hypothesis is undermined.

Hummel, Gee, et al. speculate that sauropod diet was based on these high-nutrient plants and suggest that study of coprolites might corroborate this. They go on to estimate daily food intake based on various metabolic rates, claiming the contrast between contemporary herbivores and sauropods is not as extreme as first thought:

When compared with the dietary requirements of elephants, a 70 t sauropod would have to ingest 4.3 times the amount of dry plant matter necessary for a 10 t elephant and 5.6 times the amount necessary for a 7 t elephant. However, in regards to the actual ingestion of foodstuffs, this might not have posed much of a problem for the sauropods, since adaptations such as the lack of oral food comminution of the plant matter, a wide mouth opening and the lack of cheeks would have facilitated a high intake capacity. (2008, 1019)

Hummel, Gee et al. have not disregarded every aspect of Midgely et al.'s picture: Sauropods are still, in effect, enormous stomachs. However, this is not an adaptation for consuming foods with low nutritional content.

Ruxton and Wilkinson's vacuum analogy becomes more tempting: a 70-ton sauropod would eat around 237 kilograms of ferns, conifers, and gymnosperms per day, unchewed and sent straight to the great vat of its gut. Theories of sauropod digestion are linked to theories of neck length: Explanation of sauropod gigantism requires the integration of a variety of complex, diverse components. Moreover, Hummel, Gee, et al.'s appeal to lab studies of living reptiles suggests that claims about the efficacy (or otherwise) of experimental methods in historical science must be re-examined (see Jeffares 2008).

McNab (2009) approaches size restriction as a function of resource availability and energetics. Maximal size is predicted by contrasting required energy costs (keeping warm, locomotion, and so on) with available resources. These variables provide a measurement of an organism's *field energy expenditure* (FEE), which tracks maximal size. Most lizards, for example, have low FEE due to their low-cost thermoregulative systems. Models constructed via extrapolation from these examples predict a tonnage three times heavier than actual sauropods. Your standard lizard, then, is a poor analogue in this respect. The varanids, large predatory lizards such as the Komodo dragon, have higher FEE due to their size and active lifestyle. Models based on these lineages better match both sauropod and mammal size. In calculating FEE, McNab makes use of biogeographical information and comparisons to contemporary lineages:

African elephants are known to move long distances, especially in relation to the onset of rainfall, whereas most large dinosaurs appear to have been geographically restricted to comparatively small areas, which also implies low energy expenditures in dinosaurs (compared with a mammal standard), although large species in the Late Jurassic were less provincial. (2009, 12185)

And so, taking into account assumptions about available resources, as well as energy costs such as food processing and travel, McNab predicts maximal sauropod size. His results further undermine adaptive hypotheses like Midgely et al.'s, which claim that the costs of increased body size could be outweighed by digestive benefits:

The rapid swallowing of coarse food is unlikely to increase k_h [maximal expenditure], because the limiting factor on food consumption then would be the rate of fermentation in the gut, which is reduced by swallowing unmasticated fibrous food, the time required for fermentation increasing with food intake and body mass. Thus, the huge abdominal masses of sauropods were undoubtedly large fermentation vats that may not have completely compensated for the absence of buccal process-ing of food. So, it is unlikely that k_h was appreciably higher for dinosaurs than for mammals, either because of greater food abundance or because of a higher efficien-cy in processing food, and thus could not account for their larger masses. (2009, 12186)

The role of digestion in sauropod size is, then, contentious. The physi-ological case has been made: large bodies make for large guts that can sup-port large bodies. However, whether selection targeted gigantism for its nutritional and digestive benefit is up for grabs.

Work on sauropod diet builds on our picture of sauropod phenotype and evolutionary history. Once again, we see many of the opportunistic, "scaf-folded" aspects of historical investigation that I will highlight. Midgely et al.'s hypothesis didn't make it but was nonetheless very productive. It was by virtue of that (ultimately false) hypothesis that new facts revealed by new experimental and theoretical work were evidentially relevant. Without the thought that low nutrition plants could be an evolutionary driver of increased size, Hummel, Gee et al.'s experimental work on plant fermenta-tion wouldn't be relevant to our knowledge of sauropods. Let's consider three final aspects.

2.1.4   Thermoregulation, Reproduction, and Respiration

Sauropods present a metabolic dilemma. First, if large size is selected to provide protection from predation, then as we've seen hatchlings had bet-ter grow quickly. And indeed, histology indicates that sauropod growth was extraordinarily fast. Speedy growth rates are seen in endothermic, not ectothermic, lineages. Animals that regulate their own body temperatures ("warm-blooded" or endothermic organisms) grow faster than those that rely on external temperature ("cold-blooded" or ectothermic organisms). However, endothermia limits possible body size as heat dissipation becomes problematic (hence the enormous ears of elephants).[5] So were sauropods endotherms or ectotherms?

Gillooly, Allen, et al. (2006) argue that in ectothermic lineages heating becomes less costly with increased size. Engineering and planetary sci-ences predict that massive bodies conserve surface heat due to bulk. This effect, "thermal inertia," is also seen in extant ectothermic giants; what is true for celestial objects sometimes holds for terrestrial animals. Large reptiles have an easier time regulating temperature than their diminutive cousins. The central idea, then, is that an ontogenetic scaling effect leads to "gigantothermy": as body size increases, regulating body temperature becomes less problematic as higher bodyweights retain heat. Kim Shaw-Williams has pointed out to me that this could allow sauropods to remain active (and grazing) into the cool of night, when other ectotherms might be less active.

Gillooly, Allen, et al. construct a model that predicts the effect of body size and temperature on maximum growth rate, in an attempt to identify sauropod thermoregulative systems. Body temperature is inferred from a lineage's body size and growth rate. An ectothermic organism's body temperature conforms to external temperature, while that of an endothermic organism does not. Comparing body temperature to external temperature (inferred from paleoclimatology) allows Gillooly, Allen, et al. to infer the thermoregulatory system of a target lineage. Sauropod growth rate is inferred using both standard body size estimation and bone histology. The model's effectiveness is confirmed by tests on contemporary analogues. This should be a recognizable pattern by now. Paleontologists identify a correlation, in this case what thermoregulative system you have, and the contrast between your internal temperature and the temperature of your environment. They then find ways of estimating those values in their targets, as well as constructing a targeted model to exploit the regularity. In chapter 10, I shall call these "inference tools." They are tested via extant analogues.

Gillooly, Allen, et al.'s results show that sauropod thermoregulation diverges from that of smaller dinosaurs. Diminutive reptiles had body temperatures matching external temperature, suggesting ectothermy. Larger dinosaurs, however, had higher temperatures. In fact, body temperature in dinosaurs increases logarithmically against body size. The model suggests that temperature sets an upper limit on how big sauropods could be:

If we extrapolate the model ... up to what is perhaps the largest dinosaur species (~55,000 kg for an adult *Sauroposeidon poteles*), the estimated body temperature at the mass of maximum growth is approximately 48 C, which is just beyond the upper limit tolerated for most animals (~45 C). These findings suggest that maximum dinosaur size may have ultimately been limited by body temperature. (2006, 1468)

As the temperature of endothermic lineages does not dramatically increase with body size—endotherms keep relatively stable temperatures—by Gillooly, Allen, et al.'s model sauropods were ectothermic. Testing the model against ectothermic crocodiles again reveals a logarithmic curve between body size and temperature. Overall, this supports the hypothesis that sauropods were ectothermic, benefitting from thermal inertia. However, it has been questioned whether even thermal inertia is sufficient for the high levels of activity sauropods engaged in (those levels themselves inferred from trackways; see Clauss 2011, 5). Moreover, it leaves unanswered how an ectothermic organism could grow at sauropod rates:

> The strongest argument for sauropods having been endothermic is the high growth rates recorded in the histology of their bones... There seems to be no way for giant sauropods to reach a body mass of >50 metric tons in a reasonable lifetime without having—at least partly during their life span—a high resting metabolic rate comparable to or even higher than that in mammals. (Ganse et al. 2011, 108)

So, ectothermic sauropods would have no issues with size, but would not grow fast enough. Endothermic sauropods could grow at high speeds, but are thermoregulatively implausible at larger sizes. One suggestion (Wedel 2009; Ganse et al. 2011) is that the birdlike air sacs and pneumatized (hollow) skeleton of sauropods, in addition to aiding respiration (as we will see), helped with heat dissipation.

A fascinating suggestion from Farlow (1990) and mentioned by Sander and Clauss (2008) is that sauropods led dual metabolic lives. Starting out as endotherms, their metabolic rate dropped once they reached larger sizes, switching to an ectothermic system. I'm not sure what kind of mechanism would explain such a switch. Perhaps the active processes of endothermy could be "switched off," but I'm not sure how metabolically plausible this is. Regardless, sauropods faced unique metabolic challenges due to their terrestrial way of life and enormous size, and would have required specialized adaptations. I discuss sauropod thermoregulation in more detail in Currie (2016b), where I draw on Sander and Clauss's discussion to illustrate the importance of "coherence testing"—how our picture of the past hangs together—in historical reconstruction (I discuss this in chapter 6).

Philosophers often emphasize the explanatory and heuristic functions of models. However, in studying sauropod metabolism, historical scientists use models to actively test hypotheses (see chapters 9 and 10). We also see the incorporation of theories from outside paleobiology, such as the geomorphological theory of thermal inertia. Historical scientists are nothing if not opportunistic—to use the language I will adopt in chapter 6, they are "methodological omnivores."

Two further factors important for answering phenotypic questions about sauropods concern respiratory adaptations and egg-laying.

Large animals depend on oxygen reaching their lungs at a sufficient rate and capacity for cellular respiration. The extended tracheas of sauropods present a physiological problem as it is hard to imagine how they could accommodate the required volume of air. Sander, Christian, et al. (2011) summarize the consensus that has emerged, based on an avian analogy. Birds have a complex respiratory system characterized by a pneumatized skeleton, flow-through lungs, and large air sacs in body cavities, ensuring a steady rate of oxygen dispersal. This adaptation is necessary to meet the respiratory demands of flight. Sauropods possessed an analogous system that solved their ventilation issues in a similar manner. Sauropods had a pneumatized skeletal structure, and this is taken to signal an avian-style system.

Maximal mammal size is limited by population dynamics. Typically, larger individuals require more resources, and so larger individuals will lead to lower populations and slower population growth. A small population is, all things considered, a bad thing: It is less resilient to environmental fluctuations. A particularly bad year, a nasty virus, or some climatic cataclysm might wipe out a small, slow growing population when a larger group would bounce back. Large mammals are typically "K-selected" (see MacArthur and Wilson 1967): They adaptively respond to smaller populations by investing more in individual offspring. This places a selective constraint on maximal size, since small populations will be insufficiently robust to withstand shocks. The strategy can be compared to "r-selection": have a large number of cheap offspring and hope a sufficient number will be lucky enough to make it into the next generation.

How did sauropods mitigate the population-level costs of high individual size? Sander and Clauss (2008) argue that egg-laying is key. Sauropods were oviparous; fossil remains of eggs containing prenatal titanosaurs have been found. Evidence suggests a lack of parental investment—it is hard to conceive of how a 40-ton sauropod could help, much less interact with, its tiny offspring. Eggs are restricted in size to 25 centimeters in diameter and 51 centimeters in volume, which is, as Sander, Christian, et al. rather drily put it, "extremely small compared to an adult sauropod" (133). Of course, adult sauropods could provide protection by their presence alone (perhaps akin to crocodilian egg protection). If sauropods laid bountiful egg clutches that received minimal investment, then despite individual enormity, their population could retain a resilient r-selection strategy. The continual input of cheap offspring into populations would ensure that sauropod lineages did not become extinct due to stochastic events (while also, as suggested earlier, providing a bounty of baby sauropods for hungry theropods).

Analogies are an important empirical resource for historical scientists, as seen in the appeal to bird respiratory systems as a model for that of sauropods. Defending such "surrogative" evidence will be one of my major concerns. The discussion of population-level trade-offs driven by population genetics is another example of how complex and wide-ranging the explanation of sauropod gigantism is.

2.1.5 Why So Big?

An explanation of sauropod evolution is emerging. The explanation can be characterized by what I have previously called a "complex" narrative explanation (Currie 2014a). In complex narratives, an event is explained "as if" it is unique: that is, explanation does not proceed by fitting the target into a single, unified model. Sauropods are not presented as an example of gigantism, a general phenomenon, for instance. Rather, many complex, interacting aspects are drawn upon and weaved together. The sauropods' evolutionary history is a particular, peculiar, and complex causal cascade. It relies on basal sauropod traits:

1. Long necks, which probably increased grazing efficiency;
2. Air sacs (possibly), which made for efficient respiration and may have reduced temperature;
3. Oviparity, which allowed for fast population recovery;
4. Lack of mastication, which allowed for minimal head size and maximal food intake.

These basal traits were supplemented by new adaptations driven by increased predation:

6. Increased predation in the Jurassic drove increases in adult size and growth rate;
7. Further development of air sacs and a pneumatized skeleton, allowing for extremely efficient respiration;
8. The evolution of very high basal metabolic rates (in youth at least), enabling high growth rates;

9. Gigantism itself, which mitigated the lack of gastric mill and mastication, and provided protection from predators;
10. Some way of solving metabolic challenges;
11. Particular adaptations to mitigate circulatory issues (perhaps, and particularly in head-raising sauropods).

As for the phenotype question, our picture of sauropod physiological viability has been filled out, and indeed many of the same features prominent in the evolutionary narrative crop up. Investigation of phenotype and evolutionary history are coupled (Currie 2015b). Sauropods were a unique lineage, leaving often infuriatingly ambiguous hints about their lifeways and evolution for us to puzzle over. Infuriating and ambiguous, but not empirically intractable: The scientific investigations of sauropods has progressed and deepened over the last decade, and it is beholden of an account of historical reconstruction to accommodate this and similar success.

We will hear more of sauropods later, but for now we should shift to another example.

2.2 Snowball Earth

My other major case is the Neoproterozoic "snowball" episodes. This draws us further backward in time, to 550 ma at least. I first identify the *explananda*, strange features of Neoproterozoic rocks, before moving to various attempts at explanation.

2.2.1 The Strange Neoproterozoic

Glaciers are rivers of ice. A glacier forms when the spring thaw fails: one season's snow is packed atop that of the year before, compressing it. Eventually, the glacier begins to flow across the landscape. Geologists detect previous glaciations from contemporary traces. Figure 2.3 shows the Franz-Josef Glacier, which has carved its way through New Zealand's Southern Alps. Once the glacier retreats, its path through the mountains will remain as a glacial valley. Past glaciers, then, are detectable by the formations they leave. In addition to valleys, these formations include *drumlins*, small hills in large clusters; *arêtes*, sharp ridges; *alluvial planes,* formed by melting glacial ice. Glaciers also transport rocks great distances and deposit them in unusual settings. It was these glacial erratics that first led geologists to propose the recent Cenozoic ice ages.

Figure 2.3
Franz-Josef Glacier. (Photo by author.)

Glaciers from the Neoproterozoic are far too old to leave such obvious clues. However, some ancient rocks still bear subtler, but nonetheless telling indications of glacial origin. Research into signs of Neoproterozoic glaciation reveals an unsettling pattern: Traces of glacial formation are ubiquitous. This in itself suggests that glacial periods in the distant past were more extensive. As we shall see, the Neoproterozoic hosted tropical glaciers, suggestive of global super ice ages, or "snowballs." The number of glaciation events is disputed (see Peltier, Lew & Crawley 2007), but the final episode is the main target, in part because (as we will see) it could be important for understanding early animal evolution. Which geological methods grant us access to these events?

To confirm the synchronicity of global glaciation events, geologists need to clock the Neoproterozoic rocks. *Biostratigraphy* uses fossil remains to date strata. Obviously, these are scarce in the Precambrian. *Radiometric* tests date formations based on naturally occurring radioactive materials. Such materials are few and far between in Neoproterozoic formations.

Lithography is the study of brute geological appearance and its links to the past. Neoproterozoic glacial formations are topped by distinctive dolostone "caps." Dolostone is a sedimentary carbonate rock that, in this context, is indicative of warm climates. The ubiquity of the caps suggests the geological events were synchronous. Had the events occurred at different locations at different times, then the conditions for the caps' preservation would have varied. They also indicate a fast turnover from glacial to balmy (although the number of such events, and thus the frequency of turnovers, is contentious). As Hoffman and Schrag summarize, the dolostone caps "have long been considered paradoxical because they suggest that an abrupt transformation from glacial to tropical conditions took place virtually everywhere" (2002, 142).

Earth's magnetic field affects the mineral components of rock during formation. This records both shifts in that field and the location of the rock's birthplace. *Paleomagnetism* exploits this information stream. Stratigraphers are able to measure the age of rocks against shifts in Earth's magnetism and identify the latitude of the original formation. Remagnetizing can destroy the original traces, but tests have been developed that control for biases (see, for example, Sohl et al. 1999). Evans (2000) surveyed sixteen paleomagnetic studies from different sites and found that many of the deposits formed within ten degrees of the equator.

Geologists can estimate levels of biotic activity in the Neoproterozoic using radiometric carbon decay. Sediment carbon occurs in three forms: ^{16}C & ^{17}C, which are stable, and ^{13}C, which is unstable. Photosynthesis destabilizes carbon and so the amount of ^{13}C in deposits tracks biotic activity. The more ^{13}C, the more photosynthesis has occurred, and thus the more biotic activity. They found global anomalies in carbon isotopes directly before the glaciation events. Specifically, there is a swing to negative ^{13}C values directly before the glacial period: Biotic activity collapsed.

Fossil discoveries indicate that the latest snowball event occurred relatively soon before the Cambrian explosion, the (geologically) sudden rise of complex metazoans. As we shall see, this means that certainly single-celled life, and possibly simple metazoans, survived the glaciation events.

Reading Neoproterozoic rocks, we discover the following surprising features:

1. In the Neoproterozoic, there was ubiquitous, synchronous glaciation;
2. A dramatic warming occurred quickly after the glaciation;
3. Directly before the glaciation there was a dramatic collapse of biotic activity;
4. Relatively soon afterward there was a dramatic radiation of metazoan life;
5. There were multiple global glaciations in the early and late Neoproterozoic, but none since.

We know so much about the Neoproterozoic because it has left a bounty of "traces"—downstream causal descendants—which we can identify and link to the past. Chapters 3, 4, and 5 are concerned in the main with understanding this practice and its limits. Further, geologists exploit a wide range of evidence streams to reconstruct the Neoproterozoic, from stratigraphy to paleomagnetism. This feature is important for understanding the license of the methods historical scientists employ. I will make this explicit in chapter 6. Moreover, the various techniques rely on sophisticated background ("midrange") theories that link traces to the past—another crucial cog in the machinery of historical reconstruction. Let's look at some explanations of the Neoproterozoic glaciation.

2.2.2 Snowballs, Slushballs, and Wobbly Globes

Williams (1975, 2000) accounted for low-latitude glaciation by appeal to an anomalous orbital obliquity (a tilt in the angle between Earth's axes of rotation and its orbit) during the Neoproterozoic. The moon was formed by an enormous impact early in Earth's history. Williams proposes that this caused chaotic variation in Earth's obliquity, which stabilized only after the Neoproterozoic. The hypothesis paints a terrifying picture of Earth's climate during this time:

With an obliquity > 54°, mean annual temperatures are lower around the equator than at the poles, making glaciation more probable at lower latitudes. Summer isolation is so large in the polar regions (the sun staying high throughout the diurnal cycle) that surface temperatures over land areas might exceed the boiling point of water. (Hoffman & Schrag 2002, 132–135)

With this angle of obliquity the sun's energy would be directed at the poles, leading to (extremely) "tropical" poles and "polar" tropics. The correction of Earth's obliquity would explain why no further events occurred after the Neoproterozoic, as well as the equatorial glaciation. However, Hoffman and Schrag disregard Williams's theory on three grounds.

First, even when the moon's formation is taken into account, models of the early Earth do not predict an anomalous obliquity. It must be forced into the model. That is to say that although such models do not show that Williams's scenario is impossible, it does suggest that some major unknown cause is required to explain how it could have occurred. As we saw with sauropod metabolism, model-based investigations are used to provide empirical knowledge: Geophysical models do not suggest a "wobbly" Neoproterozoic Earth. I will defend this use of modeling in chapters 9 and 10. Second, the sudden shift from a cold climate to something balmier would not occur given a correction to obliquity, as such changes would happen more slowly. A more positive force is required. Third, this theory predicts a much wider spread of carbonates in Neoproterozoic rock than we see. We will return to Williams's theory in chapter 4.

A competing explanation of the glacial events during the early and late Neoproterozoic is the "Snowball Earth" hypothesis.[6] In 1992, Joseph Kirschvink suggested that the glaciers in the Neoproterozoic are indicative of a worldwide freeze. The late Neoproterozoic was a time of continental dispersal: The supercontinent Rodinia broke up, and the megacontinent Gondwana began to form (Hoffman 1999). During glacial periods most continents clustered at the middle and lower latitudes. Kirschvink suggested that these clusters would cause a feedback loop leading to a global freeze: Snowball Earth.

Both land and ice caps have a high *albedo*—they reflect more of the sun's energy than water. Landmasses in the tropics, where more sunlight hits the earth, will increase global albedo, and their warm, moist climate increases silicate weathering (the absorption of CO_2). The effect of land clustering around the tropics, then, is an increase in albedo and a decrease in greenhouse gases, which lowers the earth's temperature, particularly at the poles, where the growth of ice sheets would lead to a freezing feedback loop: "If more than about half of the Earth's surface area were to become ice covered, the albedo feedback would be unstoppable ... surface temperatures would plummet, and pack ice would quickly envelope the tropical oceans" (Hoffman & Schrag 2002, 135).

The explanation in its broad outline is captured in the figure 2.4:

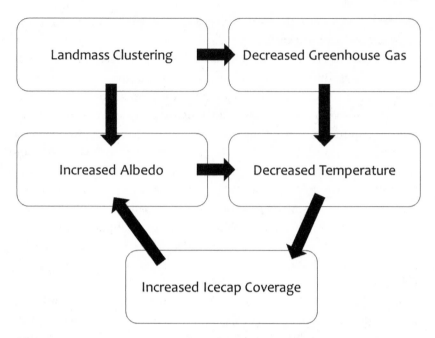

Figure 2.4
The Big Freeze.

Landmass clustering in the tropics sets off a positive feedback loop between lowering temperatures, larger icecaps, and higher albedo. Earth freezes over.

The snowball, thankfully, contains the seeds of its own destruction. The usual carbon-emitting processes (volcanic activity) would continue pumping CO_2 into the atmosphere, but the usual carbon sinks (photosynthesis and silicate weathering) would be inactive due to the freeze. An opposing loop arises: As greenhouse gases accumulate in the atmosphere, more energy is trapped, and the temperature rises. Increasing temperatures would lead to retreating icepack, and thus more liquid water, further lowering albedo (figure 2.5). The thaw would be relatively sudden, and, if global geography has shifted, the trigger for further snowballing would be removed.

This explanation accounts for the ubiquitous and simultaneous glaciation revealed by lithographic and paleomagnetic evidence (1); the sudden warming indicated by the dolostone caps (2); the collapse in biotic activity detected by carbon radiometric tests (3); and the event's rarity (because of the rarity of the trigger) (5). We still need to know whether the hypothesis

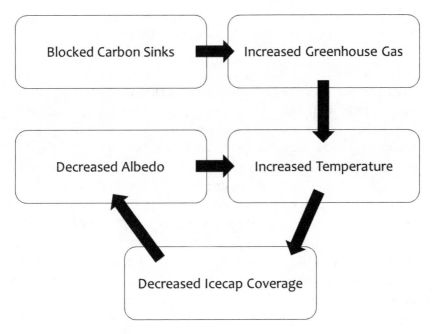

Figure 2.5
The Big Melt.

can allow for the survival of eukaryotes (4). We also might wonder whether such climatological events are likely to occur. Geologists approach these two questions using paleoclimatological models. I will discuss a few.

Hyde et al. (2000) simulate climate and ice sheet in Snowball Earth scenarios. They investigate whether snowball events could include open spaces of water, forming refuges for living organisms to ride out the freeze (I return to their model in much more detail in chapter 10). They predict ice-sheet cover based on temperature fluctuations and ice-flow models, and temperature against atmospheric fluctuations. Feeding in paleogeographical information and CO_2 concentration (as well as other factors) allows them to predict ice-sheet extent. The model is tested against various modern and Carboniferous ice ages successfully.

The simulations showed:

An ice-covered Neoproterozoic land mass is predicted by our coupled climate/ ice-sheet model with only two significant changes in boundary conditions from present values—a solar luminosity decrease consistent with estimates from solar physics, and a CO_2 level within about 50% of the present value ... an area of open water in the equatorial oceans—which could have allowed for the survival of metazoans—is consistent with the evidence for equatorial glaciation at sea level. (Hyde et al. 2000, 428)

The model reassures us that eukaryotes, even metazoans, could have survived Snowball Earth. I will expand on this in chapter 6, where we will see that the relationship between snowball episodes and the eventual radiation of these eukaryotes—the Cambrian explosion—is itself an evidential resource historical scientists exploit. As I have presented it, then, the theory ticks the geologists' boxes. However, Hoffman and Schrag are quick to admit epistemic issues: "the central difficulty with the hypothesis ... is our limited conception of a snowball event itself. With postulated conditions lying far outside familiar parameter space, there is danger that the hypothesis we question is a caricature of snowball earth" (2002, 145). In chapter 8, I will discuss the epistemic problems arising in trying to understand apparently "unique" events, or at least events for which we have poor analogues.

In addition to the importance of equatorial flood basalt emplacement (large amounts of volcanic basalt would decrease silicate weathering; see Donnadieu et al. 2004) and a methane-based greenhouse (see my discussion in chapter 10), most recent debate about Snowball Earth targets the episode's extent. Typically motivated by the biological concerns that Hyde et al. (2000) attempt to head off, some geologists argue that the earth "slushballed" rather than "snowballed." The main difference is that under snowball conditions the entirety, or near entirety, of the ancient oceans are coated in ice, whereas in slushball conditions significant amounts of open sea surround the (mostly, but not entirely) frozen continents.

Peltier, Liu, and Crawley argued in 2007 that "full" snowball conditions did not occur. In their model, decreasing temperature increases the absorption of atmospheric O_2 into the world's oceans. Increased oxygen reacts with oceanic carbon, causing a jump in atmospheric CO_2. This increase in greenhouse gases would stop an "icehouse" climate from developing. By this view, tropical glaciation occurs, but it is not accompanied by a worldwide freeze.

The "slushball" view, like the "snowball" view, is backed up by geoatmospheric simulations. Peltier, Lew, and Crawley use a coupled model of climate and carbon cycle in the Neoproterozoic. The model consists in three main parts. The first is a model of the carbon cycle—roughly speaking, those biotic and abiotic factors that are involved in creating carbon sinks, and those that release carbon into the atmosphere. The second component is a model of the "physical climate system." This is basically a model of worldwide temperature, given energy transfer from the sun, albedo, and icepack cover. The third component is a model of continental-scale glaciation. They couple the models by making the variables of one model dependent on the outputs of another. This method, investigation via modeling, receives a lot of attention in chapters 9 and 10. It is also striking that the models paleoclimatologists use and develop are basically the same as those that are used to predict the climate's future. I will draw on this in my instrumental defense of investigating the past (see chapter 12).

Whether a full "snowball" or a (still decimating) "slushball" Earth occurred is still very much a live issue. However, both pictures generally agree on Kirschvink's basic model. The difference is that, in the slushball scenario, the feedback between higher albedo, temperature decrease, and icecaps is broken by an increase in atmospheric CO_2.

So, although the basic shape of Kirschvink's suggestion appears to have held, there remains rich, and significant, space for disagreement about the extent and character of the Neoproterozoic climate. This disagreement has proven remarkably productive: Highly speculative hypotheses concerning Earth more than five hundred million years ago are concretized via simulation, thus opening up new tests. This "empirically driven" speculation is, I will argue, a central facet of historical science's success.

2.3 Moving Forward ...

Historical scientists approach different targets differently: *Obdurodon tharalkooschild*, the Neoproterozoic glaciers, and sauropod gigantism each generate a variety of investigative strategies. What unifies them, however, is the creativity and richness of investigation. As they outrun the epistemic traction granted by relatively unambiguous lines of trace evidence, such as that connecting *O. tharalkooschild*'s molar to its platypushood, much more indirect, ambiguous connections are exploited.

Throughout the book I will refer back to our snowballs and sauropods. In chapter 6, the relationship between the Neoproterozoic climate and the Cambrian explosion will illustrate the importance of exploiting dependencies between past events. In chapter 4, we will reexamine reconstructions of sauropod size and weight, delving deeper into the remarkably complete *Dreadnoughtus* specimen. Chapter 10 will examine Hyde et al.'s simulation in significantly more detail. And so on. The main point of this chapter, however, has been to get us relatively comfortable with the same set of scientific investigations and to help us appreciate the diversity and success of historical science. As will become clear, my defense of optimism about our epistemic reach into the past depends crucially on this diversity.

With the case studies anchoring us, we can now venture into more abstract, philosophical terrain.

3   Traces

Our primary window into the past is through traces: the rock, bone, and ruin that remain after time's destructive work. This destruction is lamentable, but even these fragments reveal worlds that are both alien and disquietingly familiar: the Mesozoic's giants and the Neoproterozoic's frozen oceans, for instance. Such worlds demand investigation because they are part of our story, because they are fascinating in their own right, because they have lessons to teach us about the present day and the future, and because uncovering them is challenging. I am interested in the *challenge*. How deeply can we cast our eyes into past worlds? How do we rebuild ruin as temple, clothe bone with sinew and skin, and see riverbeds, mountains, or glaciers within the rock?

At a first pass, our capacity to uncover the past relies on the availability or otherwise of *traces*, downstream causal descendants. After all, how the world *is* depends upon how the world *was*, so understanding how the contemporary world could come to be provides access to the past. In this chapter I analyze the notion of "trace" and provide an account of the epistemic work traces do. This is important for understanding the methodology of historical science; after all, most accounts emphasize the role of traces, though little attention has been paid to what traces are.

I will start by articulating two accounts implied in the literature before stipulating my own. I prefer an "evidential" story, that is, I take "trace" to be an epistemic concept (as opposed to a strictly ontological one). With that notion in place, I will describe the nature of their evidential license. Specifically, I will provide an account of *minimal dependence* and build on it to explain the epistemic warrant of trace-based reconstruction. I will finish by illustrating this framework in reference to *Obdurodon tharalkooschild*.

This chapter matters for three tasks in particular. First, in chapter 5, I will construct what I call the ripple model of evidence. This, in combination with the dependencies discussed in this chapter, allows me to set my target: the "unlucky circumstances" historical scientists face when trace evidence becomes thin and ambiguous. Explaining why historical science succeeds in the face of that epistemic situation underwrites my argument for optimism. Second, in chapter 6 I argue that methodological accounts of historical science have placed far too much emphasis on traces, that dependency relationships between events in the past matter as well as dependency relationships between the past and now. This is especially so in unlucky circumstances. Third, in much of the book (chapters 6 through 10), I argue that historical hypotheses gain support from non-trace sources.

To see this, we had better first understand what a trace is. Let's turn to that now.

3.1   The Notion of a "Trace"

Over 500 million years ago, the tropics housed glaciers, which melted in a sudden climatic reversal. The geological formations of the present day bear witness to these events. Rocks that formed during the Neoproterozoic have glacial lithology and distinctive, dolostone caps. The ubiquity of the caps suggests that geological events were synchronous, and the layer's thinness suggests a fast turnover from glacial to balmy. Moreover, paleomagnetic studies can map the magnetic properties of contemporary rocks to shifts in the ancient planet's magnetic field, allowing us to infer the positions of the rocks during formation. These rocks, it turns out, were born in the tropics.

Traces test hypotheses about the past. As we have seen, Williams (1975, 2000) argued that those distinctive Neoproterozoic formations could be explained by appeal to geophysical anomalies. Specifically, if the angle of Earth's orbital obliquity were irregular, then the sun's energy would be directed at the poles rather than the equator. This would turn the familiar arrangement of cold polar regions and hot equatorial regions on its head. The enormous impact that formed the moon is cited as a possible cause for the "wobble." An important mark against this hypothesis is its failure to account for contemporary mineral deposits. If equatorial glaciation was caused by an anomalous obliquity in Earth's orbit, the carbonates would have been widely dispersed, rather than forming caps. The clustering of carbonates in Neoproterozoic remains undermines a geophysical explanation of the glaciation.

Figure 3.1 allows us to visualize the traces of those long-ago glaciations. It represents modern geological features (dolostone caps, warm lithography, tropical latitude paleomagnetism) as "traces" of the Neoproterozoic; they bear some relation to past events (tropic glaciation coupled with rapid climate change) by virtue of which they are revelatory of them. Scientists come up with hypotheses to account for these Neoproterozoic traces, and the traces in turn test those hypotheses. Williams's account fails at least in part because it gets the traces wrong—if his hypothesis were true, we would see a different carbonate distribution than we do. An account of "trace" explains what type of relation contemporary structures have to past structures, by virtue of which such evidential roles can be played. That is, an account of traces tells us what the arrows in figure 3.1 signify. Philosophical analysis of this concept has been minimal. I will discuss two possible approaches before shifting to my own analysis. As we will see, both

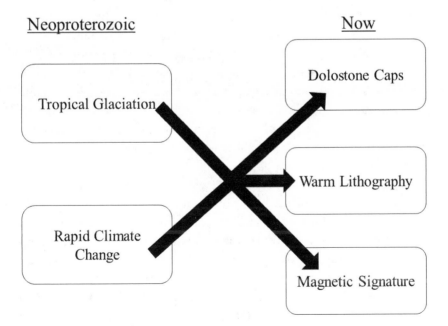

Figure 3.1
The traces of Snowball Earth.

accounts (to some extent) require metaphysical commitments to notions of "causality" or "information," and I consider it a virtue of my account that we need not make particular metaphysical commitments but rather rely on whatever notions are in scientific play.

One way of understanding traces is via causation. There are causal relationships between the tropical Neoproterozoic glaciers and the features of current rocks in Neoproterozoic strata (their "warm" lithography and "tropical" paleomagnetism). A causal account of traces, then, identifies a contemporary structure as a trace of some past event by virtue of the contemporary structure being causally downstream of the past event. In short, x is a trace of past event y by virtue of y causing x. On this view, "tracehood" is a causal relationship between a contemporary structure and a past event. Carol Cleland sometimes talks about traces in this kind of way: "an investigator observes puzzling traces (effects) of long-past events" (2002, 480).

By this account, the arrows in figure 3.1 indicate relationships of cause and effect between events in the past and contemporary phenomena. This is a schematic for a cluster of possible views, generated by cashing out different notions of "cause." And philosophy provides a smorgasbord: Woodward-style interventionist accounts, Lewisian counterfactual accounts, various species of Salmon-style accounts, and so forth.[1] Do we, for instance, consider causation is a special kind of counterfactual relationship? Or does it depend on certain kinds of energy transferal between objects? Specifying a causal account of "trace" faces tough questions. For instance, although correlation is not causation, correlations could play a role in uncovering the past and so—presumably—should be included in the notion of "trace."

Alternatively, we could conceive of traces informationally. Via processes of stratification, information about the location of Neoproterozoic glaciers is encoded in geological formations. Some contemporary structure x is a trace of some past event y just in case x contains information about y. Similarly to the causal case, the specifics of such accounts depend crucially on what we take information to be. I will illustrate this with two examples. We start with Turner (2007, 18–20), who appears attracted to an informational view. He uses Horwich's (1987) discussion of the time-asymmetry of knowledge.

Horwich envisions an ideal "recording system." A recording system can be in a series of states, and which state the system is in depends on external conditions. A thermometer's state, for instance, depends on external temperature. Ideal recording systems can be in a set of mutually exclusive states $(S_1, S_2, S_3, \ldots S_n)$, which each map uniquely onto a member of a set of external conditions $(E_1, E_2, E_3, \ldots E_n)$. When a particular external condition holds, say E_1, then the recording system will be in a corresponding state, say S_1. Under those conditions, we could say that S_1 in the recording system encodes information about E_1. Stratification is an imperfect recording system. The formation of dolostone caps (S_1) encodes information about fast climatological changes in the past (E_1) by virtue of those conditions correlating with the presence of particular geological features. On this kind of account, the arrows in figure 3.1 indicate *robust coding*, the retention of information through time.

Peter Kosso (2001) also thinks of traces in informational terms. He bases his account of historical evidence on "the flow of information from the object of observation to the observer" (41), drawing on Dretske's (1981) analysis of information. Dretske was analyzing the semantic content of signals. His starting point is to reject largely statistical accounts of information on the grounds that they track only the *amount* of information, as opposed to what the information is *about*. Merely knowing that I have an ideal recording system, for instance, only tells me that I have a lot of information about the past—a large quantity. I need to be able to understand which external states stand in relation to which states of the recording system in order to "translate" said information. I might know that the properties of geological formations map onto past states in a relatively fine-grained way, but unless I can work out which states matter, and for which past states, this is not particularly helpful. Dretske overcomes this problem by understanding information as a two-place relation: It turns not only on states of the world, but also on states of the knower: "A signal r carries the information that s is F if and only if the conditional probability of s's being F, given r (and k), is 1 (but given k alone, less than 1)" (57).

Let's unpack this a little. r denotes the signal—say, the paleomagnetic properties of a modern rock that formed in the Neoproterozoic. The possible content is "s if F": Say, "This rock formed in the tropics." In order for r to contain the relevant information, it must be that r, in combination with k—the knowledge of the observer in question—guarantees the truth of that content. In other words, if I know the paleomagnetic properties of the rock in question and I know the connection between those paleomagnetic properties and the location where it formed, then it can only be the case that the rock formed where its paleomagnetism indicates. Notice that although Dretske's account is relative to the knowledge of the subject, it is nonetheless highly stringent. Nomic necessity is required (this is for various technical reasons).

Although both Turner and Kosso think of traces in terms of information, then, their approaches come apart depending on the details of Dretske and Horwich's accounts of encoding. Where Horwich's is largely statistical, and moreover does not require nomic necessity, Dretske's is knower-relative and requires nomic necessity. So, different judgments about what counts as a trace will be made by each view. On Turner's approach, the right mapping between states is required. Insofar as Kosso follows Dretske, those mappings must be nomically necessary, and we must stand in a knowledge relation to them. What we take information to be matters.

Causal and informational accounts of trace are related. After all, it is surely by virtue of causal processes that encoding takes place, and presumably discovering facts about causal processes allows us to decode traces. They do, however, come apart. For instance, causal processes do not necessarily retain information. Calgary's C-train runs past the university either from Tuscany in the northwest or the central city in the south. Starting from either direction, the train will pass the university. If a passenger is walking out of the university train station, that information alone does not tell us whether they came from the northwest or the south. No matter what the starting location, they end up in the same place (this is a case of "convergence," which I discuss in much more detail in chapter 8). Although it looks like information is not encoded, there is a causal relation between passengers getting on the train at whichever station they left from, and their arrival at the University stop. Causes, then, are not necessarily information-preserving, and so causal and informational accounts are not equivalent. However, my purpose here is not to adjudicate between such views. First, I doubt there is an objective answer to whether a causal or coding view of trace is right. I'm not even sure what it would mean to have a right answer to this question (which is to say, I'm not sure what the success conditions would look like). Second, I am interested in an account of traces that is useful for understanding their evidential role, and so I want to move on to a view that is tailored to that task.

Recall that we are interested in understanding the epistemic situations historical scientists face. It will be useful, then, to have an account of trace that reflects its evidential function. I suspect that some versions of the causal or informational approaches could be made to work here, but I would rather build the epistemic status of traces directly into the account. Historical scientists have limited access to traces, and some evidence about the past can be better or worse. Presumably there is a lot of information about the past encoded in the present, but our incapacity to decode this undermines their evidential worth. In short, I want to include the *evidential relevance* of traces into my story of what it takes to be a trace. A thought experiment provides a useful backdrop.

Imagine that a Hadrosaur (a class of large, bipedal, sometimes duck-billed dinosaurs), let's say a *Corythosaurus*, has been separated from its herd by a flock of *Albertosaurus* (carnivorous, *T. rex*-like theropods). They chase her across a beach, where the predators give up at the sand's edge. Over time, the tracks of *Corythosaurus* are preserved as the beach is transformed to sandstone. These trackways are eventually discovered by paleontologists, who exploit them as inroads to *Corythosaurus* behavior and physiology. The scientists do not, however, see the trackways as traces of *Albertosaurus*—how could they? However, on both the causal, and some informational, accounts, the trackways are traces of *Albertosaurus*. After all, if the *Albertosaurus* had not given chase, the traces would not be present; they are in part causally responsible and thus (at least potentially) have left an informational path. The scientists are right not to treat the trackways as relevant to *Albertosaurus*. After all, there is nothing about the tracks that suggests the flight behavior indicative of a predator's presence. If we want "trace" to reflect the epistemic situation historical scientists face, we ought to take the trackways as a trace of *Corythosaurus*, but not *Albertosaurus*. Certainly by the causal account, and possibly by the informational account, the footprints are traces of both. However, if we want "trace" to reflect scientists' epistemic situation, that is not the answer we want.

By virtue of what are the trackways evidence of *Corythosaurus* but not *Albertosaurus*? It's not that there isn't information encoded in the trackways about *Albertosaurus*, or that there isn't a causal relationship between them. Rather, scientists lack the *background knowledge* required to make the requisite connection. Recall my discussion of Neoproterozoic traces. Theories of lithology and paleomagnetism, in combination with the traces, told us about the geographical location of the Neoproterozoic rocks during formation. It is by virtue of such theories that the rocks count as evidence. Turning back to the *Corythosaurus*, consider ichnology. This is a body of theory explaining how nonbody remains such as burrows, coprolites, and footprints are preserved in sedimentary rock. It provides a mechanistic or procedural account of how, under the right conditions, *Corythosaurus* footprints can be preserved. The evidential relevance of the trackways to the *Corythosaurus* relies on theories of ichnology. Peter Kosso, following Lewis Binford (1977), calls these "middle-range theories," explaining: "The material remains are meaningful evidence ... only with an understanding of the flow of information from the past to the present ... middle-range theories give meaning and relevance, that is, evidential relevance, to the material remains" (Kosso 2001, 52).

These "midrange" theories, then, track the passage of time from present traces to the past. The *Corythosaurus* trackways count as traces of *Corythosaurus*, but not *Albertosaurus*, because they meet two conditions. First, the material remains are downstream of the *Corythosaurus*. Second, they are evidentially relevant by virtue of midrange theory. This is the essence of my account of trace:

Some contemporary phenomenon, *x*, is a trace of some past state of affairs, *y*, if and only if *x* is downstream of *y*, and *x* is evidence for *y* according to some justified midrange theory.

From here on in, unless otherwise specified, this is what I mean by "trace." There are two ways for a contemporary phenomenon to fail to be a trace. First, it might not be *downstream* in the relevant sense. By "downstream" I do not simply refer to temporal ordering, but to a "descendent/ ancestor" relationship. That is, the two are connected by some historical "line." Although I turn up later in time than my mother and David Bowie, I am downstream of the former in a way that I am not of the latter. For (as far as I know) my mother did, and Bowie did not, play a direct causal role in my coming to be. In chapter 2, I mentioned that Hyde et al. (2000) tested their Neoproterozoic climate model by using it to predict the extent of Pleistocene ice ages. Although these events are more recent than the Neoproterozoic, they are also not "downstream" in the relevant respect. The Pleistocene ice ages are not (prima facie) descendants of Snowball Earth, because they lack the relevant dependencies. Counterfactual changes to how the Neoproterozoic went will not make a difference to how the climate behaved in the Pleistocene. Or if they would (and surely some changes to the Neoproterozoic would make a difference to the Pleistocene!) that is not relevant for Hyde et al.'s evidential use of Pleistocene events. This distinction will matter, and will be clarified, in chapter 7.

Second, we might lack justified midrange theory. In the absence of some theory linking the presence of *Albertosaurus* to the footprints, they are not *Albertosaurus* traces. If it was discovered that (say) *Corythosaurus* only crossed beaches when chased by *Albertosaurus*, then there would be midrange theory linking remains to predator, and thus the trackways would be a trace of *Albertosaurus*. The Neoproterozoic geological formations are evidence of past events by virtue of geological background theories such as stratification, paleomagnetism, and lithology. These together explain how past states of affairs could lead to those formations. My account makes the epistemic accessibility of traces explicit: In order to *be* a trace, a contemporary phenomenon must be linked to the past by a justified midrange theory.

By this account, there are two routes to uncovering traces. We can *discover* traces by making new finds that, by our midrange theories, tell us about the past. Or, we can do more with the remains we have. That is, we can *refine* midrange theory: New traces are generated by the application of new technology and theories.[2]

The appeal to *justified* midrange theories serves to limit what counts as a trace. Not just any story linking contemporary states of affairs to the past will do. What is it for a midrange theory to be justified? Although I cannot give a full account here, the idea is pretty straightforward. Midrange theories are empirical claims about how contemporary structures are formed and shaped by past events and processes. As such, they are justified based on the evidence we have for them, and (depending on one's account of justification) what use they are being put to. Ben Jeffares (2008) describes how experimental work on analogous critters can be used to generate and support theories that distinguish, for instance, marks made by humans from those made by animals. Robert Chapman and Alison Wylie (2016, chapter 2) describe the extensive development and calibration required for systematic, successful archaeological fieldwork. Further on in this book, we will meet similar processes of calibration in, for instance, the use of LIDAR technologies in Mesoamerican archaeology. Suffice to say, a midrange theory is justified to the extent that it is underwritten by firm empirical support.

By my account, then, the arrows in figure 3.1 are evidential and theory-relative. This doesn't commit us to any particular view about the nature of causation or information. My thought is that whatever concepts are implicitly or explicitly appealed to in midrange theory are the ones in play. This is not to say that causation—or information, for that matter—is not an important issue in understanding historical science. Indeed, later in this chapter I shall develop an account of "dependencies" that plays this kind of metaphysical role. Rather, my claim is that in understanding the nature of historical reasoning, we shouldn't be beholden to *particular accounts* of causation and information—if anything, the relationship should run the other way. It also takes accessibility seriously in a way amenable to philosophical investigation of the historical sciences as they are practiced; a contemporary observation counts as a trace when justified midrange theory tells us this is so. The *Corythosaurus* tracks are traces by virtue of ichnology; without midrange theory linking them, the footprints are not *Albertosaurus* traces. Although, like Kosso's, this account is relative to the epistemic states of the scientists in question, these states are about evidence, rather than knowledge, and are not tied to Dretske's rather stringent account of information. There may be accounts of information that are theory or agent-relative that do the same work as my epistemic approach. These would depart further from causal accounts, and would perhaps be suitable. I will leave identifying and evaluating such views for another day.

You might feel disquiet about the theory dependence of my account. It seems odd to claim that until ichnology was articulated preserved footprints were not traces of the animal that left them. And strange indeed to say they "became" a trace once the theory was articulated. Moreover, when midrange theory is abandoned, we would not say, "this phenomenon *was* a trace, but has *ceased* to be so"; rather, surely, we would say, "we were *wrong* that this phenomenon was a trace." In short, "trace" looks like an ontological success term: It refers to objective relations between objects in a non-theory relative way. When I say that fossilized trackways are traces of extinct organisms, I am making a claim about the world that is true or false irrelevant of which midrange theories I happen to have. You might insist that the structure of the world decides what is and what isn't a trace, not our theories.

Fair enough. Happily, I don't think too much rides on this. Although I will speak of traces in a theory-dependent fashion, one could easily translate my term "trace" to talk of "*potential* trace" or better "*accessible* trace." Moreover, although the account is theory-dependent, the world is certainly not left out of the equation. Justification, in my view, depends at least in part on the relationship between theory and world. And so, "justified" midrange theories will often be *true* midrange theories. A final point: I am not attempting a traditional conceptual analysis of "trace" here. I doubt the "folk" (or scientific) conception of "trace" is particularly robust, and I doubt that it deserves much philosophical respect. My definition is pragmatic and tracks an important feature of historical scientific evidence and reasoning. When scientists think about traces, they do so in terms of their epistemic prowess: the theories they have linking contemporary remains to the past. Moreover, as we will see, this account does not depart from my interlocutors' use in damaging ways.

Okay, so we have an account of "trace." But what can traces do for us, epistemically speaking? In the next section, I will continue the abstract tenor of the discussion to outline a notion of "dependence" that answers this.

3.2 Dependency Relationships

Much of this book is about "evidence." And evidence, like all central scientific notions, is a tricky notion philosophically. However, my focus is not on understanding the ultimate nature of evidence, or providing an account of what it takes to be evidence. Thus, the conceptual machinery I co-opt and construct is justified on instrumental grounds. In this section, I will explain some machinery (see Currie 2016b) that underlies discussion throughout the rest of the book.

In the last section, we saw that traces are contemporary phenomena that are downstream of past phenomena and count as evidence by virtue of justified midrange theories. But what do midrange theories pick out—how do they do their work? They pick out what I will call *dependency relationships*. My aim is to provide a minimal notion of dependency, and then build on this to understand different *kinds* of dependency relationships. This will matter in chapter 5 when I articulate a notion of "unlucky circumstance," and it will give us a framework within which to understand the epistemic warrant granted by midrange theory.

Midrange theory captures what is going on at an epistemic level in trace-based reconstructions of the past. But what is happening ontologically? In the last section, I avoided committing to notions of information or causation, but here I want to provide an ontological, probabilistic notion of "dependence." The basic thought is this: How the contemporary world is depends in various ways on the way the world was; and the way things were is restricted by the ways the world is. There are *dependencies* between the present and the past that midrange theories allow us to exploit. If there had been no glaciation in the Neoproterozoic tropics, the chances of there being glacial lithology in the relevant strata now are very low. Moreover, if there were not abundant rocks with glacial lithology nowadays, it would be less likely that the Neoproterozoic glaciations would have occurred— some geological processes would have to have wiped the record of that time.

Here is a thin notion of dependence:

Some variable, v_1, is *minimally dependent* on another variable, v_2 just when v_2 taking a particular value, or range of values, effects the probability of v_1 taking a particular value, or range of values.[3]

At this point, I will restrict dependence to holding between *present* and *past* variables. In chapters 6, 7, and 9 I will argue that this trace-centrism, which marks most accounts of historical evidence, is a mistake. For now, I will focus on dependencies of this stripe. Recall my discussion in chapter 2 of the difficulties in establishing the length of the biggest sauropods: Incomplete remains necessitated reconstructions whose results depended crucially on our choice of contrasts. Lacovara et al.'s (2014) discovery of *Dreadnoughtus schrani*, an enormous titanosaur, is exciting not because of the new lineage's size alone (*Argentinosaurus* is larger) but because of the completeness of the specimen. Where most titanosaurs are known only by fragments, approximately 70 percent of Lacovara et al.'s specimen (excluding the cranium) is available. They estimate *D. schrani* at 26 meters long and just under 60 tons. One remarkable aspect of the find is that the animal had not stopped growing—it was not "osteologically mature at death" (6)! Let's illustrate the notion of minimal dependence by looking at one feature of the find that suggests osteological immaturity.

In Lacovara et al.'s specimen the scapula (shoulder blade) and coracoid (a further structure in the shoulder of all vertebrates except for placental and marsupial mammals) have not fused. Schwartz et al. (2007) describe this feature in another remarkable sauropod find, this time a juvenile diplodocus under 200 cm long. Homologous structures in reptiles and birds also indicate that unfused scapula indicates osteological immaturity. Consider two variables. The first is located in Lacovara et al.'s specimen: the fusing or otherwise of the scapula and coracoid. The second is a feature of the *D. schrani*, dead 77 million years, whose remains we are concerned with: its osteological maturity. Consider the counterfactual situation wherein the specimen's scapula is fused. Presumably, this would affect the likelihood of whether the long-dead organism was still growing when it met its maker. "Fusing," the present variable, then, is minimally dependent on the historical variable "osteological maturity." Next, consider the counterfactual situation wherein our *D. schrani* was osteologically mature: It is likely that her remains would have fused scapula and coracoid—and so dependency runs the other way as well. This is an example of the dependencies that underlie such reconstructions of the past.[4] Midrange theory works by describing these dependencies. Another example is the length of *D. schrani*'s remains, say its femur, and the organism's overall length. Had the animal 77 million years ago been larger or smaller, surely the length of its remnants would also be different—and vice versa.

The relationship between the past and the present is not always simple. In their journey from the organic to the inorganic, bones are transformed, scattered, and broken. Reconstruction is very rarely as simple as measuring a clearly identifiable piece of anatomy and then arranging it with others to create a complete specimen. The inferential chain between a description of the remains and a hypothesis about the past organism, then, can be very long. I mentioned that the significance of the *D. schrani* specimen was its relatively completeness. The length of *Argentinosaurus* has had a variety of estimates from somewhat below 30 to nearing 40 meters (Paul 1992; Sellers et al. 2013). A major contributor to this discrepancy is incompleteness. As we saw in the last chapter, the lineage is known from a single fibula and six pieces of vertebrae. Although these are truly colossal (recall the 1.55-meter-long fibula!), the dependencies between the length of a fibula and the length of the total animal are not sufficiently known to get a good estimate without a wide margin of error. Such dependencies undoubtedly exist, of course. We simply do not have the requisite background theory, or sufficient remains. Midrange theory operates by picking out dependency relations between the past and the present. Our capacity to exploit traces relies on this.

By considering different *kinds* of dependency, we can begin to explain how and when traces underwrite rich knowledge about the past. I like to think of dependency relationships along three dimensions. First, *embeddedness*, or how many dependent variables there are between entities. Second, their *informativeness*, or how specific the information potentially provided is. Third, their *strength*, or how much support they potentially provide to hypotheses. Let's look at each in turn. The first notion is relatively straightforward; the second two require a bit more machinery.

Current phenomena and past states of affairs can have different numbers of dependent variables—be more or less *embedded*—and this matters epistemically. Consider the dependency between the length of the fibia of Lacovara et al.'s specimen and the past animal's size. These, of course, are minimally dependent. But the length of the fibia also has mutual dependencies with the animal's weight. As we have seen, a common method for estimating the weight of sauropods is via an examination of gross morphology. Chances are, then, the length of the fibula is also minimally dependent on *D. schrani*'s bodyweight. Fibula length in the specimen, then, is embedded with *D. schrani*'s properties insofar as at least two properties of living *Dreadnoughtus* are dependent on those properties of the specimen. This matters because information about the trace will thus potentially be more informative of the target; information about fibula length will likely translate into better reconstructions of both body length *and* weight. Moreover, the relationships can be tested against one another (this will be particularly important in chapter 6 when we consider relationships between events, entities and processes in the past). I will explore how evolutionary and other historical processes promote embeddedness in chapter 8.

Dependencies between traces and the past can carry more or less information. One way of capturing this notion draws on Woodward's (2010) notion of *causal specificity*, which itself is closely related to Lewis's "influence" (2000). Causal specificity tracks how systematic the relationship between two variables is. The easiest way to get the idea is to consider a *maximally specific* dependency. Recall Horwich's ideal recording system. Such a system can be in a set of states, each state mapping onto a discrete external state. So, for system state S_1 there is a corresponding external state E_1, for S_2 there is E_2, and so on. Here, Horwich has described a maximally specific dependency. There is a one-one mapping between external states and states of the recording system. Less specific relationships will diverge from this ideal. For instance, the recording system might be in S_1 for either E_1 or E_2 (but be in other states, say S_3, for corresponding state E_3). In these circumstances, learning that the recording system is in S_1 is informative: it tells us that the external state was not E_3; however, it does not discriminate between E_1 and E_2. Moreover, a highly specific relationship not only will have relatively strict mappings, but there will also be many fine-grained mappings between external and system states.

Another way of understanding the difference between more or less specific dependencies is to contrast dials from switches. My old stereo had an on/off switch and a volume dial. Let's idealize somewhat, pretend that the stereo was always supplied with power, and ignore the relationship between the switch and the dial. Learning the on/off switch's position tells us something about the volume of the stereo. If the switch is off, we know the stereo volume is at zero. The switch being on, however, corresponds with a large number of possible stereo volumes. If we took the switch's position as a signal from a recording system, it would have two states—"off" and "on." "Off" would map onto a single external state—"volume: 0"—while "on" would map onto a wide range of states—the range of volume that the stereo is capable of. By contrast, learning about the position of the volume dial tells you much more about stereo volume, since there is a mapping between positions on the dial and how loud the stereo is.

Specificity, then, is a measure of redundancy; highly redundant dependencies are like switches. There is a variety of possible states in the dependent variable given a state in the other. Highly specific dependencies, "dial-like" dependencies, allow less wiggle room: A state in one variable will confine states in the other to a narrower range. The specificity of a dependency relation can be understood as how closely it approaches Horwich's ideal recording system.

The specificity of a dependency matters epistemically. Compare the relationship between the fibula length of the *Argentinosaurus* specimen and features of living *Argentinosaurus*. It seems plausible that the dependency between overall body length and fibula length is more specific than that between fibula length and body weight. This is because body weight is also dependent on a host of other factors: as we saw in chapter 2, the overall weight of a sauropod is a function not simply of its size but also of its composition. If, for instance, sauropods have highly pneumatized skeletons and birdlike "flow-through" respiratory systems, then they will be significantly lighter, proportionately speaking, than mammals. Chances are, then, that changes in fibula length will map onto changes in body size more so than they will changes in body weight. Specificity matters because it determines how informative a trace might be about the past. If my hunch about the specificity of the fibula is right, then getting a better idea of fibula length in *Argentinosaurus* (by, for instance, finding more specimens) will be more informative of animal size than weight. Switching examples, on the face of it, discovering that *D. schrani*'s scapula is disconnected gives us good reason to think that its long-dead owner was still growing. However, it doesn't obviously tell us at what speed, or for how much longer she would have grown. In this sense, information about the scapula is like my stereo's on/off switch: it provides relatively unspecific information.

Finally, dependencies can come in different *strengths*. Minimal dependence requires but a miniscule statistical relationship between variables, but such relationships can be more significant. In addition to the number of dependent variables, and how informative the dependencies are, dependencies can increase the probability of a variable taking a value, or range of values, to a greater or lesser extent. Imagine that sauropod size depends upon the nutritional content of Mesozoic flora, but that this dependency is relatively uninformative (that is, low in specificity) and unembedded (that is, it doesn't make much of a difference to other sauropod properties). One way such a relationship could be instantiated would be if enormous body sizes required a certain lower limit of nutritional content for viability, but that after that threshold is reached, further changes make little difference. In such circumstances, learning that sauropods are enormous will give us reason to believe that Mesozoic flora had a nutritional content above a certain range, but no information about how *much* above that range. In some low-specificity circumstances, strength can be high. That is, although the dependency only ensures the nutritional content falls above a limit—it makes *damn sure* that it does. In other circumstances, there could be a variety of ways that sauropods could manage such sizes, and some of those do not involve highly nutritious flora. Strength, then, is a probabilistic measure of the dependency.

"Probability" is, of course, a tricky term (we will return to this in my discussion of Cleland in the next chapter). Do I mean our epistemic states, our "credence," or do I mean some objective "chancy" sense of probability? Neither. Aidan Lyon (2011) presents a non-chancy, yet objective, notion of probability that tracks what I have in mind here.

Lyon is concerned with scientific theories that appear to be true in deterministic worlds (so are not "chancy"), but nonetheless posit objective probabilities (which therefore cannot be understood as subjective "credence"), citing classical statistical mechanics and evolutionary theory as examples. In both cases, the theories produce predictions and explanations of their target's behaviors (the behavior of gases for the former, changes in gene frequencies in populations for the latter) that are (apparently irreducibly) probabilistic. Lyon's solution to this problem is to articulate a class of probabilities whose role is "conveying a certain type of counterfactual information in explanations" (427). The counterfactual information in question is what Jackson and Pettit (1992) call "modally comparative" information; that is, we are interested in unifying the actual world with relevantly similar worlds. We could understand estimates of *Argentinosaurus* size as explanations of the physical properties of fossil finds. We could envision the explanation in modal terms: Were a 35-meter sauropod's fibula to fossilize, its remains would match the description of the specimen. Such explanations are in the business of maximizing the number of possible circumstances described. In the extreme circumstance, the explanation is necessary; that is, it captures all possible relevant circumstances. Imagine that the production of a 1.55m fibula was possible via sauropods of varying lengths, say 30–35 meters. In that case, the explanation would be relatively weak. In other cases, the modal scope will be smaller, at minimal *one* possible way of producing a 1.55m meter fibula is via an organism of, say, 33 meters' length. As Lyon puts it,

Sometimes we use probabilities to express modally comparative information about a system, which is a certain kind of counterfactual information about the system. Such probabilities are objective since they express objective facts about the system in question. ... I call these probabilities *counterfactual probabilities*. Counterfactual probability is not probability in some counterfactual situation: rather it is a measure of how *robust* a proposition is under a class of counterfactual situations. (429)

Naturally, I am interested primarily in our capacity to draw on traces to reconstruct the past: an evidential rather than an explanatory task.

However, I think Lyon's conception of counterfactual probability applies here. First, I am not sure how clean the distinction between reconstruction and explanation is; after all, one way of conceptualizing reconstruction is as *explaining traces*. Second, and more important, there are two senses of probability at work in reconstructing the past. There is our *credence*, that is, how confident we are in our background theories and our evidence. But there is also the *counterfactual probability*, the robustness of the dependencies between traces and past states of affairs. Naturally, these two senses of probability are related—for instance, presumably, under the right circumstances the higher the counterfactual probability, the higher our credence in hypotheses relying on that probability ought to be.[5]

Consider (just once more!) the relationship between the detached scapula of the *D. schrani* specimen and the proposition that the organism was not fully grown at death. If the counterfactual probability is high—that is, if the dependency is highly robust—then having a detached scapula basically guarantees a still-growing animal. The "detached scapula" worlds are almost all "osteologically immature." If the counterfactual probability is low, then only *some* detached scapula worlds are osteologically immature. Perhaps in some worlds the detachment is due to, say, pathology. The *strength*, or level of counterfactual probability, of a dependence relation matters epistemically. Compare a weak dependence to a strong dependence. If the relationship between detached scapula and osteological maturity is weak, then there are many other live options to be considered. If it is strong, then there are fewer. Strong dependencies (when picked out by justified midrange theory) grant epistemic confidence.

By considering how enmeshed, informative, and strong the dependencies between a trace and the past are, we can get a handle on how much work a trace can do for us. Let's illustrate that now.

3.3   *O. tharalkooschild* Redux

As we saw in the introduction, Pian et al. (2013) discovered a new species of extinct platypus, *Obdurodon tharalkooschild*. They established that (1) it was a kind of platypus, (2) it was a new species, (3) it was around a meter long. They also speculated (4) about its ecological role and (5) about its evolutionary history. And all this on the basis of a single molar.

In this section I will provide a taste of how this chapter's machinery operates by explaining how one molar could be so revelatory of *O. tharalkooschild*. This counts as an articulation of the "lucky" epistemic situation paleontologists reconstructing Platyzilla found themselves in. First, though, it is worth stating the role of the machinery in the book overall.

My account of "trace," and dependency relations, will feature prominently in chapter 5 when I describe the ripple model of evidence. One thing to notice about my discussion of traces is that it suggests a kind of "evidential atomism." That is, we examine the evidential capacity of traces by breaking them down along various dimensions. As we shall see in chapters 5 and 6, this evidential atomism is misleading when we consider many historical hypotheses. This is because the relationships between evidence—their consilience and their coherence—plays an important role in supporting historical reconstruction.

A related property is my discussion's *trace-centrism*. As we shall see in chapter 6, most accounts of evidence in historical science emphasize dependency relationships of one type—that between now and then. However, I argue that dependency relationships *between past variables* are extremely important, particularly in "unlucky" circumstances. And as we shall see in chapters 7 and 9, there is evidence—surrogative evidence—which outruns traces.

But for now, let's return to the question that opened this book. How are paleontologists able to draw such a rich picture of "Platyzilla," *Obdurodon tharalkooschild*, from the remains of a single tooth? And what is different between this case and more difficult circumstances, such as reconstructing the history of sauropod gigantism?

Traces are not all equal. Highly embedded, informative, and strong dependencies make for robust reconstructions. This is just what we see with mammalian teeth.

Obviously the *O. tharalkooschild* molar is a trace of the animal. It is downstream of a real, living Platyzilla. Midrange theories of taphonomy provide a sequential explanation of how a bone tooth could trade organic material for inorganic simulacra. By virtue of these facts, the molar is evidentially relevant to the extinct lineage.

Now consider how enmeshed the trace and the extinct animal are. The molar's morphology has dependency relations with *O. tharalkooschild*'s being a platypus, being a new type of platypus, and ecological niche. The upshot of this is that increased information about tooth morphology will yield rich results about a variety of propositions concerning Platyzilla.

The molar is also highly informative. The ratio of tooth size to overall size in mammals is highly stable. There is a fairly specific mapping, then, between the size of a tooth and the gross body size of the tooth's bearer. As tooth size changes, proportional changes are likely in body size and vice versa. This allows us to estimate *O. tharalkooschild*'s size to a relatively fine grain.

Moreover, the dependencies between the molar and facts about *O. tharalkooschild* are strong. For instance, paleontologists take the symmetry between talonid and trigonid to unambiguously indicate membership of the platypus clan. This one feature, so far as paleontologists are concerned, guarantees platypushood. This relies on the ideas that (1) only Ornithorhynchids have been observed with such dental features; (2) the possibility of convergence on this highly specific feature is unlikely. Talonid-trigonid symmetry is a *diagnostic character*: Rich background knowledge about its evolvability coupled with empirical data underwrite the inference. Part of paleontological confidence in Platyzilla's taxonomic affiliation is not simply that their credence in the background theory linking symmetry to platypushood is high. It is also that the theory is unambiguous. There is (in effect) only one way to have talonid and trigonid symmetry: Be a platypus.

My account of traces and dependencies lets us begin to articulate the sense in which mammalian paleontologists are lucky. First, we have a rich, well-supported body of theory about that fossilization process, and about the dependencies between teeth and other mammalian properties. Second, those dependencies are enmeshed, informative, and strong. Third, paleontologists are also lucky because teeth, by comparison with other anatomical features at least, are likely to be fossilized. This final feature will be captured in the ripple model of evidence in chapter 5.

However, in other circumstances historical scientists are not so lucky. Uncovering evolutionary facts about *O. tharalkooschild* is difficult—for instance, was its size increase an example of a driven or a passive trend? Questions about sauropods are difficult, too: It is unclear whether they were endothermic or ectothermic, or what the evolutionary function of their long necks was. Figuring out whether the Neoproterozoic freezing events were snowballs or slushballs is another difficult task. There is no magic bullet for these questions, no one trace or investigation is going to do the work. This is because of a lack of remains that have the kinds of properties that make mammal molars so superb for reconstruction. However, as we have already seen in chapter 2, a lack of such traces has not stopped progress in historical science; rather, scientists opportunistically exploit whatever lines of evidence are available, and often this evidence does not come from traces. Later on, I will defend such non-trace evidence. But for now, we should ask whether there is anything systematic we can say about the relationship between the present and the past that is epistemically relevant. This is the topic of the next chapter.

4   Over and Under

Historical scientists reconstruct the past by drawing on downstream remnants of their targets, causal descendants of long-ago events. We have been considering such relationships: between "traces"—the way the world is now—and the deep past. In later chapters I will argue that overemphasizing this aspect of historical evidence leads to an impoverished picture of our access to the past. In this chapter I will ask whether there is anything systematic we can say about the relationship between the present and the past that bears on our capacity to reconstruct the latter.

Recently, two such claims have been made. Carol Cleland has argued that the present *overdetermines* the past. The way the world is more than guarantees the way the world was. Derek Turner argues that the present *underdetermines* the past. The way the world is coheres with several possible pasts. In a sense, I will be ecumenical: Some of the time there are more than sufficient traces to establish the occurrence of some past event, but sometimes there will be insufficient traces. However, both claims fail to be systematic. That is, on their own, they do not explain the relationship between the past and the present. Their applicability is crucially context-dependent. The relevant contexts will be captured in the ripple model of evidence in the next chapter. And so, both Cleland and Turner have important insights about the nature of trace evidence that we will carry forward. I will take each position in turn.

4.1   Overdetermination

Carol Cleland (2001, 2002, 2011, 2013) has argued that the present systematically overdetermines the past and that this underwrites a distinctive "historical method." We will meet the method itself in chapter 6, but here I want to focus on overdetermination. My strategy is to drive a wedge between the facts about the world that Cleland tries to establish and the epistemic lessons she draws (Turner 2004 takes a similar strategy). Strong dependency relationships—even overdetermined ones—do not immediately translate into good evidence, or "overdetermining" traces. This is because the "availability" of traces matters. Overdetermination may be an ontological fact, but it does not follow from this that we are able to access the relevant facts in order for this to pay epistemic dividends. So, what is overdetermination? Let's start with more common philosophical treatments. As we will see, Cleland's sense of "overdetermination" departs from these, and the contrast is instructive.

The concept of *causal* overdetermination will be familiar from the philosophy of mind, particularly Jaegwon Kim's work (Kim 2000) and the metaphysics of causation (e.g., Lewis, 1973). Some event is causally overdetermined just in case that event has more than one sufficient cause. Thelma and Louise both throw rocks at a window, simultaneously hitting it and causing the window to break. Both Thelma and Louise's rocks are individually sufficient to break the window: Had Thelma not thrown hers, Louise's alone would have done the job, and vice versa. Thelma and Louise's rocks overdetermine the breaking of the window.

In metaphysics, such cases make trouble for counterfactual accounts of causation. These accounts claim some event is a cause of a later event just in case, had the earlier event not occurred, the later event wouldn't have either. Overdetermination cases are problematic because, on the basic counterfactual view, neither Thelma nor Louise's rocks would count as causes of the window's breaking—surely an absurd result. Had Thelma not thrown her rock, the window would still break (due to Louise's throw) and vice versa. In the philosophy of mind, overdetermination is used to argue that folk categorical propositional attitudes, such as beliefs and desires, are not causally efficacious. Thelma goes to the fridge to grab a beer. By one story, her desire for a beer and her belief that the fridge contains beer causes her subsequent action. But by another story, her action is caused by a series of neurophysiological events culminating in her grabbing a beer. Both narratives cite causes that, on the face of it, are sufficient for her going to the fridge. Throw in a couple more premises and (hey presto!) "beliefs" and "desires" turn out to be causally inefficacious.

In metaphysics and philosophy of mind, causal overdetermination plays a problematizing role. That is, it is used to challenge intuitive views about the nature of folk psychology on the one hand, and causation on the other. This challenge has motivated several flourishing research programs. Cleland's use of overdetermination is quite different. Rather than providing a problem, it presents a solution—an explanation of the remarkable success of historical science. As we will see, if her claims of overdetermination between the present and past are right, there should be rich epistemic resources available for uncovering the deep past.

But what does it mean for the present to overdetermine the past? The central difference between Cleland's conception and causal overdetermination is that whereas (in causal overdetermination) causes overdetermine effects, for Cleland effects overdetermine causes. Cleland is making a claim about events, from which she draws epistemic lessons:

Events leave widespread, diverse effects (such as pieces of glass, particles of skin, etc.). Any one of a large number of contemporaneous, disjoint combinations of these traces is sufficient (given the laws of nature) to conclude that the event occurred. One doesn't need every shard of glass in order to infer that a window broke. (2002, 481)

Cleland's arguments aim to convince us the world is a certain way. They are, then, ontic as opposed to epistemic, although she wants to draw an epistemic conclusion. It makes sense to read her as shifting from *ontic* overdetermination to *epistemic* overdetermination. A weaker reading, on which the present does not overdetermine the past in ontic terms, but nonetheless certain facts about the world guarantee epistemic overdetermination, will be discussed later. But what would an ontic conception of the present overdetermining the past be like? It would be somewhat odd to suggest that the relationship is of *causal* overdetermination. After all, causes usually precede effects, not the other way around; an organism's death and its subsequent fossilization is a cause of the subsequent fossil, not an effect; the window's breaking didn't cause Louise to throw her rock.[1] Contrary to my reconstruction of her argument, Cleland sees this as reason to take the overdetermination as epistemic:

The overdetermination of causes by their effects, on the other hand, is (strictly speaking) only epistemic. Although one may *infer* that the window broke from any one of a large number of subcollections of shards of glass, those shards are not part of the *cause* of the breaking of the window; they are its effects. (2002, 488–489)

Cleland seems to think that because the direction of causation does not run from the present to the past, the relationship she tries to establish must be epistemic. This is a mistake, as there are more options available than "causal" or "epistemic"! Cleland's arguments (which I explain later) do not obviously establish an epistemic overdetermination. By conflating these, Cleland's argument looks simpler than it is.

There are ontic relationships that are not causal. Here is one (which we shall meet briefly again in chapter 6): *constitutive* relationships (see Craver 2007 for discussion). *Dreadnoughtus schrani*'s (magnificent) full length is presumably determined by, for instance, the size and number of its vertebrae, the size of its skull, the length of its tail, and so forth. However, the length is not *caused* by those factors; rather, they *constitute D. schrani's* length. Constitutive and causal relationships differ. For instance, constitution is synchronic, occurring at a time, while causal relationships are diachronic, in temporal sequences. However, Cleland's notion of overdetermination is also diachronic, so constitution is not the notion we are after.

Happily, we have a suitable ontic notion already in hand. In chapter 3, I provided an account of *minimal dependence*, a relationship between two variables, which holds when changes to one variable make it likely that the other variable will also be different. Unlike causal and constitutive relations, this notion is indifferent to temporality. This minimal notion of dependence (1) is ontic and (2) can hold between the present and the past, and so is a good basis for understanding a non-epistemic notion of the present overdetermining the past. Here it is.

Take three present states, S_1, S_2, and S_3. These all depend necessarily on some event in the past, P (see figure 4.1). That is, if P did not occur, then neither would have S_1, S_2, or S_3—P is necessary (but insufficient) for the occurrence of each state. P is required for S_1–S_3 to occur, but does not guarantee them alone. However, these relationships are *overdetermined*, that is, S_1, S_2, and S_3 are each individually sufficient for P's occurrence. This of course follows from P's necessity: If P is necessary for S_1, then S_1 is sufficient for P. Further, make the relationship one of causation: P caused S_1, S_2, and S_3. Ontological overdetermination works like this: Intervene on the causal path between P and (say) S_1 such that S_1 does not occur. Given the presence of P, S_2 and S_3 would still occur—and moreover still guarantee P's having occurred. That is, because each state is alone sufficient to guarantee P's occurrence, S_1, S_2, and S_3 overdetermine P. That is, P's occurrence is only

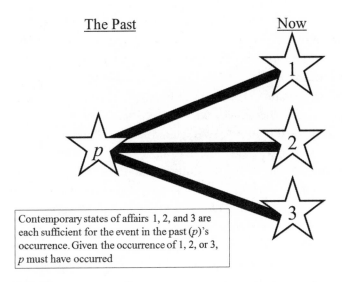

The Past   Now

Contemporary states of affairs 1, 2, and 3 are each sufficient for the event in the past (p)'s occurrence. Given the occurrence of 1, 2, or 3, p must have occurred

Figure 4.1
Ontic overdetermination.

minimally dependent (if at all) on S_1, S_2, or S_3. However, those states are maximally dependent on P. Let's put some clothes on this schema.

Dolostone caps are found in Neoproterozoic deposits all over the world, from Greenland to New Zealand. Let's take the various caps at their various locations to be the states S_1 to S_3 in the abstract example. According to geologists, the dolostone signals a fast climatic turnover from chilly to balmy conditions. After all, midrange theory says, in effect, that this is the only way the phenomenon can be produced: Quick climatic change is necessary to produce the caps. The climatic turnover is not sufficient for the present occurrence of the caps, however. These also needed to be recorded in strata and not destroyed by countervailing geological processes. The presence of dolostone caps, then, "necessitates"[2] a rather dramatic climatic switch (even if the climatic switch doesn't necessitate dolostone caps). There are many such caps, and even if it were the case that some of them didn't make it into the present, the others would still necessitate the Neoproterozoic climate. Present states of affairs, then, are more than sufficient to guarantee the climate changes in the Neoproterozoic. The present overdetermines the past.

Epistemically speaking, in this case, if I were to discover that I was in a world containing S_1, S_2, or S_3, and I knew the dependency between P and those states, that would be sufficient for my knowledge of *P*. With the right midrange theory in hand, just discovering a few dolostone caps in the relevant strata is sufficient to know that there was a quick temperature turnover in the Neoproterozoic.

On this reading, Cleland's argument needs two steps. First, convince us of the ontic claim: that the present does in fact overdetermine the past. Second, convince us that epistemic overdetermination follows from ontic overdetermination. Let's examine these in turn.

Cleland has two arguments for the ontic claim, one (2002) relying on metaphysical argument, specifically David Lewis's (1979) discussion of the direction of time's arrow, the other (2002, 2012) on physical theory.

Lewis is interested in an asymmetry between the past and the future. He starts with the intuition that the future is open, yet the past is fixed. Lewis captures this asymmetry in counterfactual terms, thus: If the present were different, so too the future; however, if the present were different the past would still be the same. That is, the future counterfactually depends on the present, while the past does not.

If the present were different, the future would be different; and there are counterfactual conditions, many of them as unquestionably true as counterfactuals ever get, that tell us a good deal about how the future would be. ... Not so in reverse. ... It is at best doubtful whether the past depends counterfactually on the present, whether the present depends on the future, and in general whether the way things are earlier depends on the way things will be later. (Lewis 1979, 455)

Cleland builds on this thought. She vividly illustrates the difference between erasing traces—and thus intervening on our access to past, and intervening on current events—thus intervening on the future. She asks us to compare covering up a crime to preventing a crime:

Footprints, fingerprints, particles of skin, disturbed dust, and light waves radiating outward into space must be eliminated. Moreover, it isn't enough to eliminate just a few of these traces. Anything you miss might be discovered by a Sherlock Holmes and used to convict you. Finally, each trace must be independently erased. ... In stark contrast, preventing a crime from occurring is easy: Don't hit the baseball; don't fire the gun. In other words, erasing all traces of an event *before* it occurs is much easier than erasing all traces of it *after* it occurs. (Cleland 2002, 487–488; emphasis in original)

Let's begin with Lewis's claim that there are counterfactual dependencies between the present and future, but not the present and the past. Lewis's reason for thinking this is an asymmetry in *causal dependence*. Events in the past have more dependencies than events in the future. Note that Lewis takes this to be a contingent fact—the world could have been different, and like Cleland he appeals to laws of entropy to ground this.[3] On the face of it, though, the idea that there are not counterfactual dependencies between the present and past looks just false: It appears to conflict with a very common mode of scientific reasoning. Recall our grounds for rejecting Williams's theory of Neoproterozoic glaciation from chapter 2. He suggested that irregularity in Earth's obliquity would account for tropical glaciation, because under some conditions the sun's focus would shift to the poles. However, this would also lead to relatively evenly distributed carbonate deposits—which is not what we see. Rather, we see dolostone caps. Here, there is a counterfactual dependence posited between a current state and a past state. If there *were* evenly distributed deposits, then it would be more likely that the Neoproterozoic freeze was caused by irregularities in Earth's orbit. This kind of trace-based reasoning (which, as we will see, Cleland has the most developed view of) seems to rely on "backtracking" counterfactuals, which Lewis explicitly does not include in his analysis (such counterfactual follow different rules, see Lewis, 1973). It could be claimed that such counterfactuals run forward, not backward. Scientists claim that had Earth's obliquity been different, the deposits would be also. And indeed, the *causation* runs in that direction. But I fail to see why the counterfactuals must also. For my purposes at least, scientific practice trumps metaphysical speculation. There is a lot more to be said about Lewis's views vis-à-vis counterfactual semantics and time asymmetries, but sadly this would take us even further from our aim of getting to grips with the epistemic situations historical scientists find themselves in. Suffice to say, if Lewis's asymmetry does not include backtracking cases such as "if there weren't dolostone caps now, then there would not have been tropical glaciation in the Neoproterozoic," and given that such counterfactuals play a central role in scientific reasoning, then it is entirely unclear what epistemic conclusions could be drawn from the position.[4]

Now, Cleland's example quoted above is different to Lewis's. It turns on our capacity to *intervene on* the past or the future. The example is, I think, misleading. Our capacity to causally interact with the past and future is at least partly independent from the ontological or counterfactual relationships that the past and future have with the present. The example, then, is best read as an illustration of the tendency for traces to *spread* over time: Causes breed effects. A cause's effects frequently spread out from it and each other. To remove traces prior to an event, then, I need intervene only once: Stop the event from occurring. To remove the traces post the event, I must intervene on a plethora of pathways. This is, I think, an important insight, which I develop in chapter 5. However, it does not appear to guarantee the present's ontic overdetermination of the past (or, as we shall see, its epistemic overdetermination).

Regardless, whether or not the present overdetermines the past is surely an empirical matter—not to be decided by armchair reflections on the asymmetry of time or our capacity to intervene on it (Tucker 2011; Turner 2007, 38–39). Our question, then, concerns the typicality of the kind of situation that Cleland highlights. That is, how often should we expect to find relationships like that expressed in figure 4.1? Why not stronger dependencies running in the other direction (for instance, P could necessitate the occurrence of the Ss)? Or, for that matter, anything else among the wide range of possible dependency relationships? It is clear that *on some occasions* ontological overdetermination occurs, but how often? Cleland's later argument, which appeals to physical theory, may help us here.

Particularly in her 2012 paper, Cleland shifts from a metaphysical argument to a thoroughly empirical one. There, she claims that two physical theories, thermodynamics and wave theory, between them guarantee that "all physical phenomena above the quantum level are subject to the asymmetry of overdetermination" (21).

Thermodynamic theory concerns the behavior of physical variables like temperature. A useful way into the theory is via comparison with statistical mechanics. They differ insofar as the former is "macro-level" and the latter is "micro-level," as follows. Imagine water in a beaker with a Bunsen burner underneath. As the water's heat increases, it boils. A thermodynamic explanation of the event will understand, say, the water as a body with energetic properties. The increased energy transferred from the Bunsen's flame is partly converted into kinetic energy, transforming the liquid into gas. A statistical mechanical explanation would envision the water as a set of particles and explain boiling in terms of the particles' behavior and interaction. On the one hand, the water is understood as a body that interacts with the increased heat, on the other hand, it is understood as a collection of particles. It is generally taken that statistical mechanics "explains" thermodynamic laws. And so, we can understand thermodynamics as a body of theory with very wide applicability that explains and predicts the behavior of "macro-level" (that is, not particle-level) physical processes.

Cleland is right that there is an important sense in which thermodynamic systems are temporally asymmetric, due to the second thermodynamic law. This law states that over time differences in energy "even out," that is, the energetic structure of the world dissipates over time. "Entropy" is the name given to this "evening out" process. So long as new energy isn't fed into a system, entropy will increase over time. If the Bunsen burner is turned off, eventually the temperature of the water will return to its previous temperature relative to the room. There have been various attempts to ground time asymmetries in the second law of thermodynamics. Philosophers who think that time has an objective direction, for instance, will sometimes appeal to increases in universal entropy (Callender 1997; Dowe 1992).

However, I do not see how appeal to entropy can underlie the kind of overdetermination that Cleland needs. Global entropy does provide an asymmetry. However, this asymmetry entails the *destruction* of traces, not the present's overdetermining of the past. If the universe is inevitably moving from a state of low entropy (that is, high structure, heterogeneity in energy states) to a state of high entropy (homogeneous energy states, low structure, or "disorder"), then this implies that no matter what the past states are, the same end-states will be arrived at. Entropy suggests overall convergence of states, which destroys past signals, not the preservation required for overdetermination. (I will return to notions of convergence in much more detail in chapter 8.)[5]

Wave theory is a set of equations that conform to the behavior of "waves." It ranges over a plethora of entities: light, sound, and radio waves, electromagnetic waves, the evolution of quantum systems under superposition, as well as more familiar phenomena such as the vibrations of stringed instruments and waves in water. We need not bother here with the technicalities, but the equation describes how, over time, an initial disturbance will spread out and dissipate. Thus, wave theory is also time-asymmetric, since waves radiate from their sources.

Later, I will draw on Cleland's discussion of wave theory to build my picture of the present and the past.[6] For now, let's assume Cleland is right and physical theory—wave dynamics, or some other physical theory, let's say—grants us reason to believe that present states ontically overdetermine past states. Here is one worry: Is such overdetermination at the right scale to matter for the sciences we are concerned with? Even if, on the level of global entropy, the present overdetermines the past, it isn't at all obvious that this means that, say, the present overdetermines the past in terms of Earth's history. More important, it isn't obvious that these ontological seeds make epistemic hay.

Recall my account of traces: Some contemporary structure is a trace of a past event just when it is downstream of it, and is evidentially relevant by virtue of justified midrange theory. My rationale for taking this theory-dependent notion of "trace" is to bring philosophical discussion closer to the epistemic situations that scientists face. It is clear that ontological over-determination of the sort I have described does not guarantee that *traces* overdetermine the past. To make this explicit, let's get clear on *epistemic* overdetermination.

An epistemic notion of overdetermination isn't problematic in the same way causal overdetermination is for metaphysicians or philosophers of mind. An agent having more than sufficient justification, or reasons for belief, or whatever, does not prima facie undermine standard accounts of knowledge. To be relevant to this discussion, we should focus on *evidence*. I take it that evidence is scalar: That is, it comes in degrees; one can have more or less evidence for a hypothesis. Cleland appears amenable to this: "Although Lewis characterizes the asymmetry of overdetermination in terms of sufficiency, it could turn out to be a probabilistic affair, with the ostensibly overdetermining subcollections of traces lending strong, but, nevertheless, inconclusive support for the occurrence of their cause" (2002, 490).

Although Lewis is talking ontically, clearly Cleland means something epistemic. But it is odd to claim that the provision of "inconclusive" support is overdetermining. Given that our credence isn't raised to one, isn't this *underdetermination*? The best reading, I think, is this. Given each event in the past, there is more than enough available evidence to provide "knowledge" of the event, or to ground justified belief in it. The Neopro-terozoic tropical glaciation is epistemically overdetermined in regard to paleomagnetic and dolostone traces just in case, were a rational agent to know about either paleomagnetism or the caps (and had the relevant background knowledge), that agent would form a justified belief that there were tropical glaciers in the Neoproterozoic. Such an agent, with both pieces of evidence, would have evidential redundancy. They could afford to lose evidence and retain their belief's epistemic justification.

And so, I interpret Cleland's claim as follows: For any (or most?) event(s) in the past, there are more than sufficient contemporary upstream states of affairs to provide justified belief in those events. Ontic overdetermination leads to evidential redundancy. Cleland's view about the relationship between the present and the past supports optimism about historical science, and moreover it has methodological upshots. If we have good reason to think that there is bountiful evidence available, this should motivate research into midrange theories and hunts for traces. However, for Cleland's claim that the past is systematically, and epistemically, overdetermined by the present to go through, she needs to argue that it follows from the past's overdetermination by the present that we can know many facts about the past via examining the present. So, can this epistemic conclusion be drawn from Cleland's ontic arguments?

Getting a grip on the *availability* of traces is essential for establishing the connection between the past's overdetermination by the present (the ontic question) and whether or not *traces* overdetermine the past (the epistemic question). Availability is determined by three features: *midrange theory*, *retention*, and *discovery*. If we lack the relevant midrange theory (/technology), or those theories lack support, then present states cannot be linked to the past; they fail to be evidentially relevant. In fact, by the account from chapter 3, they fail to be traces. We could potentially know that the way the world is now overdetermines the way the world was, but still not have the relevant know-how to actually retrodict those past states. Moreover, downstream effects of past causes must be retained over time, as traces degrade and disintegrate. In the next section, when we turn to Turner's views on the nature of historical evidence, we will examine this point in much more detail. Suffice to say, the dolostone traces must have survived various travails to make it from the Neoproterozoic to the Cenozoic. Finally, the traces must be found. There might be great discoveries hidden in the depths of the earth, but this only helps if we actually recover them.

In short, even if the present overdetermines the past, this only translates into epistemic overdetermination if we have the right midrange theory, and if the traces are retained, and we discover them. Surely, much of the time, these conditions won't hold. Cleland's epistemic conclusion does not follow, as there is a difference between the causal relations in the world and our epistemic access to them. One cannot shift easily from claims about the way the world is to claims about our capacity to know about it.

So, the systematicity of Cleland's claim fails. However, as we shall see in chapter 6, her methodological account can be separated from her claims of overdetermination. Moreover, Cleland is certainly right that *sometimes* there is more than sufficient evidence for past events, and that causes breed effects: I will build this into the ripple model of evidence in the next chapter.

In ascertaining how overdetermined the past is, we should take note of how *heterogeneous* traces can be. The hypothesis that there were tropical Neoproterozoic glaciers relies on different traces doing different jobs. Paleomagnetic traces speak to the geological position of glaciers, while dolostone caps tell us about the speed of climate change. Not only are the traces of the Neoproterozoic widely spread and of quite different natures, they also provide different information that is jointly required to uncover the nature of the past event. Additionally, how we describe and "carve up" events seems to matter. For example, there are many more traces of the general hypothesis of there being an impact at the Mesozoic's close than there are of the more specific hypothesis of the impact occurring at a particular time and place. Although traces do collaborate on the same task, much of the time there is a division of labor: They are relevant to different aspects of the target. When tasks are divided between traces, their availability or otherwise becomes more pressing, since there may be insufficient redundancy to mitigate signal decay. In short, then, there being bountiful downstream traces of an event doesn't guarantee its overdetermination: The particular dependency relations, and our knowledge of them, make a difference. The details always matter.

Certainly, some historical hypotheses are epistemically overdetermined. That an enormous extraterrestrial impact occurred at the K-Pg boundary is evidenced by various traces. The boundary contains shocked quartz, which forms only at extraterrestrial or nuclear impact sites, suggesting something big happened; the boundary has a distinctive layer of iridium, a mineral rare on earth but common on some asteroids, suggesting something extra-terrestrial; the layer marks a major extinction event—suggesting that *something* nasty happened. This evidence, it seems to me, is sufficient to justify belief in the impact. The addition of a crater of the right pedigree (size and age) off the Chicxulub peninsula was not necessary for justification. The combination of iridium, shocked quartz, extinction, and crater are more than sufficient. However, it is unclear at this stage how common we should expect this epistemic situation to be. How often are there dependencies between the present and the past that guarantee the past's being a certain way, and moreover how often these are available, that is, can be translated into traces, are open questions.

Let's take stock. Carol Cleland has argued on metaphysical and physical grounds that the past is overdetermined by the present. Even if she is right that the world is systematically overdetermined, epistemic overdetermination does not follow. Event overdetermination alone does not entail trace overdetermination. Moreover, the existence of an overdetermined past does not guarantee reflection in scientific practice. However, Cleland's insight that effects spread and multiply from their causes, and that this is exploited by historical scientists, is important, and we will return to it in chapter 5.

4.2 Underdetermination

In stark contrast with Cleland, Derek Turner has argued that hypotheses about the past are systematically underdetermined (2004, 2005, 2007, 2009a). That is, there is insufficient information available for us to recon-struct much of the past. Turner's strategy compares the fortunes of experimental sciences—the paradigm case being experimental physics—with those of the historical sciences, primarily paleontology. He points to two asymmetries. First, we are able to intervene on and manipulate experimental systems, but not historical ones. This means that historical scientists are unable to actively improve their epistemic situation by generating new evidence. I will put this asymmetry aside; it will get an outing in chapter 5, and I will respond to it in chapter 9. The second asymmetry, and our focus here, concerns the role of background theories. Turner argues that reflection on midrange theories in historical science should lead us to expect underdetermination to be rife.

As with Cleland, I will argue that Turner's claims fail to be systematic but nonetheless have important insights that will be carried through to the next chapter. Let's start by getting a grip on underdetermination and zeroing in on Turner's conception.

In the sense concerning us here, underdetermination occurs when evidence cannot decide between two competing hypotheses. One way of capturing the notion more formally uses the *law of likelihood* (Hacking 1965), which I find to be a handy way of thinking about evidence (it will crop up again in chapter 9). The law tells us what it takes for an observation to be evidence for one hypothesis over another. The requirement is just that the conditional probability of the observation given one hypothesis is higher than the conditional probability of the observation given another hypothesis. For instance, the observation that Lacovara et al.'s (2014) specimen has a detached scapula is evidence that the 26-meter *D. schrani* that died 77 million years ago had not finished growing. This is because the detached part is much more probable in a world with a still-growing *D. schrani* than one with an osteologically mature *D. schrani*. The law is as follows:

Some observation O supports a hypothesis H_1 over another hypothesis H_2 just when $P(O|H_1) > P(O|H_2)$

Underdetermination is a relationship between at least two hypotheses and a set of observations $\{O_1, O_2, O_3, \ldots O_n\}$. The basic idea is that the set of observations that concern us fail to evidentially discern between the relevant hypotheses. For instance, had the *D. schrani* not died, would it have reached osteological maturity in one year, or two years? The observation of a detached scapula does not appear to tell us. The hypotheses are underdetermined by the evidence. Different versions of underdetermination can be generated by (1) shifting evidential burden or (2) changing the observation set.[7]

We'll start with the former approach. A "pure" case of underdetermination would require that the observations are *equally likely* conditional on the hypotheses:

Underdetermination between H_1 and H_2 occurs when $P(\{O_1, O_2, O_3, \ldots O_n\}|H_1) = P(\{O_1, O_2, O_3, \ldots O_n\}|H_2)$

I would be inclined to think that if the set of observations provides only miniscule support for one hypothesis over another, say H_1 over H_2 (and my priors are about even), this isn't sufficient for me to believe H_1 over H_2. Any epistemic principle that demands that we believe the best of a bad bunch ought to be rejected. So this "pure" version of underdetermination is not attractive. Underdetermination is just too difficult to generate on such terms. Stronger requirements might ask that, for instance, both H_1 and H_2 be *possible* given the relevant set of observations, that is, at least some probability be assigned to both H_1 and H_2, given the observations. A falsificationist might think something like this: Because the observations do not deductively expel H_2 (that is, falsify it), the observations do not decide between it and H_1. Unsurprisingly, I find setting the bar that high to be unattractive as well. Less extreme variants will allow more or less wiggle room. I don't want to trip down the rabbit hole of evidential standards here, but regardless there will be a range of "sweet spots," as it were, between the extremes of requiring absolute equality and absolute certainty. One way of generating different types of underdetermination, then, is via adopting different evidential standards. The next—changing the scope of relevant observations—is more relevant to Turner.

Traditional philosophical study of underdetermination has taken the relevant set of observations to be *all possible* observations. That is, there is no observation or set of observations, of all possible observations we could make, which could decide between the hypotheses. This version is typically tied up with discussion of universal epistemic skepticism (see Goodman 1955; Kukla 1996). On the face of it, all of our best scientific theories are just as empirically adequate as the hypothesis that God, or some other mischievous yet powerful sprite, created the world three seconds ago. Turner and I are after something more mundane (2004, 2007 chapter two 37-57,). I will call this *local* underdetermination.

Underdetermination is local when the relevant observations are the set of *available* observations, that is, those that have in fact been made (Godfrey-Smith 2008; Laudan 1990; Sklar 1977; Stanford 2009). Two hypotheses may not be empirically equivalent (that is, they have different observational consequences) but nonetheless may be locally underdetermined. Recall the Corythosaurus footprints from the last chapter. Imagine competing hypotheses about the biological sex of the Hadrosaur that left the footprints. These could come apart in principle—a male Corythosaurus may leave distinctive footprints contrasted with a female Corythosaurus, or there could be other possible observations that could decide either way—but such midrange theory or observations are not available. In this case the hypotheses are not empirically equivalent, since there is a set of observations that (in principle) would distinguish between them.[8] However, they are epistemically indistinguishable given the evidence we in fact have.

Let's distinguish further between two types of local underdetermination. Sometimes the underdetermination will be *resolved*. That is, we have reason to expect future observations will dissolve the underdetermination, deciding in favor of one hypothesis over the other. Pick your favorite critical test from the history of science as an example.[9] In other cases, the underdetermination will be left *unresolved*; that is, the relevant observations will not be made.

Aviezer Tucker (1998) has a contrasting notion of underdetermination that focuses on the *uniqueness* of events. Briefly, this underdetermination occurs when an event's significant properties are either not shared by any other events or are too complex to be compared to other events. Tucker argues that such unique events generate a kind of *explanatory* underdetermination. Explanation often involves situating a case against contrasts, that is, categorizing an event as a token of a type. If the event is unique in Tucker's sense, then such contrasts are not empirically available. No matter how much knowledge we have of the event itself, then, we cannot discriminate between different possible ways of categorizing it, and thus different explanations. This might seem somewhat obscure, and so it should: At this stage we lack the theoretical machinery to compare Tucker and Turner's conceptions of underdetermination. I will return to Tucker's notion in chapter 8, when I discuss the evidential relationship between historical contingency and our access to the past.

Turner argues that the historical sciences frequently face situations where available evidence fails to decide between hypotheses (local underdetermination), and that midrange theories give us reason to believe that the relevant evidence will not be forthcoming (the underdetermination will be unresolved).

It is important to note that Turner does not deny that we do sometimes know things about the past. That there was glaciation in the Neoproterozoan tropics, that sauropods had a faster growth rate than most reptiles, that there was an extinction event 66 million years ago, and so on, are all things we surely know. Turner's arguments involve a comparison between experimental and historical science; however, it suits me to instead talk about unlucky historical scientists. In some circumstances historical scientists face local underdetermination, which we have good reason to think will not be resolved. His argument requires a distinction regarding background theory.

According to Turner, background theory can play two roles. First, it can *limit* our knowledge by giving us reason to think that underdetermination problems will not be resolved: "Background conditions serve (or should serve) as a check to the epistemic ambitions of historical researchers" (2007, 58).

In such cases, background theory provides a council of despair, citing reasons to believe that transient underdetermination will be unresolved. Second, background theory can *amplify* our knowledge by suggesting new ways of discriminating between theories: "Experimental designs always depend heavily upon background apparatus that will enable them to manipulate certain test conditions. This experimental manipulation then gives them a way to test new theories and hypotheses which, if confirmed, may provide new clues for future experiments" (2007, 58). Turner argues that midrange theories are often limiting, and so we should expect unresolved local underdetermination to be prevalent. Let's look at his arguments.

Turner's positive argument for the limiting counsel of midrange theory is by case study, of which he provides four. I am not convinced by some of these, and I am uncertain of their representativeness. I will discuss two representative examples, which will lead into my argument for being more optimistic about midrange theory.

One of Turner's cases is the competing obliquity and snowball explanations of Neoproterozoic glaciation. As I discussed in chapter 2, there are fairly strong arguments against the obliquity explanation. Geophysical modeling does not support obliquity of the required angles; a correction in obliquity would not account for the sudden climate shift; carbonate deposits would have been different. Turner may insist that the obliquity explanation has not been killed—insofar as we should assign some amount of credence to it. And indeed, it is true that my credence in Williams's obliquity hypothesis is not zero. But there are two things wrong with this. First, such an argument requires that we set a very low bar in terms of evidential standards. If all underdetermination requires is that one of the hypotheses still has *some* support, even if only a little, then generating underdetermination is too easy. I am inclined to think that the evidence is sufficient to speak against obliquity. Second, that evidence in favor of snowball (or slushball) scenarios has built over the last two decades seems to matter here. That is, we have seen significant progress in our understanding of the Neoproterozoic as new technologies, theories, hypotheses, and discoveries come online. This should at least give us some confidence in the resolvability of the relevant underdetermination.

A trickier example of Turner's is the relationship between specimens concerning taphonomy, fossilized bone and the like, and specimens concerning ichnology such as trackways and burrows. Attempting to reconcile different trackways with different lineages is extremely difficult, especially considering both the incompleteness of the records (what I call "gappiness"), and their differing resolution. In short, it is really hard to tell who made what footprints.

One problem is that the parataxonomy based on fossil footprints is coarser-grained than the taxonomy based on skeletal remains. Since background theories of taphonomy tell us that the conditions most conducive to the preservation of skeletons in the fossil record are completely different from the conditions most favorable for the preservation of footprints, nearly every fossil trackway poses an underdetermination problem: How can we tell which sort of animal made this particular set of tracks? (Turner 2007, 50)

This is an excellent case for Turner. If we had the right collection of finds—perhaps an extremely complete bunch of fossils with a large variety of corresponding trackways—we could conceivably tell who left what. The underdetermination, then, is local. However, theories of taphonomy and ichnology suggest that the chances of such finds are vanishingly small.[10] I have two responses. First, the epistemic gains of identifying footprints to particular low-level taxa categories (species and genus) seem minimal, given the purpose to which paleontologists put them. Typically, trackways are used to reconstruct aspects of dinosaur behavior: flocking patterns, gait, and so on. For this kind of task, matching parataxonomies at a fine grain of resolution isn't necessary, since those traits tend to be retained across ancestral groups, and leave telltale signs in morphology.[11] For instance, Sellers et al. (2013) simulate *Argentinosaurus* gait. Their simulant's gait would produce tracks strikingly similar to actual sauropod finds. In this case, it isn't relevant if the matched trackways are from *Argentinosaurus* or some other titanosaur, since there is reason to think that there isn't that much variation in titanosaur gaits. The second point follows directly: This picture is too trace-focused. There are ways of uncovering relationships between bodily and nonbodily remains that does not rely on uncovering new traces. Sellers et al. connect the traces to titanosaurs by virtue of the closeness of the match between their simulated gait and the trackways (Turner, 2009b, discusses similar studies of theropods). In chapters 6, 7, 8, 9, and 10 we will hear my defense of these non-trace sources of evidence.

Turner points to circumstances where our main line of evidence regarding some fact is unsuitable. Lewontin (2002) discusses a more general version of the phenomenon. As he points out, often in historical reconstruction (he focuses on adaptationist hypotheses) we want dynamic information—data about a process over time—but we only have access to static information. As I will argue, such cases are often resolved by realizing that we need not rely on a single source of evidence: As is especially explicit in chapter 6, the power of historical reconstruction (particularly in "unlucky" circumstances) lies in generating and utilizing a variety of lines of evidence—and this can overcome mismatches.

Finally, it is unclear how to assess the kinds of bets Turner asks us to make. As Patrick Forber has said,

We cannot be in an epistemic position *now* to assess whether two incompatible rivals are empirical equivalent relative to all *present and future* evidence ... we can make no reliable inferences about how our epistemic position will persist into the future. The leaves sufficient space to doubt that incompatible rivals will remain empirically equivalent after the accumulation of more data, the development of new technology, and the innovation of theory. (2009, 255)

However, as we saw in the introduction, betting on the future outcomes of scientific research is an important task for historical scientists. Moreover, optimism itself involves some kind of positive attitude toward future science—so we shouldn't simply *dismiss* such bets, or so quickly deny our capacity to make them (Turner 2016). In chapter 11, we will discuss future bets on historical science more carefully.

Turner has perhaps identified some plausible cases of local underdetermination, but these are debatable, it is hard to weigh the warrant of bets regarding the resolvability or otherwise of local underdetermination, and more important, I am unsure of their representativeness. Overall, it is difficult to ascertain what to make of the proposition that there is systematic underdetermination (of Turner's "local but probably unresolvable" brand) in the historical sciences. After all, geology, paleontology and archaeology have made great strides in the last twenty years—the case studies in chapter two are testament to this. Having said this, I am interested in *unlucky circumstances*, cases where our reach into the past is tenuous, and so surely we should expect some underdetermination to be stubbornly unresolvable. However, there are two further problems with Turner's pessimism: his account of evidence and his view of midrange theory. The former will be dealt with in future chapters, in brief; Turner fails to appreciate the epistemic resources historical scientists have at their command. However, it is time to turn to the latter.

Laudan and Leplin (1991) have argued that inferring underdetermination from a lack of deciding evidence is unjustified because of the fluid, progressive nature of background theories. Recall that there are two ways of generating traces: via discovery or via refinement. Find new stuff, or find new ways to understand the stuff you already have. Laudan and Leplin refer to the latter strategy. Two hypotheses might appear underdetermined given some set of observations, but this is only against a set of background theories. My characterization of underdetermination, then, is missing a vital ingredient: The two hypotheses must be comparatively likely in light of (1) a set of observations and (2) a set of background theories. As background theory develops, the same set of observations can resolve the underdetermination. Even if today's midrange theory is limiting, providing a council of despair, tomorrow's midrange theory may not (see also Jeffares 2010).

Turner responds by pointing to the robustness of background theory in historical science. Taphonomy, he claims, is a remarkably stable theory. Indeed, it would be surprising if it turned out we were wrong about fossilization processes. And it is this stable theory that tells us that it is highly unlikely for any particular lineage to be fossilized. I have two rejoinders. First, I am not convinced that taphonomy is as stable as Turner paints it. Second, he misses the force of Laudan and Leplin's point: It is not that currently existing midrange theory is likely to be revised, but that *new* midrange theory is likely to be added.

Taphonomy, the science of fossilization that tells us about the conditions under which organisms fossilize and where they might be located, has developed significantly. In the past, it was considered extremely unlikely (if not impossible) for the soft tissue of ancient terrestrial animals to be preserved. However, in 2005, Mary Schweitzer and her team dissolved the fossilized remains of a 70 million-year-old *T. rex* thigh bone in acid and found what they identified as blood and cellular structures. They speculated that the bone, deeply mineralized and condensed, formed a protective casing for the delicate inner structures. It turns out that the soft tissue is more likely to be preserved than we first thought.[12] This is an example of taphonomy being revised, such that its message about our capacity to know the past becomes a more optimistic one.

Moreover, our stock of midrange theory is growing. The introduction of sophisticated technology allows both new observations and new tests. As we saw in chapter 2, Hummel, Gee, et al. (2008) challenged Midgely et al.'s (2002) claim that Jurassic flora was low in nutritional content. They used in vitro fermentation to test the nutritional content of ferns and other representative Jurassic plants. Here, a "stomach" is simulated in the lab and measurements of the subsequent heat are taken as a proxy for energy production. Prior to the development of sufficiently sensitive thermometers, controlled labs, and theoretical understanding of fermentation and its relationship to thermodynamics, Midgely et al.'s claim looks hard to test. Of course, it turned out to be eminently testable, but not because existing midrange theory was modified. Rather, new knowledge was added. An example with far-reaching consequences was the addition of molecular, phylogenetic methods to preexisting morphological methods of establishing ancestral relations between organisms. Developing the techniques and midrange theory required to incorporate this stream of evidence into phylogenetic reconstruction is ongoing, but it has increased our capacity to uncover ancestral relations between extant organisms enormously (see Ayala 2009).

Another way of increasing our stock of midrange theories is through incorporating theories from other disciplines, and in this regard historical scientists are master scavengers. Gillooly, Allen, et al. (2007), for instance, draw upon "thermal inertia" from engineering and planetary science in their examination of sauropod metabolism. Thermal inertia is a scaling effect between heat retention and increased size: Larger bodies retain heat for longer. If this applies to ectothermic animals, then larger reptiles would have an easier time regulating temperature than their diminutive cousins, since their thermal batteries will be more efficient. The incorporation of "gigantothermy" puts existing evidence of sauropod gigantism in a new light: The fact that high body weights are more efficient in ectothermic lineages could provide clues about the evolutionary causes, and the lifeways, involved in sauropod gigantism. Moreover, such epistemic tools can be *tailored*: They are configured to the specific epistemic context at hand, allowing knowledge-generation to be sensitive to idiosyncratic context (see also Chapman & Wylie 2016, chapter 4 143-202).

To highlight the transformative power of new background theories, technologies, and techniques, I want to turn to archaeology—in particular an example from studies of Mayan sites. The Mayan culture was spread across Central America (southeastern Mexico, Guatemala, Belize, etc.) and had a long history stretching from around 1000 BCE to the end of the "Classical" period around 1000 CE. During the "middle" preclassical period (say, 600 BCE) the Mayans appear to have developed a distinctive urbanized social organization, including monumental architecture and a particular script. A major preoccupation of Mayan archaeology is getting a handle on the development and nature of this urbanized culture.

Asking such culture-level questions requires large-scale understanding of the organization of Mayan geography, a target that traditional archaeological methods are not well suited for. Archaeological understanding is often generated at a relatively small scale: Archaeologists dig careful ditches into particular sites and analyze the soil, artifacts, and architecture found therein. By contrast, the Mayan civilization covered large spaces, and these spaces are now coated in dense vegetation, hostile to traditional survey methods. Mayan buildings are identified by stone platforms that once supported structures made of less durable materials. These platforms are lower than the usual vegetation height, and so flyovers would, at best, only locate the largest monumental architecture, and checking things out on foot would be impermissibly costly and labor-intensive. Happily, the incorporation of technologies from other domains—carefully tailored to the Mayan context—provides ways around this.

The development of lasers in the 1960s bought with it new opportunities and techniques for probing the world. In 1963, they were used in meteorology to map the structure of clouds: so-called LiDAR, which is both a portmanteau of "light" and "radar" and an acronym (Light Detection and

Ranging). As in radar, the basic idea of LiDAR involves sending out a signal and detecting the returning "echoes". LiDAR can emit a variety of different wavelengths, tailored to the various objects it is trying to detect. Very roughly, as light passes through different mediums it "scatters," its properties change. Light of a certain wavelength bouncing back from a cloud of a certain composition, will have different properties to that bouncing from one of a different composition. LiDAR was quickly co-opted for geographic surveying practices—now standardly used by NASA for extraterrestrial mapping. It also began to be used in archaeology from the mid-2000s, allowing for precise aerial maps despite tree cover. Let's look at the application of LiDAR to the Mayan context (Hutson 2015; see also Chase et. al 2014; Fernandez-Diaz et al. 2014).[13]

LiDAR mapping isn't as simple as pointing beams of light at your target and recording the results; the results of surveys are highly sensitive to what you're interested in, what you're looking at, how you represent the results, and which wavelengths are used in data collection. In the context of Mayan ruins, the trick is to penetrate the tree cover and to distinguish between man-made stone platforms and natural forms such as outcrops. This involves distinguishing between various visualization techniques and calibrating results. There is no well-understood body of background theory that tells you how to distinguish between an outcrop and a ruin. To do this, you need boots on the ground, what is called "ground-truthing": "One cannot know whether or not a topographic anomaly visible on a LiDAR imagery represents a cultural feature until an archaeologist has inspected a number of such anomalies in the field" (Hutson 2015, 255). Hutson and his team conducted two kinds of ground-truthing. First, they conducted an extensive pedestrian survey, "in which teams walked across the landscape with 10m spacing between each team member and recorded every visible feature with GPS, tape, and compass" (256).

They covered just under 14 square kilometers in this painstaking way. The ground survey was then overlain with the LiDAR results using various visualization techniques. This allowed them to pick the technique that best highlighted the features they were after (distinguishing outcrops from platforms, for instance). Second, they took LiDAR results from a region lacking pedestrian surveys and hypothesized a group of man-made structures. These were then eyeballed to confirm whether they were indeed human structures. This allowed them to "characterize the visual signature of Pre-Hispanic platforms" (256). These calibrations provided a sense of both the error rate of such studies and how the error rate related to different land-scapes (for instance, vegetation height is not as important a source of noise as vegetation density). Using the calibrated technique, Hutson identified a further 438 platforms revealing Mayan geography, which is exactly the kind of evidence that archaeologists need to answer questions about the development of urban Mayan culture.

The incorporation of LiDAR into Mayan studies illustrates the features of background theories in historical science that push back against Turner's pessimism. First, the new evidence has been generated by the incorporation and development of new background theory in archaeology, not by the transformation of existing knowledge. Second, the knowledge generation process is idiosyncratic: It is tailored to the circumstances at hand, and thus sensitive to the requirements of the particular epistemic situation. Given this idiosyncrasy and unpredictability, arguing that local underdetermina-tion will not be resolved is foolhardy.[14] In chapter 6, I will argue that this opportunism is a central aspect of the methodology—and the success—of historical science.

Thus, even if Turner is right that our existing theories linking traces to the past are unlikely to change, he misses that our stock of midrange the-ory can, and most likely will, grow. The basic issue with his argument is that even if our background theory about some line of evidence provides a council of despair, this does not mean that *different* lines of evidence will not open up. New techniques such as LiDAR come online and are calibrated to local conditions, allowing new information to come to light. There is reason to be optimistic about our access to the past.

Turner is right, however, that midrange theory leads us to expect traces to take us only so far. As we have seen, Cleland underestimates how much time's passage dilutes information. Even though Turner has not convinced us that local, unresolvable underdetermination is *systematic* in the histori-cal sciences, it would be foolhardy to think that we will know everything about the past (In chapter 11, I will discuss whether we should bet on *which* facts!).

This chapter has been largely negative. I have responded to both Cleland and Turner's claims about systematic connections between the present and the past, which are relevant to our knowledge. However, there are lessons that I will take into the next chapter. In my discussion of Cleland, we saw that as the time between the present and a past event increases, so too does the number and heterogeneity of traces. With Turner, we saw that time degrades traces. The spread of causes over time can breed evidential redundancy, but signals decay over time as other processes disrupt and erase them. I will draw on these two insights to construct the ripple model of evidence, which will allow us to identify the "difficult" cases—the unlucky circumstances—in historical science. Let's turn to that now.

5 Ripples

In the last chapter, we identified two insights from Carol Cleland's and Derek Turner's views on the relationship between the present and the past. As time passes, events breed descendants of increasing number and heterogeneity. This underlies Cleland's belief that there is an overdetermination relation between now and then. However, these descendants—traces—degrade as time passes. This is a big part of why Turner thinks the past is often underdetermined. Neither of these positions systematically describes the epistemic situation historical scientists face, but, as we will see in this chapter, they can be drawn on to construct a conceptual model of trace-based evidence. The *ripple model*, as I shall call it, sets us up for the remainder of the book. It allows us to identify the "unlucky circumstances" in which scientists investigating sauropod gigantism and Snowball Earth find themselves, and those studying *Obdurodon tharalkooschild* do not. At the chapter's close I will identify three claims that underlie pessimism about unlucky circumstances. In later chapters I will undermine these claims, and in doing so I take myself to motivate an optimistic attitude toward the epistemic capacity of historical science.

5.1 The Ripple Model of Evidence

Although traditionally associated with electromagnetic radiation (light, radio waves, etc.), the radiative asymmetry (as it is sometimes known) characterizes all wave-producing phenomena, including disturbances in water and air. It originates in the fact that waves (whether water, sound, light, etc.) invariably spread outwards, as opposed to inwards, as time progresses, which means that the effects of a cause become increasing widespread in space. (Cleland 2012, 21)

Imagine I throw a pebble into an otherwise undisturbed pond and take snapshots of the resulting disturbance at set time intervals. Earlier times will have a smaller area of effect than later, and the disturbance may become more pronounced as time passes; however, in later snapshots the clear patterns generated by the pebble will distort and fade. This is the basis of the ripple model of evidence. In essence, the ripple model represents the epistemic tension between two processes: the spread of disturbance (which, as we shall see, has an epistemic benefit) and their degradation (which, unsurprisingly, comes with an epistemic cost). I will construct the model by incorporating Cleland's insight and then building in Turner's.

5.1.1 Dispersal

So again, imagine I throw a pebble and take snapshots of the resulting time intervals. Spreading from the initial event (t_0), at each proceeding interval (t_1, t_2, t_3) the *area* of disturbance will increase. We can represent this pictorially as in figure 5.1.

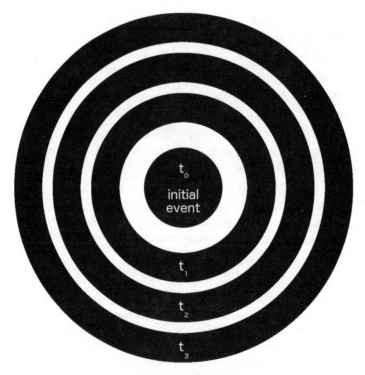

Figure 5.1
The basic ripple model. (Image by Leonard Finkelman.)

This is an unorthodox but illuminating way of thinking about causation and time. By this representation, the closer to the center, the earlier the time. The figure does not chart magnitude of effect, but rather scope, or the area of effect. For any given event, the size of the subsequent area of effect will change through time. Recalling the discussion of traces and dependencies in chapter 3, we can understand these changes in two ways, *epistemic* and *ontic*. Epistemic changes (changes to the relevant trace set) involve changes to midrange theory, the uncovering of new evidence, and so forth—that is, the action of scientific investigation. *Ontic* changes involve changes to the objects themselves, the downstream effects of the event's occurrence. Dependency relations change over time: specificity, strength, and embeddedness may increase or decrease. The ontic states (the dependencies) underwrite the epistemic states (the traces). The ripple model is concerned with this ontic notion. At t_2, the event at t_0 (the pebble hitting the pond) has a wider area of effect than at t_1. There are, then, more potential traces at the later time. Hence, there are more dependencies between t_0 and the events at t_2 than there are at t_1. Moreover, different events can have different sized areas of ontic effect. Just as throwing different sized pebbles, or throwing with differing force, will create ripples of different magnitude, different causes have more or less dispersed areas of effect.

So far, this picture emphasizes the number of downstream effects, but as time goes by the *heterogeneity* of effects also increases. For example, as we saw in chapter 2, signs of Neoproterozoic glaciation are dispersed in space: Neoproterozoic strata, with their distinctive dolostone caps, are found in many locations. However, such signs are also heterogeneous. Dolostone caps are one kind of trace; paleomagnetic traces are another. As effects breed further causal descendants, the types of potentially exploitable dependencies increase. We could represent heterogeneity with different images within the lines, as we see in figure 5.2.

Just as the number of downstream effects have increased between t_1 and t_2, so too has the number of *types* of effect. If it is right that, usually, effects spread from their causes over time, then we should see more types of effects at later times than earlier. This is crucially important: If the number of exploitable streams increases, often so will our evidence. Let's call this phenomenon *dispersal*.

Dispersal: Some event is more or less dispersed depending on the size and heterogeneity of its set of causal descendants.

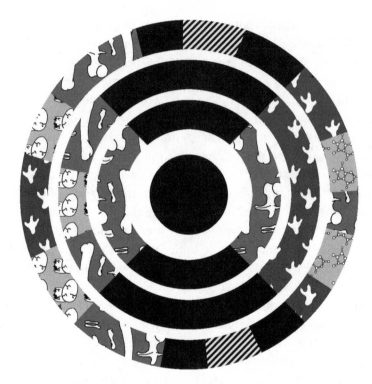

Figure 5.2
Ripple model with heterogeneity. (Image by Leonard Finkelman.)

Now consider the following two ceteris paribus claims:

1. Ceteris paribus, the dispersal of an event increases over time;
2. Ceteris paribus, the more dispersed an event, the more potential traces it has.

Figure 5.2 represents claim 1: Later times have more effects, and more types of effects, than earlier times. It is plausible because causation is transitive—or near enough. If causation is transitive, then when some event *A* causes another event *B*, and *B* causes another event *C*, then it follows that *A* causes *C*. It is probably false that causation is always transitive—that *B*'s causing *C* guarantees *A*'s causing *C*, since there are potential counterexamples. Consider one from Michael McDermott (1995). A right-handed man detonates a bomb by pressing its switch with his left forefinger. The previous day a dog bit his right hand, which is why he used his left forefinger. In these circumstances, we have the event of the dog bite, *A*, which is causally related to the left-forefinger switching—*B*—which is further causally related to the detonation of the bomb, *C*. If causation is transitive, it follows that the dog biting the hand is *causally responsible* for the subsequent explosion, which is surely a mistake. I am not completely convinced by these kinds of cases, since they seem very sensitive to how we describe the relevant causes (Lewis 2000). Regardless, claim 1 doesn't require that causation is *always* transitive (for transitivity to be constitutive of, or part of the "logic" of, causation); just that it *often* is.

Let's switch to another toy example. Imagine releasing a pinball in a standard pinball machine. This will have downstream effects: The ball bounces off various triggers, which in turn cause various lights to flash, points to accumulate, electricity to be expended, and the player to react in various ways (flipping the paddles, shaking the machine, feeling frustration or elation, etc.). As time goes on, the ball's release accumulates both effectual scope and types of effects. This is because it interacts with more entities as time goes on, and the causal consequences of *those* interactions themselves have further downstream effects. That the effects of an event tend to accumulate doesn't require that causation be transitive *necessarily*, but does make it plausible that dispersal increases through time. Claim 1, then, is at least prima facie plausible. Let's turn to the second claim about dispersal: that increased dispersal leads to increased potential traces.

Recall that to be a *trace* of a past event, some contemporary state of affairs must be both downstream of it and evidentially relevant to the past event by virtue of some justified midrange theory. In chapter 3, I provided a story about evidential relevance: Midrange theories identify *dependency relations* of various stripes between past events and their traces. If I were to change the velocity, angle, or timing of my throw, the various ripples caused by the pebble would be different. Make subtle changes to the power of the pinball's release, and further changes will likely begin accumulating downstream. Perhaps the ball is going slightly slower, and so its trajectory doesn't lead it straight between my desperately flailing flippers. At t_1, when the event is of low dispersal, there is a relatively small set of events that have dependencies with our target at t_0. At t_2, dispersal has increased. There are now more events with dependencies (and as heterogeneity has increased, there will be different types of dependencies). As dispersal increases, then so too do the number and type of dependency relations. The more dependency there is, the more opportunities there are for exploitation. Assuming midrange theories are available, and the other conditions for accessibility, retention, and discovery (from chapter 4) are met, it is possible for events to be *more* accessible further downstream than they were comparatively near (temporarily speaking) to the event!

And so, holding signal decay fixed, we can see that dispersal can cause our epistemic situation to *improve* between t_1 and t_2, due to dispersal. There are more potential traces, as there are more dependency relations between states of affairs at t_2 and the event at t_0 than there are at t_1. That is, claim 2 is true.

Consider figure 5.3. In this case, the dashed circles represent samples or scopes of enquiry. They are data sets, that is, they are traces that scientists have accessed. We can imagine the dashed circles as "readings" of the ripples emanating from the event at t_0. By virtue of their dependencies

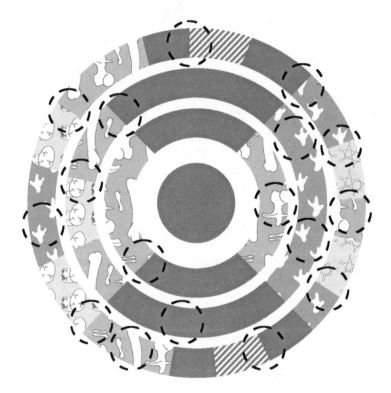

Figure 5.3
Ripple model with samples, demonstrating "overdetermination." (Image by Leonard Finkelman.)

with the event at t_0 and midrange theories that capture those dependencies, these can be drawn on to reconstruct the event. Basically, because t_2 has more data—more traces—it provides more insight to t_0. At this stage, the model does not include signal decay, so all of the potential traces at t_1 are retained at t_2. Insofar as the data from t_1 is sufficient to learn about t_0, and t_2 includes more data than t_1, then t_2's data is more than sufficient. Increasing dispersal creates evidential redundancy. Perhaps, as Cleland suggests, it epistemically overdetermines the initial conditions. t_2 has more evidential redundancy than t_1. Moreover, the increased heterogeneity of the trace set increases access to the past (this is discussed in more detail in chapter 6).

Dispersal improves our epistemic situation in two ways. First, the size of the potential trace set increases. There will be, then, a larger data set to draw inferences from, which potentially provides redundancy, protecting against signal decay. Moreover, presumably a larger potential trace set increases the chances of discovery. Second, a highly dispersed event leaves heterogeneous traces. The sauropod lineage did not merely leave bodily remains as fossils. Its members also left nonbodily remains: footprints, coprolites, eggshells. More ephemerally, they may have influenced the evolution, behavior, and morphology of lineages around them. Although I have found no research on this, one would think that animals requiring such prodigious amounts of food as sauropods would have had an evolutionary impact on Mesozoic flora. Additionally, some lineages (although sadly not sauropods) have descendants still among us. These different traces can all be leveraged and bought to bear in reconstructing the past. Given my discussion in chapter 6 about the importance of utilizing multiple streams of independent evidence in confirmation, a highly dispersed, and so heterogeneous, trace set provides significant epistemic advantage.

This is, of course, only half the story. By the model so far, increased temporal distance between a target event and researchers is epistemically advantageous. Put simply, it is better to be at t_2 than t_1. But that can't be right: If we're taking the analogy seriously, the ripples at t_2 would have interfered with one another, dissipated, and faded. It is time to draw on Turner's insight about the degradation of traces.

5.1.2 Faintness and Gappiness

Although it is right that (ceteris paribus) the effects of an event become more numerous and diverse the further downstream we go, they will also degrade. I shall understand this in terms of two concepts: "gappiness" and "faintness." Let's start with the former.

A nice way into "gappiness" involves considering phylogenetic reconstruction. Phylogenies (of large, sexually reproducing animals at any rate) are hypotheses about the ancestral relations between various lineages. They are basically family trees, representing who is more closely related to whom. In more ambitious contexts, this includes estimating the timing of various splitting events between lineages. One way of testing a phylogenetic reconstruction is by comparing it to the appearance and disappearance of lineages in the fossil record. For instance, one hypothesis is that two families of sauropods, *Diplodocus* and the titanosaurs, are sister-clades, splitting from a common ancestor sometime in the early to mid-Mesozoic. Such a claim makes predictions about which sauropods will be found in various Mesozoic formations. If the titanosaurs are basal (the original form), for instance, then we should not expect to see *Diplodocus*-like specimens earlier in the era. However, neither the fossil record nor phylogenies are trustworthy. In the face of some mismatches, particularly negative results, paleontologists suggest there are "ghost ranges": geological periods where some lineage probably existed but happened to leave no trace. As I will expand upon later (especially in chapter 7), patterns in the fossil record can be read in two ways: as geological signals, revelatory of biases in the processes of fossilization; or as biological signals, revelatory of which lineages existed. Gaps in the former sense involve ghost ranges; they explain the incomplete fossil record. Although it is sometimes assumed that such periods are more frequent further back in life's history, M. Wills (2007) has shown that they are also frequent in more recent strata—processes of fossilization and stratification leave an incomplete record of the past. A ghost range is a postulated gap in a lineage's downstream effects: Those particular ripples have disappeared.

I'm not sure how to capture what a "complete" record regarding some event in the past might be. Perhaps the "complete fossil record" would be every organism preserved in state at the moment of its demise,[1] or one of Horwich's perfect recording systems matched up to each property of every organism. However, we can get a handle on how sets of traces might be more or less complete. Consider again the Neoproterozoic glaciation. The dolostone caps are found in numerous sites, spread throughout the world. That the glaciation was global is inferred from their bountiful number, and their dispersal. If there were more sites, the set would be richer, less *gappy*; if there were fewer, it would be more impoverished, *gappier*. The more gaps, the less data (of less variety!), and thus the weaker the base from which to draw inferences. To take another example from chapter 2, *Argentinosaurus* length, which is estimated from a few vertebrae, is significantly more contentious than *Brachiosaurus* length, for which more complete skeletons are available (see Soinski, Suthau, & Gunga 2011).

Here is a highly abstract, counterfactual way of thinking about gappiness. A potential trace set is gappy to the extent that it is smaller than the trace set there would have been, were signal decay to not have occurred. That is, compare a model that contains only dispersal (that is, no signal decay) to the potential trace set we in fact have. The more the former diverges from the latter, the gappier it is.

So far, the ripple model has two components: dispersal and gappiness. Dispersal tracks the increased number and heterogeneity of downstream effects over time. Gappiness is a measure of a potential trace set's incompleteness. However, it is not the end of the story about signal decay. Dependencies between a past event and the contemporary world can be more or less *faint*. Although dispersal and gappiness are features of the world, understood in terms of dependencies, it is helpful to understand faintness in terms of traces. I understand faintness as the combination of two epistemic features. These are the stability or otherwise of midrange theory and the difficulty of trait discovery.

In chapter 3 I explicated different types of "dependency" relations. Dependencies could be more or less *enmeshed*, that is, a contemporary structure can have different numbers of dependency relations with a past event's variables. They could be more or less *informative* or specific, that is, have differing mappings between states of the trace and states in the past. They could have differences in *strength*: the extent to which the trace makes the event likely. As time passes, these dependency relations change. The relationship between, say, the pinball's release and the first trigger it hits is pretty direct. Changes in the velocity of the release will make fairly straightforward changes to the angle with which it hits the trigger: The dependency between release velocity and the triggering event will be highly informative. However, further downstream, more events with their own causal influences would have occurred. Say the pinball slips between the flippers. This later event would depend not only on the angle and force of the original release, but the player triggering the paddles at the right times, and so forth. As time goes by, the dependencies between the initial event and its downstream descendants will get weaker and less informative. The lines of information connecting the past with the present will become less direct, more distorted. More sophisticated and complex midrange theory will be required to make sense of it.

This indirectness breeds opportunities for mistakes and necessitates wider margins of error. Paleobiologists are very confident of *Brachiosaurus* length—relatively complete skeletons are available, and missing parts can be reconstructed on the basis of robust regularities about sauropod morphology. However, there is much less certainty about *Brachiosaurus* weight. Reconstructing weight requires more than a ruler. It requires assumptions and judgments to be made about overall density—a property difficult to access from fossil remains. For instance, if it is right that sauropods have a birdlike respiratory system and are highly pneumatized, then it is likely that they were significantly less dense proportionately to mammals. But we have no experience of terrestrial animals anywhere near that size. By virtue of this, estimates of *Brachiosaurus* weight are much less stable and confident than those of length, because the traces are so much fainter (Sellers et al. 2013).

Moreover, locating and isolating traces can be tricky. Possibly the best-known paleohistoric event is the K-Pg impact, along with the ambiguously related mass extinction that sent the dinosaurs on their way. Everyone agrees that the impact happened—indeed, the impact's occurrence is surely an example of the kind of evidential redundancy Cleland highlights. As we have seen, the presence of the iridium layer, shocked quartz, and extinction event are sufficient to justify belief in the impact. Discovering an enormous crater was unnecessary for epistemic justification. However, ascertaining the relationship between the impact and the extinction event is a much trickier proposition. Was the impact a mere coup de grâce for an already collapsing biota, or did it play a major role in triggering the crisis? These hypotheses each make different predictions about the strata leading up to the extinction event. By the former, we should expect to see decreasing diversity; by the latter, Mesozoic biota ought to still be thriving right up to the bitter end. However, figuring out the patterns of biodiversity in the last million years of the Mesozoic is difficult. In addition to the fossil record's gappiness, it is often a biased and misleading recording device. Locating a signal and attaching it to the past is tricky. Ward (1983, 1990) tackled the problem by hunting for ammonites, a diverse, successful lineage of nautilus-like aquatic organisms. If ammonite fossil number and diversity drop off toward the end of the Cretaceous, this would provide reason to think biodiversity was decreasing overall. The thought is that the fortunes of a rich, successful clade like the ammonites probably reflect worldwide trends—a risky induction already. Even beyond that risk, there is further trouble: Where, of all the world's Cretaceous rocks, should we look? After all, a *lack* of finds could be a geological signal—a ghost range—and so fail to tell us anything about the biodiversity of the late Cretaceous. In earlier work, Ward hunted a Spanish fossil field and found very few ammonites. He concluded that this supported the coup de grâce hypothesis. However, later, he found bountiful ammonites in a nearby location—apparently supporting the opposite conclusion. The signal from ammonite scarcity to late Cretaceous biodiversity is doubly faint. First, the inference from ammonite richness to overall biodiversity is tricky (to say nothing of the assumption that ammonite finds from one location are revelatory of worldwide ammonite diversity). Second, the signal is difficult to uncover. Cleland (2013) rightly uses this case to illustrate the difficulties involved in testing historical hypotheses.

We can identify two further properties for the ripple model, then:

Gappiness: the relative "completeness" of a potential trace set.

Faintness: (1) the relative difficulty of linking traces to the past (quality of midrange theory), (2) the relative difficulty of *locating* traces.

I have already implied that increases in gappiness and faintness are bad news, epistemically, but it is worth spelling this out. Here are three more ceteris paribus claims:

1. Ceteris paribus, faintness increases through time.
2. Ceteris paribus, gappiness increases through time.
3. Ceteris paribus, the higher the faintness or gappiness of an event's traces, the less retrievable the event.

It is worth highlighting the sense in which 1 and 2 are ceteris paribus. We have already seen that increased temporal distance doesn't entail decreased evidence (see Sober & Steel 2014). The claim is that holding fixed the event in question—its initial conditions, as well as its subsequent dispersal—we will at least typically see increases in gappiness and faintness. Faintness and gappiness, of course, are closely related. Especially so, on informational or causal accounts of traces. As time goes by, causal connections could be stretched and diluted until they disappear (after all, remember, causation probably isn't purely transitive!). As time goes by, information about past events degrades to the point of erasure. However, on a theory-dependent account of traces, the two come apart nicely. A faint trace *is a trace* by virtue of midrange theory connecting it to the past. However, its evidential value has diminished due to the instability of midrange theory and the long reach required for the inference. A gap is *not* a trace (at least prima facie, as we will see), since it is not connected to the past. This matters, since presumably cases of "complete erasure" are rare. After all, the materials involved in causal interactions in the past have not been destroyed, at least atomically. It is at least possible that, with access to all the contemporary facts, much if not all of the past could be reconstructed.[2] However, this "in principle" discussion has very little to tell us that is epistemically or methodologically relevant to historical science. Imagine that some of the atoms that constitute me contain information about, or are causally downstream from, Snowball Earth or the sauropods. Well, a fat lot of epistemic good that would do. We can't locate the relevant atoms, and moreover lack the midrange theory required to connect them to the past. Accessibility matters.

In some contexts, highly gappy trace sets nonetheless license stable, well-confirmed inferences. A single fingerprint can ensure a murder conviction;[3] the Chicxulub crater is a single trace but is pretty unambiguous; there being only a single, tiny remnant of *O. tharalkooschild* doesn't seem to undermine her reconstruction. Notice two things about such situations. First, the dependency relationships are strong and specific, providing a clear signal to the past. Fingerprints are good evidence in criminal cases because of the strength and specificity of the dependency relations between individual humans and their fingerprints. Each of us has (in effect) a unique fingerprint pattern, and so if the mapping relationships are known, individuals can be picked out with confidence. Second, as I will expand on in chapter 6, midrange theory is "stable." Stability tracks the sensitivity of a hypothesis to midrange theory (Wylie 2011). Unstable inferences have wildly different results if midrange theory shifts. Stable inferences are more resilient. It would take some impressive theoretical changes for the Chicxulub crater to end up *not* signaling an extraterrestrial impact (although presumably the timing of said impact is less stable). In stable conditions, gappiness does not undermine inference. And so, under some circumstances gappiness is not problematic. However, this is not all that frequent. Moreover, it is certainly not available for the "unlucky" circumstances I shall soon focus on.

But wait. Ward took a *lack* of finds as evidence of a decrease in Mesozoic biodiversity. It is true that an absence of traces can occasionally distinguish between historical hypotheses, but this is tricky. Sober (2009) has a nice probabilistic discussion of just this point. In essence, whether or not absence counts as evidence depends crucially on how likely it would be for us to find a *lack* of absence in that case. Everyone agrees that a meteor strike occurred at the K-Pg boundary. However, as we saw, it is not clear what role the impact event played in the extinction. It could be that the meteor finished off already diminished biodiversity. As we saw earlier, the information we need could be in the fossil record, and uncovering this was Ward's intent in his investigations. Additionally, comparing the winners and losers of the mass extinction can be illuminating. How did such a resilient bunch as the (non-avian) dinosaurs meet their demise while apparently fragile groups— frogs, for instance—survive? However, testing via gaps in the record is difficult. As we have seen, the fossil record is not merely gappy but also inconsistent. So some time periods, and some areas, are more amenable to preservation than others. In other words, an absence of fossils can fail to be a biological signal, revelatory of life's history, but rather a geological signal. In this vein, Cleland (2011, 2013) emphasizes the idea that historical scientists make "prognostications" rather than "predictions."[4] "The problem with most so-called predictions inferred from hypotheses about long past events and processes is that they are too vague to fail," she writes. "This is in stark contrast to classical experimental science where failed predictions are viewed as a very serious threat indeed" (Cleland 2013, 6).

Paradigm physical theories are precise, and physical systems can be carefully manipulated to provide rigid tests. A diminishing of fossils near the K-Pg boundary doesn't necessarily speak to the truth or otherwise of hypotheses about the meteor's role in the mass extinction. I think this is too pessimistic. First, failure to find evidence over time should affect our credence in the hypotheses. If it were found, at numerous sites, that once bountiful fossils became rare, this would be reason to downplay the role of the impact. Second, midrange theory lends a hand. We know, for example, that some lineages are more likely to fossilize than others. Ammonites, for instance, were widespread, bony and aquatic—and so, were as likely as any lineage to leave a record. Moreover, as midrange theories tell us which conditions are friendlier to fossilization, it can tell us where to find them. Midrange theories provide expectations of the gappiness, faintness, and dispersal of trace sets—and this can tell us when to take absence as evidence.

5.1.3   The Complete Model

We are now in a position to present the complete ripple model pictorially, with figure 5.4.

Figure 5.4
Complete ripple model. (Image by Leonard Finkelman.)

The shade of the lines represents the "faintness" of dependencies between the original event and its potential traces. Gaps in the lines represent, well, gaps. Although the size and heterogeneity of dependencies increase over time, so too does the number of gaps and the faintness of those dependencies. At t_1, there are clearer connections, stronger and more specific dependencies, to the event at t_0 (hence the darker line), while at t_2, these connections have become less direct, and the line has faded. By t_3, evidence is only faintly connected to t_0, and the dependencies have become not only dispersed but also extremely gappy.

What does all this mean? By the ripple model, some event's *epistemic retrievability* (for traces) is set by the following:

1. *dispersal*, or the expected number and heterogeneity of the dependencies between now and the past event;
2. *faintness*, or (a) the ease of trace discovery; (b) the ease of trace extrapolation; (c) the stability and support of midrange theory;
3. *gappiness*, or the "completeness" of the potential trace set.
4. The informativeness, strength, and enmeshedness of the dependencies between the event and contemporary states of affairs.

When considering our epistemic access to events in the past according to this model, we should ask how dispersed, faint, and gappy their remains will be, and how enmeshed, strong, or informative the relevant dependencies should be. Events in the past with high gappiness and faintness, unless offset by high dispersal, will have a low retrievability. If dispersal is high and dependencies do not fade or become gappy too quickly, events should be easily retrieved. Midrange theory provides insight into retrievability: Given an event's occurrence in the past, what kinds of traces ought we expect? The impact event at the close of the Cretaceous would have high retrievability. Such a global calamity leaves widely dispersed traces, hardly faded in the intervening millennia. These events should be relatively easy to access. Other events have low retrievability: the fate of a single sauropod; what Napoleon had for breakfast one insignificant morning of his childhood in the late eighteenth century; whether gigantism evolved in sauropods in response to selection for gut size. Such cases are likely to be gappy: The chance of a particular sauropod fossilizing, or young Napoleon's breakfast being recorded, is miniscule. They are also likely to be faint: Whether it is plausible for selection to have targeted gut size in sauropod evolution requires some heavy lifting from midrange theory; the dietary habits of 18th Century Corsicans might not be well understood.

"Retrievability" is our expected capacity to access the past, given the ripple model of evidence. It sets what our credence ought to be, given our background theories, in discovering some past fact. Naturally, we can still have knowledge of facts with low epistemic retrievability. As we saw in chapter 2, McNab (2009) denies that selection would favor increased size for the purpose of increased gut capacity. The thought is that the cost of digesting increased amounts of unmasticated food would not be mitigated by increased ingestion volume. This argument asks much of midrange theory. It relies on comparisons of contemporary mammal and reptilian "field energy expenditure," correcting for the varanids' (monitor lizards, including the Komodo dragon) carnivorous lifestyle, the incorporation of biogeographical information, and various assumptions about sauropod lifeways and Jurassic flora. Heavy lifting indeed—but nonetheless something that generates hypotheses that we could have good reason to believe.

Retrievability differs for different events, entities, and processes. Compare the retrievability of *Argentinosaurus* length with mass. To discover the length of *Argentinosaurus*, paleobiologists had to extrapolate from very little data, relying on midrange theory, which predicted total sauropod length based on vertebrae size. As Soinski, Suthau, and Gunga (2011) document, these extrapolations are made on the basis of more complete skeletons of related sauropods. Length estimations vary depending on which sauropods are used. Because inferring length is sensitive to which contrasts are used, this contributes to faintness. Determining *Argentinosaurus* mass is fainter still. Here, in addition to length, we must estimate the density of the animal. Such estimates are even less stable. Recall how Gunga et al.'s 2008 study, which incorporated theories about sauropod physiology (in particular air sacs and pneumatized skeletal features) dramatically reduced the mass of *Brachiosaurus* from 74.4 to 38 metric tons.

The events of the Neoproterozoic were, overall, highly retrievable—at least in broad detail. A worldwide glaciation event, followed by a sudden heating would be widely dispersed. Moreover, the expected trace set is not particularly gappy, especially when compared with, say, events from earlier in Earth's history where we would expect the stratigraphic record to be destroyed by subduction and other geological processes. The geological record is relatively stable for such a dramatic event, which occurred a modest 600 million years ago. Moreover, the rocks' record is unambiguous: Lithography, radiometric tests, and so on, leave little doubt as to their signal. However, there are tricky questions about the Neoproterozoic. Determining whether episodes were snowballs or slushballs appears to outrun available evidence (although, as we have seen, the presumed survival of metazoans suggests a complete snowball is unlikely). As I shall emphasize in later chapters, geologists resort to other sources of evidence like simulations to uncover which is more likely.

From this discussion, we can identify two epistemic problems we face when dealing with past events with highly faint and gappy downstream effects. First, underdetermination: The available traces themselves are insufficient to decide between hypotheses. Second, stability: Our reconstruction of the past (particularly measurements) can be highly dependent on midrange theory, and small shifts can have dramatic consequences (see Gunga et al. 2008). It is, broadly speaking, my task to argue for a set of epistemic resources that historical scientists have at their disposal that (1) are not necessarily captured by the ripple model and (2) allow them to overcome both underdetermination and stability problems when faced with difficult epistemic targets.

It is worth pointing out that we already have good reason for optimism about many of the events historical scientists care about. They are not generally interested in uncovering every detail about the past. They are, rather, interested in important events that made a difference to history's unfolding. Mass extinctions, radiations, calamitous geological events, and the formation of geographical features (seas, mountain ranges) are important in part because of the depth and diversity of their temporal impact. Such events have high dispersal and low faintness. The events historical scientists really care about, in many situations, are events for which there will be bountiful traces and low signal decay. However, I am interested in what resources historical scientists can draw on when their target has low retrievability; I shall call these "unlucky circumstances," and we will get to them in the next section.

The ripple model also provides a new perspective on the tension between Cleland and Turner. For Cleland, the investigator at t_3 is in a better situation than the investigator at t_1 because t_3 has a more dispersed data set, and dispersal provides more opportunities for discriminating hypotheses. For Turner, the investigator at t_1 is better off because the data set is less faint and gappy. That Cleland's dispersal and Turner's degradation (faintness and gappiness) are compatible suggests that the question of whether the historical sciences overall face situations painted as Cleland does, or in Turner's more somber hues, isn't particularly interesting. The ripple model replaces a dichotomy with a graded set of possibilities, with Turner and Cleland focusing on either extreme. It is more pressing to work out how to determine which epistemic situations various historical scientists face, and ask what they can do about it. The answer is to provide a range of epistemic possibilities—a continuum between targets with high retrievability and those with low retrievability. Scientists with different research agendas target different events and processes in the past. Where such targets fall on this continuum tells us, via the ripple model, what kind of task the scientists in question face.

My interest is in scientists whose targets have low retrievability by the ripple model. After all, if we have grounds for optimism about those difficult cases, surely we have grounds for optimism about historical science generally. This shows that Turner and Cleland's dispute must be understood on a case-by-case basis. Only after examining midrange theory and a wide range of individual, representative cases, could we conclude whether historical sciences overall tend to have overdetermined or underdetermined hypotheses.

Tucker (2011) has also synthesized Turner's pessimism and Cleland's optimism. Examining Darwin's explanation of biological form and function, he argues that historical science begins by asking questions about patterns. First, a general pattern is identified. Darwin closely examined the homologous relationships between many species, noting that their forms followed distinct patterns. So, a general pattern is identified. Second, some general explanation is provided for the pattern. For Darwin, patterns of homology were explained as patterns of inheritance. Homologues exist by virtue of sharing a common ancestor. Finally, with those past states of affairs on the table, we ask questions about them. What were the common ancestors of lineages like? Tucker claims that theorizing in the second stage is overdetermined (there is bountiful evidence that two homologues share a common ancestor) and that theorizing in the third stage is underdetermined (reconstructing the actual properties of common ancestors is very difficult).

Tucker's approach is quite different from mine (although they could be complementary). Where he focuses on the kinds of questions historical scientists ask, I focus on epistemic situations and resources. I agree that explanatory interest plays a role in the evidential status of hypotheses. That an extraterrestrial impact occurred, or there was a mass extinction, is relatively easy to establish—but whether the impact caused the extinction is a tough nut to crack. One worry with Tucker's approach is his reliance on common cause explanation—see chapter 6 for my reservations about this. Moreover, I am suspicious that Darwin's example is representative of historical enquiry. It isn't clear that investigations of sauropod gigantism or the Neoproterozoic climate follow his hypothesized trajectory. Tucker views historical enquiry as shifting from the easy to the increasingly difficult. My discussions of dependency relations and the ripple model should give us traction on just what the "easy" and the "hard" cases are.

In their 2014 paper, Elliot Sober and Mike Steel ask a very similar question to mine: "How does increasing the temporal separation between present and past affect the amount of information that the present provides about the past?" (558) Their approach is, again, quite different, and contrasting our positions is instructive. They abstract from the details of midrange theory made explicit by my account of traces and (as implied by the quote) take an "informational" account of traces. They then consider a very simple branching system with four properties. Informally stated, these are (1) that the probabilities of transitions (that is, the tree branching) are constant, (2) the states are connected, insofar as there is at least some chance of returning to a previous state, (3) it is possible to remain in the same state, and (4) what they call the *Markov* property. This last property holds when the total probability of reaching a particular state at a particular time depends only on the probabilities of the states immediately prior to that time. That is, each time-slice screens off the probabilities of the one before, a kind of temporal version of causal closure. They show mathematically that if these four properties hold, the information available connecting the past variable to the present will approach zero. In effect, this is an information-based, formal way of capturing gappiness and faintness. Note, of course, that this doesn't show that the more distant a past event the worst the information we have about it (this depends on the branching events and the probabilities). They use this framework to understand, in highly abstract terms, different kinds of evolutionary processes and how the influences on them tend to be more or less information preserving. One surprising result is that under some specialized conditions information will not degrade after all.

I think Sober and Steel's work is extremely helpful and is a good step toward understanding the epistemic underpinnings of some historical inferences (particularly in phylogenetics). However, it is less helpful for my purposes for several reasons. Most obviously, they adopt an informational account of traces. Such a stance limits how much it can inform us about historical science in practice, for the reasons I have already mentioned: They cannot take the retrievability of traces into account. That is, midrange theory and the probability of the traces being discovered is not part of the story. More important, all information is treated as the same—as a single measure. This makes it difficult to distinguish between different types of evidence, and, as we will see in chapter 6, this is essential for understanding the warrant of many historical inferences, particularly those that rely on consilience or coherence. To reiterate what will become a theme: I am skeptical of the value of deeply abstract accounts of evidence or information for the purpose of understanding the epistemic situations of scientists. Much of the action depends on often messy local detail.

The ripple model is only a first pass at characterizing the factors determining the retrievability of past events. We shall build on it in future chapters. In discussing Turner's pessimism, I mentioned that historical scientists have access to a wider set of epistemic tools than he allows. The ripple model of evidence tells us about the accessibility of events in the past in regards to *traces;* in the following four chapters, I argue that historical scientists have more at their epistemic disposal than this.

So, how do the claims of this chapter so far fit with what we learned about traces in chapter 3?

In chapter 3, I stipulated and defended an account of *traces.* A contemporary state of affairs is a trace of a past event just in case, first, that contemporary state of affairs is downstream and second, given a justified midrange theory, that state is evidence of that past event. Trace-based reasoning involves the scientific exploitation of dependency relations between the present and the past. Dependencies can be understood in terms of how embedded, strong, and informative they are. Midrange theory works by explicating those dependencies.

In this chapter I have introduced the *ripple model*, which conceptualizes our epistemic access to the past in terms of the *dispersal, faintness*, and *gappiness* of dependencies. Ceteris paribus, at increased temporal distances, dependencies become more dispersed—leaving us with more, and more heterogeneous, traces. Yet those traces grant less access to the past as they become gappy (data sets will be incomplete) and faint (our confidence in the connection afforded by midrange theories decreases).

This picture is a framework for understanding historical evidence. It is not a view on whether, say, historical scientists are in some particular epistemic situation or another. It could turn out that many of the events that historical scientists are interested in are the ones that have a high epistemic retrievability. Regardless, I am now in a position to highlight the epistemic situations that interest me.

5.2 Unlucky Circumstances

"Lord if it wasn't for bad luck, I wouldn't have no luck at all."
—Lightnin' Slim, "Bad Luck Blues"

Some scientific tasks are easier than others. Discovering *Obdurodon tharalkooschild*'s size or her platypushood is relatively straightforward. Understanding facts about her behavior or evolutionary history (What was the evolutionary function of her gigantism? What did her teeth enable her to hunt?) are trickier. Many questions about sauropods and the Neoproterozoic climate are similarly tricky. Determining sauropod mass requires understanding sauropod density, and this requires knowledge of their physiology: how pneumatized were their skeletons, whether they have other avian-style respiratory features, and so forth. Such information is not easily or directly forthcoming; it must be coaxed from an often stubborn, recalcitrant past. My aim in this section is to characterize such circumstances and then provide three reasons for pessimism about our capacity to know much about them. Later, I will show that this pessimism is too quick.

In the introduction, I spent some time describing pessimistic attitudes about the historical sciences and the methodological upshots of this. Most strikingly in some strains of archaeology, but in other places as well, a general pessimism about our epistemic capacities vis-à-vis the past is linked with a kind of conservatism about method (or, a kind of free-for-all subjectivism!). These expressions of pessimism tend to rely on general, or at least systematic, claims about the past. Such general claims are mistaken: Some historical targets are quite forgiving. But sometimes things are harder. Let's call these cases "unlucky circumstances." By identifying these harder situations, and further pointing to the epistemic resources that historical scientists appeal to in such contexts, I will undermine pessimism generally. As we will look at in more detail in chapter 11, if I can motivate at least *some* confidence about our success in unlucky circumstances, it seems as if we should be much happier with the lucky ones.

There is something artificial about my use of the term "luck." "Luck" implies a certain lack of control on the part of historical scientists. However, as Turner (2016) rightly points out, historical scientists have a fair bit of epistemic access to which investigations are likely to produce results and which are not. Research targets are not selected at random (and indeed funding strategies encourage research likely to get results!). Presumably a smart historical scientist will look for untapped, "lucky" circumstances. By showing that progress can be made in unlucky circumstances as well, I similarly provide reason to conduct such research (see chapter 12).

Let's contrast some examples. First, consider the length of *O. tharalkooschild* and the weight of *Argentinosaurus*. As we have seen, inferring the length of the platypus from its tooth is kosher. This is because the dependence between them is *strong*: In mammals, tooth size is proportional to body size with very little variation (other than obvious cases, such as the upper incisors of saber-toothed predators, which we will meet in chapter 8). Moreover, it is *specific*: Changes in tooth size track changes in body size systematically. By virtue of these dependencies, *O. tharalkooschild*'s molars are firm (rather than faint). We are not so lucky when it comes to *Argentinosaurus* weight. Here, the relationship between information we have access to (the size of the animal) does not track weight in a straightforward manner. It is neither strong nor specific, since other physiological factors, pneumatization for instance, interfere. Moreover, the likelihood of mammalian teeth fossilizing, making it to the present, and being discovered is much higher than for the relevant sauropod remains being recoverable. Teeth are easily fossilized and their morphology is often retained to a remarkably fine grain. This is not so for the physiological information required for Sauropod reconstruction. *O. tharalkooschild*'s molars then, are likely to be less gappy than *Argentinosaurus* physiology (and regardless, as we have seen, due to the strong dependencies involved, a single tooth will do). Finally, in the platypus case we have the epistemic goods—the well-supported midrange theories—required to connect trace to past entity. We are not so lucky with the sauropods. There, we lack an extant animal that brings together both large size and birdlike respiration, so we don't have a good basis upon which to reconstruct. Much speculation is necessarily involved.

Now compare reconstructions of *Argentinosaurus* weight to the tropical glaciers of the Neoproterozoic. As before, we have very good reason to believe the glaciers existed, as well as in fairly fine-grained claims about their timing and their location, because of the strength of the dependencies, and the midrange theories connecting them. But I want to highlight their dispersal. There are many, and varied, traces of the ancient glaciation. The lithography of rocks in Neoproterozoic strata, their paleomagnetism, the dolostone caps, and perhaps even the Cambrian explosion (see chapter 6) are sources of information about these glaciers. The global-scale nature of the event ensured that these signs are scattered in many locations. The record is relatively ungappy, and it speaks relatively unambiguously. The dispersal of the event engenders rich, heterogeneous data sets. Again, we are not so lucky with the weight of *Argentinosaurus*. The information we have to draw on in its reconstruction is significantly less dispersed: Remains are incomplete and not heterogeneous (it is not obvious which other lines of evidence might be drawn on to support reconstructions of *Argentinosaurus* weight).

So, we can identify *unlucky circumstances* as those with low retrievability by the ripple model: scenarios involving limited trace evidence. They are (1) faint; traces are difficult to locate, and midrange theory is unstable. They are (2) gappy; the potential trace sets are incomplete. They are (3) not dispersed; there are limited lines of evidence connecting trace to past entity. Moreover, (4) the dependency relationships between trace and past entity are weak, unembedded, and nonspecific.

In such circumstances, you would be forgiven for thinking that our epistemic task is hopeless. However, in chapter 2 we have already seen that investigations under such unlucky circumstances can be surprisingly rich and progressive. And so, what might motivate a pessimistic attitude in unlucky circumstance? That is, what could motivate an empirical bet that current underdetermination and instability will not be resolved? I identify three reasons, three targets for the rest of the book.

Characteristically, basically by definition, unlucky circumstances have limited trace evidence. The dependency relationships are weak and unspecific, and the midrange theories identifying them are ambiguous and uncertain. A source of pessimism, then, could be the claim that the only evidence we have about the past is trace evidence. We could understand this as an inductive argument:

1. Our available evidence about the past is limited to traces;
2. In unlucky circumstances trace evidence is of poor quality;
3. Therefore, reconstructions of the past in unlucky circumstances are likely to be of poor quality.

To be valid, the argument requires a further premise stating that poor quality evidence leads to poor quality reconstructions. As we will see in chapter 6, there is reason for caution here. There I will argue that the role of consilience and coherence can make for rich reconstruction in spite of fragmentary evidence. Premise 2 of the argument is true roughly via stipulation—I am focusing on those very cases where traces are not able to do much work. If premise 1 is false—if there are other sources of evidence about the past—then the argument could be circumvented.

A way of strengthening the argument would be to also argue that the situation vis-à-vis traces is *unlikely to improve*—that is, *further traces are unlikely to be found*. This is the pessimistic bet Turner encourages us to make based on reflection of the "limiting" nature of midrange theory.

1. Our available evidence about the past is limited to traces;
2. In unlucky circumstances trace evidence is of poor quality;
3. Because trace signals degrade, we are unlikely to uncover further traces;
4. Therefore, reconstructions of the past in unlucky circumstances are likely to remain poor quality.

My second target, then, is premise 3: the pessimistic empirical bet about the likelihood of discovering new traces.

The third way of motivating pessimism is a logical corollary of the first and second. Specifically, *historical scientists cannot manufacture evidence*. As I will make explicit in chapter 9, the thought is that historical scientists are unable to conduct experiments to test their claims about the past. Experimentalists are able to overcome unlucky circumstances—a lack of naturally occurring evidence—by constructing their own tests; making their own luck. This puts nonexperimentalists at a disadvantage. As Derek Turner has argued:

The experimental manipulation of microphysical entities and events makes it possible for scientists to test, and in some cases, confirm new theories. We cannot, however, manipulate things and events that existed and occurred long ago. This may seem like a trivial and uncontroversial point ... However, this ... means that there is something—namely, our inability to intervene on the past—that limits our knowledge of the past. (2007, 24)

And so, let's reiterate these three pessimistic claims and point to my arguments against them.

Our available evidence about the past is limited to traces.

In chapter 6, I argue that emphasizing a trace-based methodology in historical science is impoverished because it misses the importance of dependency relations *between past events and entities*. Just as we draw on the relationship between current phenomena and the past, we also draw on relationships between aspects of the past that we know. In chapters 7 and 8 I discuss the role of *analogues*: naturally occurring surrogates of past entities. Here I argue that, under some circumstances, these can provide rich, non-trace-based lines of evidence about the past.

We are unlikely to uncover further traces.

In chapter 11, I argue that historical investigation is *scaffolded*: Progress must be made from the basis of previous hypotheses. One upshot of this is that figuring out what will be evidentially relevant prior to investigation is extremely difficult. This further suggests that we are biased toward pessimism when making educated guesses about the future success or otherwise of historical investigations. Why? Because from a scaffold, the evidentially relevant trace-pool is more likely to increase than stay the same, but because we cannot tell what will be relevant, we are likely to underestimate the total amount of evidence. Moreover, we saw in chapter 4 that there is reason to think that our access to the past will improve via *refinement*: As new technologies and midrange theories come online, our capacity to exploit dependencies between the past and present increases.

Historical scientists cannot manufacture evidence.

In chapter 9 I argue that, in principle, historical scientists can use surrogates such as simulations to make their own luck, analogously to how an experimentalist does. This position is expanded in chapter 10 by connecting this position to a wider discussion of the function of modeling in science. Where others have emphasized the heuristic or explanatory roles of models, I argue that in historical science at least they often play an evidential role. Idealization actively supports increases to our knowledge of the past.

Recall that one reason to care about optimism is that scientists themselves have to make judgments about which avenues of research are likely to be fruitful—so we want to know which targets to be optimistic about. This motivates a contrastive approach to epistemic situations. A pessimist about unlucky circumstances is likely to recommend that research time be spent on questions that are more likely to pay epistemic dividends. Lewontin does something like this in his arguments against seeking evolutionary explanations of human intelligence. By counteracting this pessimism, we see that even when trace evidence is poor, progress is to be expected. Moreover, insofar as I undermine these rationales for pessimism about unlucky circumstances, a more general optimism about our capacity to find out about the past is a result, and I will argue in chapter 12 that this matters for our reasons for doing historical science in the first place. In chapter 11, I will discuss what kind of optimism such considerations support. Let's get going!

6   The Main Business of Historical Science

What is the best way to navigate a maze? Well, assuming the walls stay stationary, here's a good technique: Place a hand on one wall and make sure that it remains in contact as you walk around. That is, always keep a wall on your left or right side. By doing this, you will be sure to exhaust the possible paths through the maze until you find the exit, without the danger of repeating or missing routes. This is a *method* of maze navigation: a way of generating successful solutions to a particular kind of problem.

This chapter is about method in historical science. That is, how do historical scientists go about generating knowledge, and why does it work? Such discussion comes apart, at least in principle, from questions about the nature of knowledge. Imagine that I am lost in a foreign city (this is not infrequent). Three methods for finding my way are to ask for directions, consult a map, or explore in an attempt to find recognizable landmarks. Which method I should apply depends crucially on context, on facts about me and facts about the city. How good am I at reading maps? Do I have reliable informants? Does the city have recognizable landmarks? Is it well designed? And so forth. A successful account of methods for finding your way in unfamiliar cities will likely include sets of instructions, which, given certain contexts, will achieve navigational goals. Given the kind of agent I am, and the circumstances in which I find myself, which knowledge-generating method should I employ? The job of a philosopher interested in method, then, is to identify and evaluate ways of generating knowledge in various epistemic and pragmatic contexts.

Accounts of the method of historical science typically provide simplified, idealized models of the "main business" of investigations into the deep past. What is the source of historical science's success, and what are its limits? Most accounts specify a particular method: a schematic paradigmatic picture of how evidence about the past is generated. Historical science is given a methodological "essence." My aim in this chapter is to argue that, in this sense, there is no "main business" of historical science. Attempts to simplify historical method miss their *opportunism*. Historical scientists are not methodological specialists, but *methodological omnivores* (Currie 2015a). This matters both for our understanding of historical science and for pessimism. As we shall see, most stories of historical inference focus on one kind of dependency relation: that between contemporary phenomena and the past. This misses the relationships between past entities, which, as we will see, is a crucial facet of historical method.

The chapter is in five parts. I start by introducing and criticizing common cause explanations. I then turn to accounts of "consilience," or the exploitation of independent evidence streams. I emphasize what we might call "coherency tests," exploiting dependency relationships between past entities. I close by arguing that there is no "essence" of historical science and characterizing the epistemic license of methodological omnivory.

6.1   Common-Cause Explanations

It is only because of natural recording systems, and because correlations usually have common causes, that we can reasonably claim to know anything about the past.
—Turner 2007, 23

Here is a mystery. One hot summer's night in Canberra, I returned home late after rehearsal for a musical performance slated for Kim Sterelny's sixtieth birthday. I noticed that the house was unusually messy—my roommate at the time, it must be said, kept a much tidier house than I. Moreover, some items didn't seem to be in their usual spots. Most saliently to me, my banjo was missing. On entering my bedroom, I found that the window had been smashed. Now, here is a bunch of unusual events: a messy house, a missing banjo, a broken window. Each event is individually unusual, but their correlation is more so. Each event's individual remarkableness was increased by the remarkableness of their co-occurrence.

But of course there was no mystery: I quickly ascertained that I had been robbed. After all, if I had been robbed, then the three events would be quite likely, since they would be unified by a *common cause*. Their individual improbability was rendered likely by virtue of postulating a single past event. This is an extremely common, and commonsensical, piece of reasoning, and one that has been identified as the main business of historical science.

Many accounts of the method of historical science appeal to common cause explanations. We should distinguish two related claims. First, a descriptive claim about what method in historical science is like. This is to say that when I inferred from the messy room, missing banjo, and broken window to a theft, I used the kind of reasoning that is distinctive of historical science. Second, a more normative set of claims about the reasoning's warrant. Not only are historical sciences appealing to common causes, but in so doing, they also do the right thing.

Regarding the descriptive claim, I say: Yes, historical scientists frequently unify traces via common causes. But this is not *all* they do, and moreover, such reasoning is less applicable in unlucky circumstances. When we examine difficult cases, examples of historical scientists departing from this method crop up.

Regarding the normative claim, I say: Yes, sometimes there is warrant for common cause explanations. However, such warrants are localized. I will argue that *even if* there is an all-things-considered reason to prefer common cause explanation, in actual practice such reasons are swamped by context. Common causes should be preferred (when they ought to be), not because of some general license but because of a set of local facts. Let's briefly canvas the literature.

The most developed account of a "common cause" methodology in the historical sciences is Carol Cleland's (2001, 2002, 2011, 2013). In brief, her picture is as follows: First, scientists identify a mysterious correlation. Say, a messy house and a missing banjo. Second, hypotheses are generated that can account for (that is, unify) those traces. Perhaps I have been robbed, or my roommate has gone on a bender, messed up the house, taken my banjo, and run off. Third, we hunt for further traces—smoking guns—that can decide between those live hypotheses. It is most unlikely that my roommate would break my window. But this is a likely entry point for a thief. And so the broken window strongly supports the former hypothesis over the latter. I was robbed.

Many historical investigations have this character. Cleland frequently appeals to the impact event at the K-Pg boundary, and with good reason.

We have an unusual correlation between an extinction event, a layer of shocked quartz, and so forth. The extinction event and the iridium could be accounted for by either an extraterrestrial impact or a period of increased volcanism. However, the shocked quartz at the boundary is a smoking gun, since it discriminates between these hypotheses. Shocked quartz is found at extraterrestrial impact sites and nuclear testing zones, not at volcanic eruptions.[1]

In chapter 9 I will spend some time making Cleland's notion of a "smoking gun" clearer, but all I require at this point is her basic appeal to common causes: "Hypotheses concerning long-past, token events are typically evaluated in terms of their capacities to explain puzzling associations among traces discovered through fieldwork" (2011, 552]).

Other philosophers and scientists also appeal to common causes. There are differences between these accounts and Cleland's, to be sure, but they all emphasize the *unification of traces*. Take Aviezer Tucker (2004, 2011), for instance:

The historical sciences are concerned with inferring common causes or origins: contemporary phylogeny and evolutionary biology infer the origins of species from homologies, DNA, and fossils; Comparative Historical Linguistics infers the origins of languages from information preserving aspects of existing languages and theories about the mutation and preservation of languages over time. (2011, 20)

Tucker takes historical reasoning to be about the unification of traces. Phylogenetic reconstruction, applied to both the tree of life and linguistics, exploits relatively robust, well-understood dependencies between present traces and the past. I will discuss phylogenetic methods in more detail below.

Klienhans, Buskes, and de Regt (2005, 2010) emphasize reliance on common-cause explanation in the earth sciences. Although they find fault in Cleland's discussion of "smoking guns," their account is nonetheless similar:

A number of hypotheses are developed which potentially explain the observations. By contrasting and testing a number of (incompatible) hypotheses, a biased attempt at confirmation is prevented ... when these various data and model scenarios all point to the same (underlying) explanation or common cause, earth scientists accept this explanation as (tentatively) true. (2010, 13–14)

Their account incorporates model testing and the utilization of multiple strings of evidence (although, in principle, so does Cleland's!), making it more amenable to points I will make later on. However, the emphasis is still firmly on common causes. And it is undeniable that positing common causes is an important aspect of historical science. As Cleland in particular makes vivid, historical inquiry typically starts with a surprising contemporary fact. For instance, much early geological investigation of erratic rocks could be characterized as being motivated by questions like, "How on earth did *that* get there?" However, we should note two things about this claim.

First, the hunt for common causes is not a good method in unlucky circumstances. The reason should be obvious: Common causes require effects—traces—to unify, but in unlucky circumstances *there are not many traces*. Moreover, for many of these cases, finding extra traces is unlikely to resolve the problem. Take, for instance, one of Turner's examples of underdetermination from chapter 4: the difficulty of reconciling body and nonbody fossil phylogenies (2007). Which animals made which tracks? To answer this, we must distinguish between different hypotheses about the taxonomic affiliation of the organism that left a set of footprints. For instance, are the footprints the remains of *Corythosaurus* or a different Hadrosaur, say, *Edmontosaurus*? Finding sufficiently rich traces to decide either way would be unlikely. Moreover, because the two sets of traces tend to preserve information at different grains, it is not obvious that finding *more* traces would help. The conditions under which body and nonbody fossils form, and the information that can be drawn from them, differ to such a degree that marrying data from the two is difficult. The challenge is fitting together two data sets at different grains, and it is not obvious how simply increasing the size of the data sets will achieve this. We need a particular kind of data: that which allows us to bridge the gap between trackways and body fossils. If historical inference is characteristically about the unification of traces and the provision of common cause explanations, then very little solace is provided in unlucky circumstances. So much the worse for optimism, we might say. However, as we will see, historical scientists are not restricted to such methods. And so, in unlucky circumstances, hunting for further smoking guns, looking for new traces, is not a good strategy. This leads us to the second point.

There are bountiful counterexamples to the claim that historical scientists are primarily in the business of unifying traces. As we saw in chapter 2, and as I will argue over the next four chapters, historical scientists do not provide only common-cause explanations. They seek and exploit independent streams of evidence, and they exploit dependencies between past entities and events. They use analogous reasoning. They construct smoking guns using simulations and the like. To claim that historical scientific reasoning is captured by common cause explanation is at best an overly enthusiastic idealization. Such views are impoverished and miss the rich, opportunistic strategies these scientists employ.

Let's move on to normative claims: the warrant for common-cause explanation. Such discussion is typically rather abstract, asking whether there is some *general reason* to prefer an explanation that cites a common cause over one that cites two separate causes. I will quickly sketch the kinds of views available before arguing that this general approach is problematic.

To begin, take two hypotheses, H_1 and H_2, which both provide explanations of two contemporary phenomena a and b. H_1 does this by unifying a and b by appeal to single past event, x, while H_2 accounts for a and b by appeal to two past events, y and z. As follows:

H_1: a and b are effects of x

H_2: a is an effect of y, b is an effect of z

Given just this information, is there reason to prefer H_1 over H_2? We need to be careful about the nature of the reasons we provide. We could appeal to a psychological, heuristic, or methodological reason. For instance, it might be that humans tend to *prefer* hypotheses like H_1 over those like H_2. Or preferring H_1 might be a good heuristic for building theories. One could, for instance, start with a simple picture of the past and add more events only as necessity requires it. I am not interested here in this kind of defense, since they do not seem to provide any kind of response to pessimism. If anything, some encourage pessimistic attitudes. After all, if the only reason we prefer H_1 over H_2 is because we find such hypotheses easier to work with, or some other non-truth-tracking reason, then a central pillar of historical method becomes worryingly detached from empirical adequacy.

No. We should seek *epistemic* defenses of common cause explanations, that is, defenses that claim that H_1 is more likely to be true than H_2. Such a defense would support something like the following:

Principle of Common Cause: Ceteris paribus, one hypothesis, H_1, is more likely than another hypothesis, H_2, just in case H_1 unifies more traces than H_2.

In other words, all else held equal, the hypothesis that posits the fewest causes in the past is the most likely. If two hypotheses have otherwise identical empirical support, the principle acts as a tiebreaker. What might a defense of such a position look like? As Cleland (2002) points out, these could be a priori and a posteriori. An a priori defense of common causes claims that the principle simply falls out of one or another account of the nature of evidence. An a posteriori defense claims that the principle is based on some worldly fact. Let's look at an example of each.

Salmon (1975), building on Reichenbach (1956), provides an a priori justification. He understands the justification of common causes in terms of how they "screen off" probabilities. Take a surprising correlation between two traces. This improbable correlation is rendered probable by demonstrating their mutual dependence on a common cause. For instance, the occurrence of dolostone caps atop glacial lithology is a common feature of Neoproterozoic strata. It would be a surprising coincidence if these different features were unrelated. By postulating a common cause—global glaciation—the events' correlation becomes unmysterious. It would be a coincidence indeed if, on the same night, my window was broken, my banjo went missing, and the police visited my house—so long as the occurrences of those events were independent. Of course they were not: I was robbed. In short, on Salmon's view, it simply falls out of probability theory that hypotheses like H_1 are more likely than H_2, because the common event in H_1 "screens off" the apparent improbability of the events.[2]

Carol Cleland provides an a posteriori justification of the principle of common cause. Recall chapter 4: There I discussed her view that present states of affairs overdetermine past states of affairs. Cleland claims that this ontological fact of the matter underwrites the common-cause methodology that she takes to be the main business of historical science:

the principle of the common cause provides a global constraint on scientific reasoning. ... The use of the principle of the common cause by historical natural scientists rests upon a substantive thesis about the nature of the world for which there exists overwhelming empirical evidence, namely, the thesis of the asymmetry of overdetermination. (2011, 20)

I object to both of these a priori and a posteriori defenses of the principle on the grounds of their generality, and their distance from actual scientific practice. Briefly, the ceteris paribus clause of the principle of common cause barely ever holds, and moreover appealing to such global principles obscures the real, local, action.

To see why, let's return to our toy case from chapter 3: the *Corythosaurus* trackways. Imagine that in addition to the trackways from the ancient beach, we find another, similar set of *Corythosaurus* prints on the remains of a neighboring beach, and imagine further that they date to the same day (this really takes some imagination—the chances of such a find are mind-bogglingly low, and our capacity to date finds so finely is pure science fiction). Consider these two hypotheses:

H_1: The footprints were made by the same individual *Corythosaurus*.

H_2: The footprints were made by two different *Corythosaurus*.

I think that H_1 is supported by the principle of common cause. Although it does not unify the footprints via a single event per se, it does unify them insofar as they have a single cause: the same individual *Corythosaurus*. H_2 unifies fewer traces, as we must postulate two different *Corythosaurus*. But should we prefer H_1 over H_2? Clearly, *Corythosaurus* facts matter here. If, for instance, we had reason to expect *Corythosaurus* population density to be low—maybe they were highly territorial, solitary animals—then such traces are likely to belong to the same individual. Unfortunately for H_1, however, Hadrosaurs were pack animals, one of the most populous lineages of the Mesozoic. We should expect the signal of any particular Hadrosaur to be weak, drowned out in the clamor. If that is right, in this context it would be a mistake to think that H_1 should be preferred due to positing less individuals.

Here is another case. Between 1989 and the early 2000s the Australian cricket team was in very good form. Accordingly, they dominated the Ashes tournament throughout those decades, winning in 1989, 1991, 1993, 1995, 1997, 1999, 2001, and 2003. In each of the years in which they won, summer was also warmer than winter. Surely the correlation between summer's relative warmth and Australia winning the Ashes does not cry out for a common cause explanation. Why? Because the correlation simply isn't surprising. The Australians fielding a good cricket team, and summer being warmer than winter, are not surprising events, and so their correlation isn't surprising either. Okay, so what? Here's the point: It is because we know about Australia's predilection at cricket, how popular it is there, how much funding it receives, and so forth, that their winning streak is unsurprising.

Moreover, we have good reason to doubt that there would be any connection between seasonal patterns as stable as the difference between summer and winter, and something as context-sensitive as who wins a cricket tournament.

At this point you might accuse me of cheating. These two frivolous examples break the principle of common cause's ceteris paribus clause, and so are irrelevant. We want to know whether we should prefer a unified or a disunified explanation for a correlation when *all else is held equal*. By bringing in our background knowledge of Australian cricket, or *Corythosaurus* behavior, I only bring noise into the equation. That is the right objection to make: These cases are in no way counterexamples to the principle of common cause. But counterexampling is not my purpose here. My point is this: When historical scientists provide common cause explanations, the ceteris paribus clause holds in only a vanishingly small number of cases, if any. A defense based on global claims, be they a priori or a posteriori, will be lost among the rich contextual details. Focusing on general defenses, then, obscures where the action is in common cause explanation.

This underwrites *local* a posteriori defenses of common cause explanation. That is, we do not prefer common causes because of some general rule, but because of knowledge we have about particular contexts. In some moods, Elliott Sober comes to similar conclusions in his discussions of cladistic inference, and it is worth sketching this to illustrate the point (see Sober 1981, 1990). I take phylogenetic trees, at least of charismatic eukaryotes, to be hypotheses about the actual ancestry of the relevant lineages: Who is more closely related to whom, in what order did the lineages split from one another, and so forth. When constructing a phylogeny, a systematist matches groups of "characters," features or traits of the lineages relevant to their relatedness. When different lineages have different characters, this represents an evolutionary divergence somewhere in the two trees. The basic question facing systematists[3] is: Which of the many possible trees is more likely? What principles can guide us? To illustrate this, let's consider a highly simplified case (I will come clean on the simplifications later in this chapter), represented by figure 6.1.

The figure represents two hypotheses about the ancestral relations between three lineages, *A*, *B*, and *C*, in reference to a single character, *t*. The character is present in *A* and *B* but is not present in *C*. According to the first hypothesis, *A* and *B* form a clade, but *C* is excluded—that is, *C* first

H1: *t* is basal, *t* is lost in lineage *C*   H2: ~*t* is basal, *t* evolves in
                                                        lineage *B* and *A*

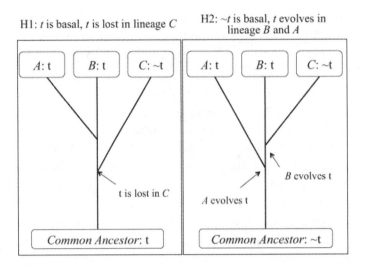

Figure 6.1
Simple cladograms.

split from *A* and *B*. According to the second hypothesis, *A* is the outgroup.
The hypotheses also differ on the presence or otherwise of *t* in *A*, *B*, and *C*'s
common ancestor. By H₁, *t* is "basal," that is, it is present in the common
ancestor. By H₂ it is "derived," that is, it is a later evolutionary development.
Notice that by H₁ there is a single evolution event—*C*'s loss of *t*, whereas
in H₂ there are two evolution events, *t* evolving in *A* and *B* separately.
"Cladistic parsimony" recommends that we prefer the tree with the fewest evolutionary events—the fewest character changes over time. And so,
insofar as figure 6.1 captures the relevant information, we should think H₁
more likely than H₂.

What justifies cladistic parsimony? The justification does not, I think,
arise from a claim about the past being "simpler" than the present, or some
similar general claim. Rather, it comes from our knowledge of living systems. Evolutionary theory should lead us to expect highly adapted systems
to have certain properties. Since high-fidelity transmission across generations is required for selection to accumulate complex traits, such systems
will have strong lines of heritability. Highly complex, evolved systems, then,
will exhibit "phylogenetic inertia": Traits are likely to be retained across
generations (Currie 2014b; Griffiths 1996, 1999; Levy & Currie 2015). Such
inertia could underlie cladistic parsimony.

Here's how this might go. According to phylogenetic inertia, we should expect most traits to be more likely to remain stable than to change over relatively short periods of evolutionary time. We should expect the high-fidelity transmission of traits. If that is right, then more often than not, a trait present in one lineage will also be present in a sister, ancestor or descendent clade. That is, ancestrally related lineages will share traits. If so, then cladistic parsimony should also be expected. Cladistic parsimony tells us to minimize the number of evolutionary events, where such events are understood as changes in character states. As a change in character state represents a partial failure in trait transmission, such events should be comparatively rare if phylogenetic inertia holds.

Note that the applicability of this defense is itself highly sensitive to our knowledge of the relevant characters. For instance, as Hall (2002) discusses, traits related to climbing are excluded from morphology-based reconstructions of monkey phylogeny. This is because, given the high selection pressure on such traits, phylogenetic inertia is likely to fail. We expect non-climbing traits to be replaced by climbing traits (for instance, in the muscular structures related to grip). So traits related to climbing are likely to often evolve in the lineage—potentially violating cladistic parsimony. Further local knowledge, then, allows us to identify exceptions to cladistic parsimony. Knowledge of the evolutionary "lability" of particular traits (their stability across generations), can trump the principle's general license. My examples here are very distant from actual scientific cases. In modern phylogenetics, large quantities of typically molecular data are used. As opposed to a single preferred tree, many possible trees are generated and resolved into several "best fit" options, with attached confidences. Many statistical processes are used to resolve these trees, cladistic parsimony among them. The processes to use and the relevant confidences are often guided by midrange theory. For instance, the "neutral" parts of genomes generate preferred data, as they are taken to be insensitive to noise from selection. Moreover, as we have just seen, climbing adaptations are excluded from phylogenies for similar reasons.

Cladistic parsimony, then, is a version of the principle of common cause that is justified on local grounds and subject to specific exclusions on the basis of contextual information. Specifically, exceptions are due to more localized knowledge about the lability of particular characters. Once we see the work that midrange theory does in these contexts, it isn't clear what role remains for a "general" defense, such as an appeal to probabilities being "screened off," or present-to-past overdetermination.

Historical scientists are rarely, if ever, in situations where the principle of common cause, taken as a general ceteris paribus claim, is an important theoretical virtue. Their epistemic status is simply too complex. In cases where common-cause explanation plays an essential role, as in phylogenetic reconstruction, a body of justifying theory exists for the practice. The prior probabilities of the hypotheses always matter—and midrange theory informs this.[4] Even if, then, there is a general (a priori or a posteriori) justification of the principle of common cause (and I think this is an interesting question in other contexts), if we want to understand the epistemic resources and situations of historical science, this is a red herring. General defenses of the principle of common cause do not help us understand the actual patterns of reasoning that historical scientists employ, since they obscure the localized, midrange-theory informed justifications they use.

Thus, common cause is a good rule when it is justified by a body of theory. Some patterns of traces, such as phylogenetic ancestry, are likely to be unified. Others, such as we saw in the *Corythosaurus* case, are not. Appealing to common cause is, of course, a very important part of historical methodology. It is, however, not a good strategy in unlucky circumstances, where traces are thin on the ground. Moreover, I shall argue that it is not *the* business of historical science. Indeed, focusing on common-cause explanation alone obscures the range of available epistemic resources.

6.2   Consilience

Some accounts of the warrant and method of historical inference emphasize the exploitation of independent evidence streams or "consilience." The thought behind the view is relatively simple. Imagine we have two methods of ascertaining some past fact. Further, imagine that the methods are independent of one another. That is, they both provide information about the same facts, but do so based on different assumptions and techniques. Here is the clincher: If both methods provide the same results, yet those results are false, this would require (1) each method to fail separately, but nonetheless (2) converge on the same result. Assuming the methods are indeed independent, the likelihood of this occurring is low. The convergence of independent results, then, can make for robust hypotheses.

Patrick Forber and Eric Griffith (2011) argue that weaving together independent evidence is the business of historical science. This is, I think, a step too far, but they do a nice job of capturing why consilience matters. Here's a traditional philosophical worry: *testing holism*. Recall from chapter 2 that one motivation for rejecting the obliquity explanation of Neoproterozoic glaciation is that an orbital obliquity would result in widely distributed carbonates, rather than the characteristic dolostone caps we see. It is tempting to see the reasoning as follows:

1. If Earth had an irregular orbital obliquity in the Neoproterozoic, carbonates in Neoproterozoic deposits would be widely dispersed;
2. Carbonates are not widely dispersed in Neoproterozoic deposits;
3. Therefore, Earth did not have an irregular orbital obliquity in the Neoproterozoic.

So far, so good. However, another way of viewing these three premises is as an inconsistent triad. Propositions 1, 2, and (the negation of) 3 cannot be all true at the same time. One premise has to be false. But why should premise 3 take the fall? We could instead insist that Earth had an irregular orbital obliquity in the Neoproterozoic (after all, it accounts for the tropical glaciation, and is possible given computer models!), and that carbonate is not widely dispersed in the Neoproterozoic deposits. From this we conclude that the conditional linking Earth's obliquity with dispersed carbonate must be mistaken. Or we could claim that our data gathering is flawed, and that Neoproterozoic-strata carbonates are not widely dispersed after all. This is an old point: For any apparent falsifying or confirming evidence, there will be other vulnerable "auxiliary premises." Perhaps our theories of how carbonate deposits form are faulty, or perhaps our ideas about how a global obliquity might affect carbonates is mistaken, or perhaps our methods of observing and determining the dispersal of deposits do not work.

In my view, testing holism does not undermine scientific inference generally (for the classic treatments, see Duhem 1954; Quine 1951). Indeed, it is problematic only if one takes evidence in an all-or-nothing sense. In practice, we do have defeasible grounds for preferring some premises to others. For instance, in the case above we have a handle on which of the premises in the inconsistent triad ought to be rejected—the third one. After all, failing to predict carbonate distribution is not the only mark against the hypothesis. As we saw in chapter 2, that the obliquity needs to be "forced" in simulation runs suggests that it wouldn't arise given the ancient impact that Williams appeals to, for instance. Moreover, we have reason to think the other two premises are solid: We certainly know much more about how carbonate deposits form than we do about irregular obliquities in Earth's orbit. Absolutely, there are still possible circumstances where we might prove to be embarrassingly wrong about such things (indeed, this has happened before!), but I am not moved by testing holism as a general worry about scientific inference. However, it is a useful context for illustrating the advantages of independent evidence.

Consider the modeling evidence against the obliquity theory (and for now let's assume that you are convinced of my arguments in chapter 9, and are happy to say that such studies produce empirical evidence!). One could reconstruct the reasoning here as follows:

1*. If Earth had an irregular orbital obliquity in the Neoproterozoic, then models would predict this behavior without forcing;

2*. Models only predict an irregular orbital obliquity if they are forced;

3*. Therefore, Earth did not have an irregular orbital obliquity in the Neoproterozoic.

As in the last example, this is also vulnerable to testing holism: We might be wrong about our model's connection to the Neoproterozoic events, or about our model's behavior. Regardless, note that in both arguments the concluding premise, 3 and 3*, are the same—they support the same result. But notice that the supporting premises, 1 and 2, and 1* and 2*, are quite different. Now, for us to deny the conclusion, we must show that both a supporting premise in the first argument, and a supporting premise in the second argument, fails. This is a more daunting prospect.

And so, independent evidence streams—in this case deposits in Neoproterozoic strata and models of Earth's obliquity—can make for well supported historical hypotheses. Forber and Griffith go so far as to argue that evidential convergence "provides the primary source of support for such historical reconstructions" (2011, 1). I will argue against this claim later on in this chapter (and will also bring up another worry about consilience's effectiveness in chapter 8). For now, I have two further aims. First, I want to further articulate the relevant sense of "independence" required for consilience, and second, I want to echo the last section's emphasis on localized rather than general justifications of such principles.

Alison Wylie (2011) has rightly emphasized the importance of multiple strands of evidence in an archaeological context: "Of necessity, evidential reasoning depends on multiple strands of arguments: it emanates from disparate elements of the archaeological record, draws on background knowledge that originates in diverse source fields, and bears on an array of conditions and events that constitute the complicated lives of the material things that make up the archaeological record" (386–387). In reconstructing the past, we draw on many different epistemic sources. Wylie has a useful distinction between *vertical* and *horizontal* independence. Briefly, vertical independence concerns cases where different lines of evidence play different roles in an inference, while horizontal independence concerns lines of evidence whose results support the same hypothesis but rely on different auxiliary premises (i.e., consilience). We will return to Wylie's notions of independence in this chapter's concluding section.

I understand vertical independence in terms of which conclusions the evidence in question supports, that is, the two streams of evidence either provide information about different aspects of the historical target, or one plays a supporting role. The classic case of a supporting role is midrange theory. Recall that Lacovara et al took the detached scapula in their Dreadnoughtus specimen to be reason to think that the animal was osteologically immature, that is, it was still growing when it died. Here, we have a body of theory about the relationship between a physiological feature—the detached scapula—and the osteological maturity of a specimen. We also have a particular specimen with this feature. The inference of Lacovara et al is licensed via the midrange theory and the specimen's features *in combination*. This is different in character to Forber and Griffith's discussion of consilience. In particular, vertical cases do not overcome testing holism. In Forber and Griffith's horizontally independent example, different premises support the same conclusion. In vertical cases different, but related, conclusions are supported. Consider another example, the glacial lithology and the paleomagnetism of the Neoproterozoic. In combination, these give us reason to believe that there were glaciers in the Neoproterozoic tropics. However, notice how they do this.

First, take lithology:

1. Rocks bearing glacial lithology formed during glacial periods;
2. Neoproterozoic rocks have glacial lithology;
3. Therefore, Neoproterozoic rocks were formed during glacial periods.

Now, paleomagnetism:

1. Rocks with tropical paleomagnetism formed in the tropics;
2. Neoproterozoic rocks have tropical paleomagnetism;
3. Therefore, Neoproterozoic rocks formed in the tropics.

It is only *in combination* that we get the claim that the Neoproterozoic had tropical glaciers. Vertical independence is not consilience. Merely utilizing multiple evidential sources is not sufficient for the epistemic benefits that Forber and Griffith highlight. You need the right kind of independence. For instance, imagine that our knowledge of the behavior of carbonate deposits and models of global obliquity were not horizontally independent. Perhaps our theories of how carbon finds its way into strata relied upon global obliquity, for instance (this is not the case, of course!). In such circumstances, horizontal independence would fail, as the failure of *the same* auxiliary premise could undermine both conclusions. In short, two sources of evidence are horizontally independent when the same conclusion is supported via different premises. They are vertically independent when different, but related, conclusions are supported.

Finally, horizontal independence is a matter of degree: Two lines of evidence can share some auxiliaries and be independent in other respects. Especially if mutual auxiliaries are well supported, it can still be more likely that two independent auxiliaries fail—and more likely still that the mutually supported conclusion is right.

This brings me to a further point: As in the case of the principle of common cause, relying on an abstract, formal account of the justification of consilience misses where the action is. What matters is the local work done by midrange theory. After all, these bodies of knowledge set how likely and stable we suspect lines of evidence to be. Failures of horizontal independence are divergences from the ideal scenario that Forber and Griffith highlight.[5] How problematic these divergences are depends on how likely the relevant auxiliaries are. I won't spend much time on this point here, since I would end up repeating much of the last section. Suffice to say, it is sophisticated midrange theory and the particular evidential context that tell us how problematic failures of independence are. Historical scientists do not often find themselves in situations as evidentially pure as Forber and Griffith describe. Focusing on failures of independence and the general worry of testing holism obscures the local action.

6.3 Coherency Testing

Consilience and common cause explanations can both be understood as different accounts of method in the historical sciences, that is, they are different stories about how historical scientists generate knowledge and why it works. Either historical scientists hunt smoking guns—they seek to unify traces using common cause explanations—or they exploit independent streams of evidence. Although these accounts might both be captured using the same story about the ultimate nature of evidential warrant, they give different accounts of method, and that is our concern here. One thing that unites the accounts, however, is their emphasis on traces.

Forber and Griffith discuss historical reconstruction as providing "the resources to successfully explain puzzling extant traces, from fossils to radiation signatures" (2011, 1). Cleland is even more specific. By her lights, historical hypotheses are judged by how well they "explain puzzling associations among traces discovered through fieldwork" (2011, 552). This *trace-centric* approach to understanding historical methodology misses the role of coherency: Historical scientists exploit not only traces but also dependency relationships between past events, processes, and entities. I will give an example and then explain how coherency can count as evidence, that is, be truth tracking, in such cases (see also Currie 2016b; Currie and Sterelny 2017).

Consider the relationship between two events in the deep past: Snowball Earth and the Cambrian explosion. According to the Snowball Earth hypothesis, at least twice during the late Neoproterozoic (say, 590 million years ago), Earth froze over. Relatively soon afterward (a mere 50 or so million years later), Earth's rocks recorded an unprecedented radiation of metazoan life known as the "Cambrian explosion." It is generally accepted that these events are linked: The ancestors of Cambrian fauna must have survived Snowball Earth. And this has consequences for our knowledge of both events. For instance, how could complex life survive a frozen planet? As Hoffman and Schrag put it, "Assuming snowball events occurred, what refugia ensured the survival of eukaryotic plankton, and early metazoans if they existed? How did the climate shocks entering and exiting snowball events impact their evolution?" (2002, 147).

In response, scientists construct simulations testing between a complete freeze (a snowball) and something less extreme (a "slushball") (see Donnadieu et al. 2004; Hyde et al. 2000). The occurrence of the Cambrian explosion gives reason to believe the Neoproterozoic freeze was incomplete.

Moreover, Neoproterozoic events are revelatory of the Cambrian. For the radiation to occur, pockets of life must have been isolated in order to diverge both phylogenetically and developmentally, without evolving complex, novel traits. Snowballs could act as pelagic filters, ensuring life remained relatively simple due to the "almost complete destruction of terrestrial biota and shallow-water, bottom-dwelling life" (Runnegar 2000, 404). This could explain how early life was phylogenetically separated and thus able to evolve divergent resource pools for evolution to exploit once balmier Cambrian conditions arrived: "Some environmental filter was required to maintain early metazoans in 'larval mode' after they had invented set aside cells. This enabled them to diversify into well-separated lineages that ultimately became the independent sources of radically different body-plans" (Runnegar 2000, 404).

Aspects of the Neoproterozoic glaciation and the Cambrian explosion, then, depend on each other. Given the explosion's occurrence, the glaciation must have occurred in certain ways. Had the glaciation been different, so too would the explosion have been different. Although the causal relationships are asymmetric (as illustrated in figure 6.2), the relationships of dependence are more symmetrical.

Not only did the Neoproterozoic glaciation leave traces in modern rocks, but it also influenced how other events—such as the Cambrian explosion—occurred. Scientists utilize both traces from the Neoproterozoic and theories about the Cambrian explosion to test and support the Snowball Earth hypothesis. They also call on events from the Neoproterozoic to explain the Cambrian radiation. In short, the dependency between the two events is exploited to further investigate them. So, geologists not only care about how the past events they hypothesize gel with traces, the modern day Cambrian fossils and Neoproterozoic deposits, but they also care about the gelling of past events—the Cambrian explosion and the Neoproterozoic snowballs.

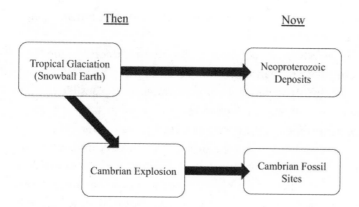

Figure 6.2
Causal relations between geological events in the Neoproterozoic, biological events in the Cambrian, and their subsequent traces.

In addition to dependencies between the present and the past, there are dependencies between and within past events. Background theories pick out these dependencies just as midrange theories do, and thus act as a kind of evidential conduit: Our knowledge of the Neoproterozoic carries over into the Cambrian, and vice versa. So, my account of the evidential warrant granted by knowledge of dependency from chapter 3 applies not to traces alone—dependencies between now and the past—but also to dependencies between variables in the past. For example, the variables of snowball events (their timing and extent) matters for variables related to the Cambrian explosion.

The notion of minimal dependence from chapter 3, then, can be used to understand both trace-based evidence and "coherency." Note that dependence relationships include both constitutive and causal relationships. Many causal relationships are *inter*-entity, process, or event; they hold between variables in different systems. Constitutive relationships are *intra*-entity, process, or event; holding between variables in the same system. The dependency between Snowball Earth and the Cambrian explosion is inter-entity, while some dependencies appealed to in reconstructing sauropods is intra-entity; sauropods need small teeth to allow for the small heads required for their long necks. Although I characterize dependence in terms of "variables," elements in a system, a derivative sense between entities is readily available. Two entities are dependent just when there is dependency between at least one of each entity's variables.

Consider the events in figure 6.2. Not only are the "traces" of the Neoproterozoic and the Cambrian minimally dependent on those events, but they are also minimally dependent on each other. It is plausible that, if one event were different, the other would be too. If, for instance, snowball events had been insufficiently extreme to act as pelagic filters, the Cambrian fauna would not have evolved such divergent developmental and evolutionary resources. The explosion, then, would have likely been less dramatic. By virtue of this, if we have either theories representing those dependencies or empirical evidence of their correlation, they will be evidentially relevant to one another.

Thus, my account of "coherency tests" doesn't rely epistemically on coherency per se. Rather, past entities can count straightforwardly as evidence in the same way traces do. We have knowledge of dependencies, and we exploit this to reconstruct the past. This avoids objections to coherency-based accounts such as Carol Cleland's:

Because much is unknown about the events in the sequence, narrative explanations have a significant fictional component, involving omissions and additions. This poses a potential problem insofar as it conflicts with the traditional emphasis in natural science on evidential warrant. The problem is exacerbated by the central role of explanation in the confirmation and disconfirmation of historical hypotheses. If the primary reason for accepting a historical hypothesis is its explanatory power and it draws its explanatory power primarily from the coherence and continuity of a quasi-fictional story, then historical natural science really does seem inferior to experimental science; in the absence of empirical warrant a narrative explanation amounts to little more than a "just-so" story. (2011, 17)

Cleland is suspicious of the evidential importance of constructing "coherent narratives" of the past, since our trace sets are so gappy. I take it that her request for "empirical warrant" is met by the account I have provided. Just as midrange theory connects the present to the past by providing explanations of how historical processes could give rise to contemporary traces, our knowledge of dependencies between past events provides evidential warrant for coherency tests between events in the past.

This point also matters for pessimism. As you will recall, one motivation for pessimism was this: *Our available evidence about the past is limited to traces.*

But this claim is mistaken. In addition to trace evidence, connections between present states of affairs and past states of affairs, historical scientists can exploit dependencies between past states of affairs. Of course, this turns on our having knowledge of those states of affairs in the first place, and of course this is going to depend to some degree on traces. However, the capacity to exploit those connections can dramatically extend our epistemic reach. We have good reason to believe that one feature that set sauropods apart from other giants was their lack of mastication. This enabled them to have small heads, and thus long necks. This is a way in which they depart from mammals, whose fancy teeth require a larger head-to-body size ratio. The connection between mastication and neck length is not based on traces. Trace evidence tells us that sauropods did not chew, and that they had long necks. It is by exploiting the relationship *between* long necks and nonmastication that this vital piece of the sauropod puzzle is grasped.

Reflecting on chapter 2 provides more examples of scientists exploiting dependencies in the past (see also Currie 2016a, Currie & Sterelny 2017). However, I now want to shift to a more general lesson about methodology.

6.4 Methodological Omnivory

We have seen that in some contexts historical scientists unify traces, exploit independent lines of evidence, and draw on dependencies between past entities. So, what is the main business of historical science? In this section I have two aims. First, I argue that there is no "essence" of historical method; rather, scientists adopt whatever methodologies are appropriate given their evidential situation. They are what I have called "methodological omnivores" (Currie 2015a). Second, I will provide a preliminary sketch about how to tell when these different methods come into play.

Of the three methods I have discussed in this chapter, is there any reason to take one as quintessential of historical science? Carol Cleland's work distinguishes between what she calls "experimental" and "historical" sciences, but she makes clear that these are archetypes: "Historical" scientists will sometimes do experiments, and "experimental" scientists will sometimes hunt for smoking guns. Ben Jeffares (2008), for instance, emphasizes the use of experimental methods to establish midrange theory. The historical sciences, then, are not entirely free from experiment. If these are only archetypes, then, what's the point? I take it that the point is *explanatory*. Such simple analyses are trying to account for some important aspect of historical science. We should understand them, then, in those terms.

Cleland's main aim is to vindicate the "historical" sciences in the face of a certain pessimistic strain that claims that the legitimate sciences rely on experiment. Her account of smoking guns, then, is supposed to provide a vindicatory contrast with experimental science. You do not have to be an experimentalist to be a legitimate, successful scientist. In this, she succeeds. Investigations that proceed by unifying traces provide a rich source of knowledge about the past that is prima facie different from how investigation goes in the lab. I am less convinced that this is what historical scientists are in the *business* of doing, however. As we have already seen and will certainly expand on in the next few chapters, historical scientists do much, much more than hunt for and unify traces.

In contrast to Cleland, I am interested in how much we can know about the past, specifically in which resources historical scientists can draw on in unlucky circumstances. In this context, we need to ask what can be done when the traces run out. It turns out that a lot can be done. Historical scientists try to find new traces both through discovery and refinement. They look for dependencies between past entities. They exploit vertical and horizontal evidence streams. As we shall see in forthcoming chapters, they also draw on analogues, simulations, and other surrogative evidence. Historical scientists are not methodological "obligates," focused on one method or another. Rather, they are opportunistic—methodological "omnivores." Just as the right method for finding my way about an unfamiliar city depends on both my epistemic features and the properties of the city in question, different methods for uncovering the past are more or less applicable in different contexts. That is, it depends on both the scientists—their background knowledge, technologies, and interests—and on their targets, as we will see.

Omnivory involves the frequent co-optation of epistemic tools and their being fit to local contexts and needs. In Currie (2015a), I describe this strategy in the reconstruction of extinct lineages. General mechanical principles (such as levers) are used in biomechanics in combination with local facts about organisms—muscle structure, and so forth—to infer the bite force of extinct critters. Such techniques are calibrated using experimental studies of living (typically related) lineages. Another example of such fine-tuning is described in Currie (2013). Primatologists have noticed a regular connection between teste size (in relation to body-mass) and mating strategy—a regularity that paleoanthropologists use to theorize about mating strategies in the past. The regularity is incorporated into a simple model mapping relative teste size with mating strategy. Exceptions are used to further refine the model. For instance, other features can increase body mass independently of mating strategy or teste size. Yellow baboons have unusually long and strong forelimbs, a trait that skews their overall body size. By attending to trunk size rather than body size, yellow baboons fit into the model—and the model is further refined. Finally, in chapter 4 I discussed the use of LiDAR techniques to locate and map ancient ruins in Central America. There, calibrating the method to local conditions involved "ground-truthing"—as opposed to using an explicit theory of how various kinds of structures (natural or cultural) will appear on LiDAR visualizations, archaeologists simply compare LiDAR results to ground-based surveys. The success of historical science often relies on opportunistically developing epistemic tools geared toward local conditions. Success is not simply about incorporating various lines of evidence, but about *generating* lines of evidence.

Methodological omnivory should be understood in terms of the "disunity" of science. This cluster of views rejects the idea that science aims for unified theories, instead seeing it as a more or less integrated patchwork of models, techniques, theories, and so forth. Omnivory has significant parallels with Wylie's (1999) emphasis on scientists' exploitation of disunity and Sandy Mitchell's (2001, 2003) "integrative pluralism." Mitchell's view focuses on how nonequivalent models can be integrated to provide sufficient but theoretically disunified explanations of local phenomena. Her view and mine are distinct. First, for Mitchell, disunity's source is the nature of modeling, the trade-offs involved in idealization, and complexity, whereas methodological omnivory is driven by incomplete data. Second, Mitchell is mostly interested in how models and theories generate explanations of phenomena, while I am interested in how evidence is generated. Methodological omnivory and integrative pluralism are both interested in how scientists exploit disunity but have different focuses. The two, surely, are complementary.

Given such an opportunistic, disunified story about method, one might ask: Well, what is a poor philosopher to do? Can we say more than just

"Well, scientists do lots of different stuff?" I think we can. The machinery I have developed thus far, dependencies and the ripple model of evidence, allows us to speculate about the applicability of various methods in various epistemic situations. I am only going to gesture here, since I am eager to get on to the remainder of the book, where I defend a more expansive account of our epistemic reach into the past.

Consider past events that are highly dispersed but not particularly gappy or degraded. In such cases, we will expect large, heterogeneous trace sets and relatively robust dependencies between traces and the past. Under such circumstances, we should expect to find many traces, and we should expect them to provide relatively fine-grained empirical delineations between hypotheses. It strikes me that under these conditions the smoking gun methodology ought to dominate. After all, the strategy relies on there being a variety of different traces able to finely discriminate between hypotheses. Large scale, dramatic events tend to have this character: Consider the K-Pg impact event, or the background radiation of the big bang. High dispersal underwrites the hunt for smoking guns.

How about when traces are faint, that is, our midrange theory has trouble connecting them to the past? Under these circumstances, we should expect consilience and coherence to come to the fore. First, if traces are faint, we would want to maximize the number of different types of traces, thus exploiting the benefits of horizontal independence. Weak dependencies can be mitigated by exploiting a large number. Second, uncovering dependencies between these events would further increase our epistemic reach.

This is only a hint, of course, at the kind of methodological work that needs to be done. In the future, we might more carefully identify the circumstances in which various methods and approaches could be successful. Such work has the potential for providing scientific prescriptions: Suggestions could be generated about which investigative practices are likely to provide dividends. Regardless, the take-home message of this chapter thus far is as follows. Part of the explanation of the success of historical science is methodological omnivory: By opportunistically exploiting a variety of methods, our epistemic reach is maximized. There is no "essence" of historical method—and this is why it works so well. Moreover, some of these methods are not "trace-centric." For instance, in some circumstances they rely on dependencies between events in the past. I will close the chapter by using an archaeological case study to tie this all together.

6.5   The Material Remains of Spiritual Life

Some scientific investigations target robust, clear lines of evidence. Consider phylogenies drawing on molecular traces, reconstructions of mammalian size from teeth, and tracking human individuals via their fingerprints. These all involve the targeted exploitation of rich veins of information. In such circumstances, there is something approaching (but, of course, only approaching) a magic bullet. Indeed, scientific communities grow up around these rich veins. However, as with laboratory science, our evidential reach in such circumstances is often limited. Mammalian teeth tell us an awful lot about their morphology and ancestry, but less about their ecology, and often vanishingly small amounts about evolutionary history. This is because the dependency relations holding between traces and the past are weaker in these other contexts. As we switch from lucky epistemic situations to unlucky ones, successful scientific strategies change. Instead of focusing on a single rich vein, a mixed strategy is employed—many lines of evidence are sought out and woven together. Following this, expertise in scientific communities becomes more diverse and the needs of integration come to the forefront. This capacity to engage in different investigative strategies and develop different methodologies depending on context is the essence of methodological omnivory. Tying historical reconstruction to any particular method—common-cause explanation, for example—would cripple investigation.

As we saw in the introduction, according to Christopher Hawkes, the trickiness of an archaeological inference turns on how "human" the target is—that is, how distant the feature is from basic, ecological subsistence activities. One of his examples is spirituality: How are we to infer the ethereal world of a culture's spiritual beliefs from scant material remnants? In this section, I will use an example of archaeologists doing something like this to illustrate the picture of historical method this chapter has focused on, one that emphasizes methodological omnivory. Alison Wylie's work (2011) is particularly useful here, and I will develop her discussion. Let's start with the archaeology—Christine VanPool's work in reconstructing the shamanistic practices of various North American indigenous groups (2009; see also Currie 2016a)—before switching to Wylie's analysis.

Archaeologists have traditionally divided *shamanistic* from *priestly* cultures. Shamanistic practices are idiosyncratic and individualized, often involving trance states. Priests are more organized, are typically parts of hierarchies, and are more explicitly linked into local power structures. We shouldn't take these categories too seriously; as VanPool says, they are "analytically useful groupings that reflect the co-occurrence of religions traits that tend to correspond with one another as the level of cultural complexity shifts" (2009, 178). That is, the distinction between "shamanistic" and "priestly" cultures is not discrete, but rather tracks a bunch of spiritual practices that tend to clump together in different cultural contexts. If pessimists like Hawkes are right, we should expect inferring details of spiritual practice from material remains to be next to impossible. Let's see.

VanPool first divides different aspects of shamanistic practices into those that are likely to be idiosyncratic and those that are likely to be more universal. Identifying the universal aspects allows her to construct midrange theories. Shamans achieve altered states—trances—through a variety of practices: taking psychoactive drugs, chant and ritual, sensory deprivation, and so forth. These different methods themselves have characteristic effects on human psychology and perception. Thus, there are potential connections between shamanistic methods of entering altered states and the altered states themselves. If those states can then be connected to material remains, we will be cooking with gas. VanPool does this first by connecting various kinds of intoxication to various visual experiences, and second by connecting those visual experiences to artistic representations.

Nicotine and peyote intoxication affect our visual systems in different ways. Nicotine effects color perception—the palate is restricted to white, black, and yellow. Peyote, by contrast, produces vivid "psychedelic" experiences. As VanPool points out, these perceptual effects are universal, but their interpretation by different shamans is likely to be idiosyncratic, and this difference sets a research program:

Both entoptic images and hallucinations are universal, but their utilization and interpretation by shamans ... are *culturally* specific. ... Understanding the cultural filter used to interpret the hallucinations encountered during SSC [trances] should be central to the anthropology of religion, given that it reflects cultural transmission between the practitioners, aspects of a culture's cosmology, and their view of the spirit world. (2009, 180)

In ethnographic records, the art associated with these different substances tracks the differences in phenomenal affect. That is, the use of nicotine tracks black and white imagery, and of peyote more colorful representations. So, we can connect the artistic imagery (rock art, for instance) with particular methods of achieving trance states—themselves potentially indicative of different social groupings and religious practices.

Further, VanPool identifies various material remains beyond artistic representation that are also related to sacred practices. These include musical instruments, the remains of psychoactive plants (macrobotanical remains of datura, for instance), tools involved with their consumption (pipes, for instance), and sacred spaces reserved for shamanistic activities. The similarities and differences between different sites provide inroads into tracking different spiritual practices in different contexts. In effect, by weaving these different traces together, VanPool develops a localized model of shamanistic practices across the relevant geographical location. Unlike in lucky circumstances, where a single line of evidence takes precedence, VanPool is characteristically omnivorous: The power of her reconstruction turns on the "confluence" of evidence. Let's turn to Wylie's work to describe this.

Wylie's entry point is the remarkable capacity of material remains to resist the "imperial" power of theory. That is, given the apparent paucity of material evidence—the underdetermination of historical hypotheses— we would expect our interpretation of material remains to be more or less determined by our preconceptions. But this is not what happens. Archaeological evidence stubbornly refuses to play theory-laden ball, transforming our conceptions of the past despite the destructive processes of time and our tendency to see what we want in evidence. Wylie's explanation of the intransigence of material remains appeals to what I call the *confluence* of evidence. What is "confluence"? It is the sum effect of various evidential sources coming together—not in the specific way required by consilience (although this is assuredly part of the story)—but in a more general sense. I understand confluence as the combination of three factors that Wylie (2011) identifies: vertical independence, horizontal independence, and stability—all of which we have already met in passing. My characterizations of these notions differ from hers, but let me give a brief one for each factor (see figure 6.3).

Horizontal independence is, in effect, consilience. We can understand this as a relationship between different dependency-relations or lines of

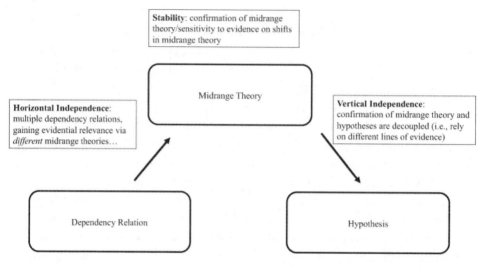

Figure 6.3
Confluence.

evidence. Two lines of evidence are horizontally independent just when they are evidence by virtue of different midrange theories. As we have seen, this matters for problems of testing holism. Different sources play the same role: testing some hypothesis.

Stability is a relationship between midrange theory and dependency relations. The more "stable" our midrange theory, (1) the better confirmed that midrange theory is; (2) the less sensitive the evidential value of the trace is to changes in midrange theory.

Vertical independence is a relationship between midrange theory and the hypothesis being tested. It holds insofar as the evidential support the midrange theory garners is different from the evidential support of the hypothesis. This matters for worries about being overly theory-laden. If horizontal independence holds, although midrange theory and hypothesis are theory-laden, they are laden by *different theories*.

The power of VanPool's reconstruction is based on confluence. First, there is horizontal independence between different lines of evidence. The connection between rock art and varying means of achieving trance state is derived from (1) the general perceptual effects of different drugs and (2) ethnographic information connecting imagery to perceptual states. The connection between macrobotanical remains and shamanistic practices is based on different midrange theories from botany and chemistry. Second, midrange theory is stabilized both by the range of evidential sources (psychology, ethnographies, and so on) and via the construction of *localized* models. VanPool's insistence that we identify the scope of the various regularities involved allows us to zero in on different cases. Third, different theories rely on different evidence—they are vertically independent. The hypothesis that some group used tobacco in its rituals is supported by the various *sacra*, or religiously laden artifacts, that remain—pipes, black-and-white-imagery, and so forth. These themselves count as evidence by virtue of midrange theories relating to perception, ethnographic information, and so forth—the midrange theory doesn't presuppose the hypothesis.

The lesson here is that, unlike inferring a mammal's size from its molar, the power of VanPool's reconstructive machinery lies not in a single line of evidence but in their confluence—how the whole picture hangs together. This omnivorous strategy maximizes our epistemic reach into the past. By drawing further links between past hypotheses, as we have seen in the relationship between Snowball Earth and the Cambrian explosion, this can be further developed and extended in a localized, piecemeal way.

Taking lucky circumstances, and successful strategies in that context, as a model for good historical science, then, is not only mistaken, but potentially blocks epistemic progress. Success in unlucky circumstances turns on adopting a different method: It is localized, speculative, and opportunistic—methodological omnivory. It is underwritten by the confluence of evidence.

7 Parochialism and Analogy

"History is merely a list of surprises," I said. "It can only prepare us to be surprised yet again. Please write that down."
—Kurt Vonnegut, *Slapstick, or Lonesome No More!*

You should not be surprised to learn that, unlike Kurt Vonnegut's narrator, I do not think that history is merely a list of surprises. Historical scientists care deeply about repeated patterns in the past. They study them for their own sake, and they use them to further our understanding of particular events. My aim in the next two chapters is to understand the "contingency" or "historicity" of deep time, as it relates to the nature of historical science. In particular, I am interested in how these qualities affect our epistemic access to the past.

In this chapter, I hope to convince you of four things. First, historical scientists care about and employ regularities. Second, these regularities are "fragile," that is, they hold only under certain conditions and are often exception-ridden. Third, "analogies" provide epistemic access to these regularities. Fourth, this use of analogies supports optimism about the historical sciences insofar as they are a counterexample to the idea that all historical evidence is based on traces.

My strategy is to begin by considering a diverse group of views that agree that biology (or historical science) is restricted in scope, and is so by virtue of its historical nature. I present two versions of this kind of view and argue in favor of the less extreme one. With this out of the way, I turn to "analogies." I contrast them with trace-based evidence. Where traces are evidence by virtue of midrange theories that connect them to the past through lines of causation, analogies are evidence by virtue of being tokens of the same type of process. Analogies are instances of a regularity or pattern and are therefore empirically relevant to those patterns and each other.

Some philosophers might have qualms about my use of the type/token distinction. Often, for instance, "types" are understood as abstract objects, while "tokens" are concrete instantiations of types. The type "Octagon" might exist in Plato's heaven, while the Octagon (Dunedin's town "square") is a physical instantiation of that abstract object. However, there are less metaphysically expensive ways of understanding my use of the distinction. By "type," you can read me as referring to "kinds," or to qualitative similarities, or to members of the same set. Personally, I find the "type/token" language useful in this context, and you should feel free to tailor your gloss to your metaphysical inclinations.

7.1   Parochialism

It is tempting to think that historical science is (let's call it) *parochial*. That is, the explanatory range and empirical power of historical investigations are restricted to local events. This thought is, to some extent, right: Our hard-won knowledge of sauropods is surely rather limited in scope. And this is particularly plausible in comparison with the grand generality of the theoretical edifices of physics and chemistry. But how limited, and by what? And, moreover, what are the consequences of this for our knowledge of the past? Such questions are our concern for the next two chapters. In this section, I start by discussing a group of views that approach parochialism in various ways. This is certainly a disparate group: I will focus on Jack Smart, Alex Rosenberg, and Ken Waters. From this I articulate two views, one more extreme than the other. The extreme view, which I call *parochialism*, takes historical scientists to be restricted to actuality, both in their explanatory interests and their evidential resources. The less extreme view, which involves what I will call investigations of "fragile systems," takes it that historical scientists often appeal to regularities: They are interested in more than the actual world. I argue against the extreme view.

Before beginning, I should mention philosophical discussion of "laws." There is much in the philosophy of science concerning laws: what they are, whether there are historical or biological (or ecological, or economical) laws, whether all laws are the same or whether some are ceteris paribus, and so on and so forth. This is, I think, tangential to my discussion. I remain silent (indifferent even) about whether my discussion of patterns and generalizations in historical science involves laws. I can point to an account, however, that does the work I need, appealing to Sandy Mitchell's work. She recommends we identify laws via their use in scientific theorizing: "The function of scientific generalizations is to provide reliable expectations of the occurrence of events and patterns of properties" (1997, 447).

Within a certain range of "reliability," this is true of much theorizing in historical science. What is midrange theory for, if not to explain how various patterns occurred in terms of the historical processes that generate them? Moreover, this account explains why scientists focus their investigative efforts as they do. For instance, sauropods are surprising in part because their size bucks our expectations about gigantism. That is, we have generalizations about maximal size in terrestrial animals, for which sauropods are dramatic exceptions. So, if you want to call these generalizations and regularities "laws" or "lawlike," you're welcome to—but I don't think the term "law" carries any special weight. As we will see, I focus on more nuanced ways of understanding regularities and their role in scientific theorizing.

7.1.1 A Common Thought

The theory of evolution and ecology are two branches of biology which are quite obviously "historical" in nature. They are concerned with a particular and very important strand of terrestrial history. No doubt there are analogous histories on remote planets, but in the theory of evolution we are concerned with the hereditary relationships of those particular species we observe, and so we are not concerned with laws in the strict sense. (Smart 1959, 366)

For Smart, biology is unavoidably terrestrial. As opposed to physics and chemistry (and ideally metaphysics), its scope is earthbound—and earthbound in two senses. First, the explanatory targets of biology are particular events, yielding "narrative" explanations.[1] Second, regularities governing other phenomena are not evidentially relevant, or at least have very limited applicability. Targets are, in a strong sense, unique, and so cannot be unified via general models (I will articulate a relevant sense of "unique" in the next chapter). If a science is terrestrial in the second sense, then the available evidential resources are impoverished. Why? Because, as we shall see, our reconstructions of the past would then rely on trace evidence alone; hence the relevance of this to our discussion. Smart explicitly links his view to biology's inherently historical character. Where other scientists seek universal, exceptionless regularities, historical explanation is restricted to particular histories. Whether or not this is problematic (it is not obvious to me why the search for universal regularities is important), it is false. The reason is two-pronged. I argue that, some of the time at least, historical explanation is not problematically parochial. Historical science is not tied to Earth and its particular history, but rather to the types of systems found on Earth and, for all we know, elsewhere. Of course, much of the time historical scientists are concerned with particular terrestrial histories, but they often rely on regularities that are not so restricted. In other words, even when they are parochial in the first sense I take from Smart (that is, they aim only to explain particular, earthbound events), they are not parochial in the second, pernicious, sense.

Why should we consider biology (or historical sciences more generally) to be restricted to terrestrial matters? An appeal to common-cause explanation could do the job. As we saw in chapter 6, on some views the prototypical method of justification in historical science is taken to be positing past events in order to unify traces, and then discriminating between those posits by searching for further traces ("smoking guns"). If that is right, then we might expect much historical inquiry to be parochial (in one or both senses), since the main methodology of historical science is accounting for and unifying observed patterns of traces. Of course, focusing on common causes does not eschew wider understanding of the system in question. After all, understanding how such systems behave under counterfactual perturbations is essential for identifying smoking guns. However, in common-cause explanation our interests are largely restricted to providing the best account of the actual distribution of traces. I argue that such views are too quick: Historical scientists are intimately concerned with system types, and they draw on non-trace, analogous evidence in investigating such systems.

For Smart, biology is terrestrial owing to the nature of its target. Where nomological sciences, physics and friends, make universal claims about homogenous structures, atoms and their ilk, biologists are concerned with individuals. "Mouse" doesn't pick out a kind, but rather a bunch of ancestrally related furry rodents. Smart argues that any attempt to define "mouse" using nonhistorical properties (such as we do with physical or chemical kinds) results in false universal claims (for somewhere in the universe there is bound to be something sufficiently mouselike that doesn't, for instance, enjoy cheese). Biologists make generalizations, to be sure, but these generalizations only concern terrestrially bound individuals. Smart's clearest mistake is to endorse the dichotomy between universal, and merely terrestrial, generalizations—as I will explain, there are varying degrees of modal scope that a generalization might cover.[2]

Rosenberg (2000, 2006) comes to a similar conclusion as Smart, but for different reasons. He takes biological kinds to be individuated by *selected function* (a wing is such by virtue of its being selected for flight), arguing that selection cannot discriminate between functionally equivalent states. A form of parochialism emerges from selection's blindness to substrate: "Since selection for function is blind to difference in structure, it is easy to see that there will be no laws in any science which, like biology, individuates kinds by naturally selected effects, that is, by functions" (2006, 138).

Why? Because a universal generalization of the form "all *F*s are *G*s," where the *F* is functional and the *G* is structural, will be false if the process picking out the *F*s cannot discriminate between the *G*s. For example, "all wings are feathered" is clearly false—consider bats, pterosaur, and insects. The universal statement fails because although feathers are one way to instantiate a wing, they are not the only way to do so. Natural selection cannot guarantee that a functional kind will always be realized by the same structure. This can be taken too far. Although selection can realize its effects across different substrates and structures, some substrates are better suited to some functions than others. There is a good reason why there are no flying, or, for that matter, arboreal[3] cephalopods. Moreover, regularities need not take Rosenberg's exceptionless logical form to play "lawlike" roles in biological explanation.

Although Rosenberg differs from Smart in delineating biological kinds by selected function rather than ancestry, they are similar insofar as biology's lawlessness is due to its historical nature. Biology, on this view, is largely natural history:

Just as the radio engineer uses physics to explain why a circuit with a certain wiring diagram behaves as it does, so the biologist uses physics and chemistry to explain why organisms or parts of organisms (e.g., cell nuclei), with a certain natural history description, behave in the way they do. ... Descriptive biology consists in generalizations of natural history, not laws in the strict sense ... biology is physics and chemistry plus natural history. (Smart 1959, 364)

Biology, it is claimed, is restricted to describing facts about terrestrial individuals, or structural instantiations of functional kinds. True explanation happens elsewhere (presumably, at some reductive "more fundamental" level). Moreover, this is due to the historical nature of biology.[4] It is my contention that biology and other historical sciences are frequently not parochial in this sense and, even when they are, their explanations draw on models of general, if fragile, biological systems.

Another recent philosopher of biology who can be read in a parochial light is Ken Waters (2007). He argues that biologists (or, at least geneticists intervening on model organisms) interested in uncovering and manipulating *actual difference makers*.

To understand the notion of "actual" difference-making, it would help to have a handle on difference-making itself. Jim Woodward has developed this notion in order to analyze both causes themselves and causal explanation (2001). In brief, some variable y is a difference maker to some event x just in case x's value changes upon an ideal intervention on y, with other variables relevant to y held fixed. My banjo is tuned to open G. To get this tuning, various pegs must be manipulated to increase or decrease the tension on the strings. Turning the peg on a particular string, say the drone string, will change its pitch and thus the tuning of the banjo as a whole. As such, if I were to intervene on the variable of the position of the drone string's peg, this would change the value of the banjo's tuning. Hence, tuning peg position is a difference maker to banjo key. Moreover, my restringing my banjo or otherwise is a difference maker for its having a bright tone. This is because, if we hold fixed the other properties of my banjo—its design, the wood it is built from, who is playing it, and so on—an intervention on the age of the strings (y) will change the banjo's tone (x). Specifically, it will make it brighter the younger they are, and duller the older they are.

Of the space of difference-makers—anything that, by Woodward's "manipulability" account, is a difference-maker—some are special. They actually vary in populations, and actually make a difference. Such are Waters' "actual" difference-makers. For instance, for the last year or so my banjo has been left unplayed under Paul Griffiths's desk in his office in the Sydney University Quadrangle.[5] Over that time, the tone of my banjo has most likely become significantly less bright as the strings age. Regarding my banjo's tone over the last year, it being restrung is only a potential difference-maker—if it *were* to occur, then my banjo's tone would be brighter than it in fact is, but this has not occurred. However, as we will see, the age of the strings is an *actual* difference-maker.

Waters's concept of an "actual" difference-maker goes as follows. Among a population, individuals have varying properties. Those difference-makers that (1) in fact vary in the population and (deep breath) (2) whose effects are invariant across variables that in fact vary in the population are *actual difference-makers*. The first clause is self-explanatory. The second clause requires that the difference-maker's effects remain relatively constant across the population. That is, the population must be relevantly homogenous in ways relevant to the difference-maker's difference-making. Much of Earth's water (populations of H_2O) freezes at $0°C$: Groups below that temperature are frozen, and those above it are liquid or gas. Temperature in fact varies for earth's water populations, and so is an actual difference-maker in regard to water's chemical state. However, the relationship is not invariant, for both salinity and pressure can change the exact temperature at which water freezes. Restricting our population to sea level, fresh (and so nonsalinized) water, salinity and pressure are merely potential difference-makers, while temperature is an actual difference-maker, to the water's chemical state.

So, Waters splits difference-makers into two categories. *Potential* difference makers are those causal factors that, were they intervened on, would make a difference. *Actual* difference-makers actually vary, and do make a difference by virtue of their variation. For example, among the population of my banjo's time-slices over the last year, restringing is only a potential difference-maker with respect to its tone. However, the age of the strings is an *actual* difference-maker. Aging has led to the strings becoming duller—and in a relatively regular way (seeing as, more's the pity, they're not being played). Earlier time slices of the banjo have a brighter tone than later time slices by virtue of actual variation among the time-slices in the age (and thus wear) of the strings.

Waters's distinction goes some way toward capturing the difference between "background conditions" and "true" causes. It seems as if my striking a match is an important cause of the match's lighting, while the presence of oxygen in the room is merely a background condition. In most circumstances, appealing to the presence of oxygen would be an odd (or at least irrelevant) addition to an explanation of a match's lighting.

For Waters, this intuition is explained because (for most populations of matches that might interest us) of their being struck or otherwise are actual difference-makers, whereas oxygen levels are merely potential difference-makers. Although oxygen supply varies at different altitudes, this variation is typically insufficient to affect whether or not a match lights.

Thus, potential difference-makers would affect the phenomena of interest if only they varied enough, while actual difference-makers do make a difference by virtue of their variation.

Waters explains biologists' apparent indifference to the lack of their science's generality by claiming that they are interested in actual (as opposed to potential) difference-makers. If biologists are in the business of uncovering actual difference-makers, then their divergence from physicists—their relative parochialism—makes sense:

Whereas physical scientists often try to establish generalizations that express causal relationships that hold over conditions that have not and will not be actualized (to their knowledge), biologists typically care only about whether a relationship holds under conditions actualized in organisms and their environments (or in the laboratory). ... Biologists are more interested in explaining actual differences than possible differences, and explaining actual changes over developmental and evolutionary time than possible differences. This is why they are more interested in identifying actual difference makers than potential difference makers. (576–577)

It should be noted that Waters couches his claims in terms of scientists being *more or less* interested in the actual, and so insofar as we disagree this is a matter of degree. Moreover, he focuses on experimental programs such as genetics that are, I think, quite different in character from what tends to go on in the historical sciences—well, at least in unlucky circumstances. Nonetheless, he provides a very direct characterization of a parochial science: Biologists care about describing the actual distribution of events and causes in the systems that interest them, rather than situating them in a broader context. Waters and I agree that biologists (and potentially other historical scientists) target systems of narrower variance than physicists—this is his main point. Waters argues, and rightly so, that the assumption that scientists always prefer broader generalizations when they are available is mistaken. However, I don't think that the "actual/ potential" difference-maker machinery captures this narrower variance in the historical cases I am concerned with. Moreover, I think this conception of biology (and the historical sciences more generally) can be taken too far. Much of the time, historical scientists are deeply concerned with potential, rather than actual, difference-makers. Moreover, the causes they cite and model are often too variant across the relevant populations to meet Waters's definition.

Smart's view that biology is parochial might be philosophically anachronistic, but we can hear its echo not only in Rosenberg's skepticism about biological laws and Waters's emphasis on the actual but also in Hull's defense of species as historical entities (1976), Griffiths's emphasis on ancestry in biological explanation and individuation (1996, 2006), and Dennett's dismissals of mere "frozen accidents" in biology (1995). I hope it is clear that (except perhaps in Smart's case) none of these philosophers explicitly endorse parochialism in the sense that I will articulate in the next section. Regardless, there is a general idea that biology is restricted in scope—perhaps quite radically so—and that this restriction has something to do with biology's historical nature. Perhaps Darwinian processes (selection, drift, and unpredictability in the generation of variation) are to blame for this parochialism; moreover, this provides an obvious way of splitting paleobiology and evolutionary theory from geology. There is something to this, but I will leave discussion of why historical targets might be so restricted to the next chapter.

The main point of the forthcoming discussion, in terms of the book as a whole, is to show that historical scientists target restricted regularities: They construct models of causal system types. While the systems they target are relatively variant, they hold only within a relatively specific set of conditions (and have frequent exceptions within these conditions); scientists both attempt to understand these systems for their own sake and draw on that understanding to gain epistemic traction on particular events. These systems play an important role in the confirmation of historical science. And so, besides the issue of parochialism, this discussion is vital for understanding the epistemic security of historical science: The support a historical hypothesis may garner depends, in part, on its scope.

7.1.2 Two Distinctions

Let's solidify parochialism into two views. This requires several distinctions. First, I distinguish between "patterns" and "causal structures," the former being the actual distribution of events and traces, the latter being the regularities by virtue of which the patterns hold. In a sense, when "patterns" are *explananda*, "causal structures" are *explanans*, since causal structures give us reason to expect the occurrence of patterns. Second, I distinguish between two ways in which an investigation's scope might be limited. Each limitation is linked to the previous distinction. Inquiry could be *parochial*, tied to particular patterns, or it could target *fragile systems*, systems that hold only under fairly specific circumstances. It is my contention that historical science (like other sciences!) is frequently concerned with fragile systems, rather than being parochial.

We will approach the distinction between "patterns" and "causal structures" via consideration of mass extinctions.

Much paleobiological work attempts to distinguish noise from signal in the fossil record: detecting a pattern and then working out which past events they indicate (see Benton 2009). Consider Raup and Sepkoski's claim that the fossil record contains periodic "mass extinctions," events wherein at least 40 percent of marine genera become extinct (Raup & Sepkoski 1984). In establishing this pattern—that the actual history of life contains such events—there are two challenges to meet. The first challenge is to respond to technical questions about patterns of traces. Which taxonomic analysis is appropriate? How precise can we be about timing? Which sorts of statistical techniques can be bought to bear? For instance, our choice of taxonomic category leads to different results. Often, an analysis based on species-level taxa provides a different picture than an analysis based on family-level taxa. This occurs because species are categorized into families, so it is possible to have significant species loss without significant family loss. This is important, since the rough grain of the fossil record (because of its faint and gappy nature) can inflate the drama of sudden drops in species number vis-à-vis the large scale patterns paleontologists care about.

The second challenge is to determine the signal's source: Do we have a *biological* signal, revelatory of patterns in life's unfolding, or a *geological* signal, due to biases in sediment formation? Geological and climatological events, as well as organisms' lifeways, affect the likelihood of their fossilization. Climatological shifts (say, an increase in world temperatures, which would increase liquid water volume, and thus increase the number of environments conducive to fossilization) can increase the number of preserved fossils without proportional changes in biodiversity. To argue that there are periodic mass extinctions, we must defend a method for identifying patterns in traces, and defend a reconstruction of a past pattern on that basis.

Thus far, my discussion of Raup and Sepkoski has focused on determining *patterns*: They detect occasional spikes in the disappearance of fossil genera in the record. They go on to argue that this pattern in the traces reveals a further pattern in the biological past: mass extinctions. To do this they must establish (1) that their analysis of the traces is legitimate and (2) so also is their extrapolation from traces to the past. I reserve the term "pattern," then, for the actual distribution of contemporary traces, or events in the past.

I contrast patterns with *causal structures*. These, in part, explain traces: They are modally robust and lead us to expect the patterns we observe. Causal structures take many forms. Some are "lawlike": They support counterfactuals that lead us to expect the occurrence of the explanandum. Midrange theories, for example, represent the mechanisms and processes connecting traces to the past. While a pattern is the actual distribution of events (be it trace or otherwise), causal structures are the regularities, mechanisms, laws, or processes that explain the patterns. The (disreputed) Nemesis hypothesis, for instance, modeled a causal structure intended to account for Raup and Sepkoski's pattern of extinctions. It claimed the sun is twinned to another star ("Nemesis") that, at regular intervals, sends a cluster of meteors from the Oort cloud hurtling through our solar system, thus dramatically increasing the Earth's extraterrestrial impacts during these periods. According to the hypothesis, the Nemesis causal structure is to blame for the earth's regular mass extinctions (Bailey 1984). If we believe in that structure, we are led to believe in that pattern. Note that causal structures are *ontic*—they are processes, mechanisms and regularities in the world. Later, I will discuss causal models that represent such structures.

And so, I distinguish between *patterns*—the actual distribution of events—and *causal structures*—the modally robust regularity, process, or law that the pattern is an instance of. The pattern concerns what is observed, while the causal structure concerns how the pattern might vary under different conditions. Scientists frequently invoke models of causal structures to account for observed patterns.

Clear examples of the distinction between patterns and causal systems can be found in paleobiological studies of the tempo and mode of morphological change at the macroevolutionary level. Consider the following quote from Brusatte (2011):

In order to test whether a certain clade radiated adaptively or underwent punctuated episodes of change, it is first necessary to quantitatively measure morphological rates of change. More broadly, it is simply interesting to establish patterns: are certain clades or time intervals characterized by more morphological change, or higher rates of change, than others? Once patterns are robustly established, they may be marshalled as evidence in favor of certain evolutionary processes or used to describe the large-scale narrative of clade history. (53–54)

To empirically investigate macroevolutionary theory, paleobiologists must first characterize patterns in the fossil record, then establish that the pattern is revelatory of biological history. With this done, the patterns can be used to evidentially ground models of causal structures: the regularities, processes, or mechanisms from which those patterns emerge.

With the distinction between pattern and causal structure out of the way, let's move to a further distinction between "parochial" investigations, and those concerned with "fragile systems." The terms of this distinction concern the interests and the evidential range available to historical scientists. Let's begin with another quote from Smart: "Biology, in contrast to physics and chemistry, is cosmically parochial ... though there is speculation about exobiology, biology is concerned with terrestrial matters" (2008, 227).

As we have seen, Smart echoes a common thought: In contrast to "nomological" sciences like physics, biology is lawless, concerned with contingent, earthbound events. Moreover, this is due to biology's historical nature: Because each lineage has a unique evolutionary history, our explanations and evidence are constrained by that history. But are historical disciplines such as biology parochial? This question is ambiguous. In light of the distinction between pattern and causal structure, there are two senses in which an investigation could be restricted. Let's draw the distinction as follows:

Parochialism: The scope of explanation is limited to a target pattern; the scientist is interested in why those particular events occurred. The counterfactuals appealed to, if any, are about that system.

Fragile systems: The scope of explanation is limited to some causal system, the scientist is interested in how systems with those parameters, or with those initial conditions, behave under different conditions. The causal system is "fragile" insofar as the conditions under which it holds are (more or less) highly specific and/or contingent:[6] They are relatively variant. Counterfactuals are about that *system type*.

Parochial investigation is concerned only with *the pattern in question*: explaining a particular distribution of events. Investigation of fragile systems situates patterns into a broader class: They are the result of the dynamics of a certain type of causal system.

Prima facie, both sauropod gigantism and Snowball Earth are examples of parochialism. The scientists are interested in explaining why that particular event occurred. However, much historical enquiry targets causal systems across deep time. Paleobiological work covering the nature of mass extinctions (Raup 1991), the nature of speciation (Gould & Eldredge 1977), and the role of natural selection at the macro-level (Gould et al. 1977; Huss 2009) are all concerned with regularities governing how life's shape unfolds, not the explanation of patterns or past events. Moreover, biologists often concern themselves with the conditions that make life possible in the first place. Some studies into early and extraterrestrial life are like this, as are attempts to understand Darwinian processes as "algorithms for creating design out of chaos" (Dawkins 1983; Dennett 1995). In what follows, I provide two counterexamples to "parochialism" as understood in this sense. I will then shift to an archaeological discussion to tie a concern for fragile systems to methodological omnivory.

7.1.3 Counterexample the First: Cope's Rule

In this section I describe a set of paleobiological examples of nonparochial investigation. According to Cope's rule, the average size of individuals in a lineage increases over evolutionary time. That is, individual organism size trends upward as a lineage increases in age. Investigations of the rule are representative of paleobiological concerns with the behaviors of fragile systems.

From his examination of Cenozoic mammals, Edward Drinker Cope (1887) uncovered a pattern: Individuals in lineages tend to increase in size over evolutionary time. He attempted to explain the pattern by postulating a causal structure: Although new species are typically small, some innate drive for largeness usually influences their subsequent evolutionary trajectory. Paleobiologists have since focused on Cope's rule, in part because its simplicity facilitates large-scale statistical studies of the fossil record, and because it is a good testing ground for discriminating passive from driven trends (a distinction briefly touched upon in the introduction—but don't worry, I'll run over it again in a second). Several papers (Hone and Benton 2005; Hone, Keesey, et al. 2005) use data from sauropod evolution to test Cope's rule. These investigations proceed by examining dyads in a phylogenetic tree—that is, by comparing sister-lineages. Sauropod taxa from the later Cretaceous are 25.7 percent longer than earlier taxa—Cope's rule, as a pattern, holds for sauropods. Let's look at how we might explain instances of Cope's rule (perhaps I should call it "Cope's pattern" at this point) in mammals.

Alroy (1998) analyzed 1,534 mammal species from the late Cretaceous to the Pleistocene. The analysis was carried out on taxonomically paired species that are treated as ancestor-descendant pairs. He took the uncovered pattern to be "overwhelming": "Newly appearing species are on average 0.0874 natural log units (9.1 percent) larger than older congeneric species, a highly significant difference ..." (732). And so, a pattern is uncovered. Alroy then placed a restriction on possible explanations of the pattern: "The only clear-cut hypothesis that predicts such a pattern is the most narrow and deterministic interpretation of Cope's Rule; namely, that there are directional trends within lineages" (732). Alroy rejected nondirectional explanations, such as diffusion from a boundary, or different rates of origination and extinction, on the grounds that such explanations cannot explain within-lineage patterns. He thought the overall trend might be due to "a balance of forces operating both within and among lineages" (733). For instance, although his hypothesized optima for large size in mammals is not observed, he suspected this may be due to higher rates of extinction or lower rates of origination.

In sum, Alroy describes a pattern and then hunts for a causal system (some force, law or mechanism) that could account for it. The pattern is the statistical correlation between size differences in ancestor-descendant lineage pairs. The proposed causal systems can be understood as competing within versus between lineage effects. Size could increase overall because new species are diminutive when they originate and larger sizes grant a fitness benefit—Alroy hypothesizes two size optima in mammals. Here, external pressures acting on the lineage do the explanatory work. Alternatively, size increases could be put down to lineages tending toward different origination and extinction rates: The effect could be due to lineage sorting. Both "internal" and "external" explanations lead us to expect the observed pattern. They shift from mere description to explanation by providing reason to expect such patterns in the fossil record.

Dan McShea generalizes the kinds of competing explanations that Alroy discusses into *passive* and *driven* trends (McShea 1994). The classic passive trend is represented by a diffusion model. There is, surely, some lower bound to the size of individuals in a species. Assuming lineages have, on average, equal chances of evolving to larger or smaller sizes, we should expect to see lineages increase in size overall (see Gould 1996). Alroy prefers a driven trend. There are some optimal body sizes for mammals and, given a tendency for new lineages to appear at smaller sizes, we should expect selection to "correct" that bias over evolutionary time. Another example of a driven trend is Hunt and Roy's (2006) suggestion that climate change drives body size evolution. Bergmann's rule states that species are on average larger in the colder parts of their range. Hunt and Roy investigate whether "Cope's Rule may simply be an evolutionary manifestation of Bergmann's Rule; species and lineages that conform to Bergmann's Rule should evolve toward larger sizes during periods of climatic cooling" (1,347). Perhaps the fluctuation in animal size captured by Cope's Rule in fact tracks fluctuations in world temperatures.[7] Focusing on a lineage of poseidonamicus, prehistoric ostracods (tiny, clamlike crustaceans), they find a strong correlation between climatic cooling and size increase and, like Alroy, take this to support a driven, within-lineage explanation.

Whether a trend is passive or driven, notice that both explanations account for some pattern of traces by appeal to a causal system—we are led to expect the observed pattern. Although we are interested in a pattern at first, investigation quickly shifts to discussion of causal systems. Such discussion concerns both actual and potential difference-makers. On diffusion models the existence and position of a lower bound is an actual difference-maker (assuming lineages in fact approach it). However, understanding the nature of diffusion models, when they apply, when they break down, and so forth, requires consideration of potential (as opposed to actual) difference-makers. In the laboratory situations that Waters examines, scientists are able to *actualize* difference-makers. Laboratory conditions ensure that what were once merely potential, are now actual with respect to the isolated laboratory populations. Historical scientists are not able to do this, and instead investigate potential difference-makers. Sets of actual difference-makers (or at least, factors hypothesized as such) are applied to explain particular instances of Cope's rule. Perhaps cooling can explain some lineage's size increase, for instance, yet this explanation might not be complete or relevant to every instantiation of the pattern. Paleontologists care deeply about fragile causal systems, and they understand potential difference makers in order to understand the circumstances under which such systems hold. Such systems do not always behave regularly enough (or we cannot force them to behave regularly enough) for the variance required to establish that actual difference-making holds. Having a good grip on the actual and potential difference-makers of such systems affords both the resources to tailor individual explanations to particular instances and to understand when they do not occur. Understanding what would happen, sometimes in rather outlandish circumstances, matters to historical science.

And so we can discern two important questions from paleobiological investigation of Cope's rule. First, does the rule in fact hold? Is the pattern we discern a biological signal, or is it merely due to biases in the fossil record? Second, when the pattern (the "rule") does occur, we ask after the underlying causal system: whether it is due to the influence of selection (a driven trend) or is passive owing to random walks from a lower bound. Finally, fitting particular cases to such models of system dynamics—or failing to—can help set the scope and direction of investigation. This is seen dramatically for the application of regularities about maximal size to sauropods.

Burness et al. (2001) compare maximal animal size across taxonomic and trophic levels and attempt to explain discernible patterns as an effect of metabolism and resource availability. Their model can be roughly captured with three ceteris paribus claims: (1) larger landmasses have more resources; (2) herbivores require more resources than carnivores; (3) ectotherms require less food than endotherms. Burness et al. quantify these principles and employ them to explain various apparent surprises. The largest varanid, for instance, the Komodo dragon, is able to survive on the small island of Flores because of the low food requirements of its ectothermic metabolism. Burness et al.'s model captures many cases of gigantism, in particular large Oligocene mammals (the monstrous marsupial *Diprotodon*, the giant hyena *Pachycrocuta*, and so forth). However, their model fails for the sauropod and theropod dinosaurs: "the top carnivorous dinosaurs were still 12 times heavier, and the herbivorous dinosaurs 1.5-3 times heavier, than predicted ... therefore, the never-since-surpassed size of the largest dinosaurs remains unexplained" (14,523).

I think this discussion is sufficient to show that paleobiological investigations of Cope's rule are surely concerned with fragile systems rather than being parochial. Paleobiologists concern themselves with how system types behave under a range of counterfactual conditions and variances. Parochial investigations are only concerned with system tokens. They may tackle counterfactuals (such explanations are causal, after all) but, crucially, such counterfactuals are taken to concern only that target instance. All the action is in the actual difference-makers. Clearly, paleobiologists examining Cope's rule are only in part interested in the pattern: They want to explain the rule. It is also not earthbound: If complex extraterrestrial life exists, we might expect Cope's rule to hold there as well.

7.1.4 Counterexample the Second: Pebbles on Mars

In the previous section I argued that historical science is not merely parochial—it is frequently concerned with fragile systems. Here I hammer that point home with a geological example.

One of the main purposes of NASA's Martian exploration is a hunt for signs of ancient water. A discovery would be momentous, dramatically increasing the likelihood of Mars having supported life, and its capacity for doing so in the future. Mars exhibits many features suggestive of an aquatic past: "deltas, alluvial fans, valley networks comparable to terrestrial river valleys, and giant outflow channels carved by catastrophic floods" (Williams et al. 2013, 1068). These can all be dated to more than 3.5 billion years ago. There are several competing theories about Mars's past. The "warm and wet early Mars" hypothesis (Carr 2012) postulates a thick atmosphere in Mars's infancy. This would enable a Martian hydraulic cycle despite its lack of a molten core. A competing hypothesis explains the signs of water-based erosion by postulating periodic, aquatic episodes nestled among a generally staid, waterless Mars. Because the formations evidencing Martian water are so old, many traces will have been distorted or covered by alluvial processes.

NASA's Curiosity rover discovered conglomerate rocks bearing features that have striking resemblance to features indicative of ancient riverbeds on Earth. This has been taken to be conclusive evidence of Martian waterways (Jerolmack 2013).

Conglomerate rock is formed when stones of various types are carried together by some force—marine, fluvial (river), alluvial (wind), or glacial—and eventually "glued" together by sand. The causes forming the conglomerate can be identified via (1) the types of rock, (2) the "roundness" of the rock (lithology) and (3) the stratification. For instance, wind erosion leaves different marks from water erosion. Alluvial conglomerates will typically have more jagged rocks, while fluvial rocks will be well rounded and sorted in strata by size. Geologists use midrange theories, which mechanistically connect formation processes to contemporary rock form, to infer from contemporary strata and lithology to past states.

Williams et al. (2013) take the discovery of conglomerate rock on Mars to show that about 3.6 billion years ago water flowed in rivers across Mars. They highlight the "roundedness" of the lithology (different from lithology at other Martian sites, which are usually put down to volcanic or alluvial processes), the sorting of the rocks (they are well sorted in terms of size) and the alternating pebble-sand nature of the stratification (pointing to transport of at least a few kilometers). This is, so far, the same evidence that would be taken as utterly uncontroversial evidence of an ancient earthling riverbed (see figure 7.1 for a comparison). Williams et al. discount various other hypotheses: Massive landslides and wind are discounted due to the nature of the deposits, and transportation and formation via other liquid substances is discounted based on Mars's chemical composition and past conditions.

In addition to the brute appearance of the rocks, larger-scale structures on the Martian surface are also suggestive of an *alluvial fan* (figure 7.2): triangular dispositions of gravel and other materials indicative of the passage of a river's path being interfered with by higher ground.

Terrestrial models, then, are transported into the Martian context with minimal tinkering: "On Mars, the elastic collisions within the flow may have had lower energy due to the reduced gravity, resulting in lower abrasion rates and longer transport distances to achieve similarly rounded pebbles" (Williams et al. 2013, 1070).

Geologists working on Martian remains transport the same criteria for reconstructing Earth's past, with only minor corrections to mitigate Martian nonconformity. At the very least, then, the midrange theories that geologists employ on Earth are not parochial. That rounded pebbles are indicative of the fluvial formation of rocks is taken as true on planets that

Figure 7.1
Conglomerate rock. The left is Martian; the right is Earthling. (© NASA; NASA/JPL-
Caltech/MSSS and PS.)

Figure 7.2
On the right, a fluvial fan on Earth; on the left, NASA's topographic map of the cu-
riosity rover's landing site (cross) and the proposed Martian fluvial fan. (Detail from
Jerolmack 2013, 1055. © American Society for the Advancement of Science.)

are relevantly earthlike. Even though geologists are interested in explaining particular patterns on Earth or Mars (rather than the regularities), such explanations rely upon regularities that move beyond the target pattern.

And so, even though it is true that NASA has parochial (if not literally terrestrial) concerns—they want to know whether water flowed on Mars—to discover this they rely on nonparochial epistemic resources. They construct a causal model linking Earthling and Martian conglomerate rock formation to the past presence of rivers.

So far I have distinguished between "causal systems" and "patterns." I have further articulated two senses by which an investigation might be restricted: "parochialism," where the scope of inquiry is restricted to the distribution of traces or past states of affairs, and "fragile systems," where investigation is restricted to fairly contingent system types. The last two sections have argued that historical science is not merely parochial. As we saw in investigation of Cope's rule, historical scientists are interested in understanding the mechanics of fragile regularities and systems under a wide range of (often counterfactual) circumstances. As we saw in the search for Martian water, historical midrange theories are not restricted to the terrestrial, but can be applied wherever relevantly similar system types occur. We can conclude from this discussion that merely being historical or having a particular target is not sufficient for parochialism in the sense I have defined it. Historical scientists are interested in system types of varying fragility, and their interests are not just limited to the actual variation of actual system types, but in counterfactual instances as well. Against Smart and Rosenberg, historical scientists employ regularities—and not merely those from physics and chemistry. Waters may well be right that in some contexts (the biological laboratory in particular) the action lies in actual difference-makers, but in historical reconstruction and explanation we rely on, and exploit, potential difference-makers. Counterfactuals and potential difference makers are a critical part of concerns and reasoning of historical science.[8]

Let's connect my discussion of fragile systems to methodological omnivory.

7.1.5 The Birth of the Sun and the Moon
Geology and paleobiology are often considered (by archaeologists at least) to be closer to the "physical" sciences, and thus less likely to be parochial

in my sense. So, I will close this section with an archaeological example: reconstruction of ritual sacrifice in the Mesoamerican city of Tikal. Here I want to emphasize the utilization of a wide variety of different models in the reconstruction of token events. Exploiting and integrating complex patchworks of fragile systems allow historical scientists to reach deeply into the past even in unlucky circumstances. Concern for fragile systems is part of the explanation of historical science's success.

The Mayan classical period begins around 200 CE and lasted for 700 years. During that time, the region that is now Guatemala was characterized by increasingly idiosyncratic, often warring, city-states. I will focus on Mazariegos et al.'s (2015) reconstruction of a human sacrificial scene from this period.

Their target is the 2004 discovery of a partially cremated double burial of two males. Here's Mazariegos et al.'s description of the basic event:

Sometime in the fifth century AD, the inhabitants of the Lowland Maya city of Tikal witnessed an extraordinary sacrificial ritual. Two individuals were thrown into a pit, especially dug for the purpose and supplied with sweltering fuel. They may have been dead or nearly dead when thrown, but it is equally possible that they died from burning. Their charred remains were left inside the pit, which was filled shortly afterwards. (187)

As should be par for the course by now, Mazariegos et al. are characteristic methodological omnivores: opportunistically drawing on a varied toolkit. Their reconstruction relies on (1) pathology and taphonomic work, which describes the circumstances of the individuals' deaths and (2) isotopic studies, which allow them to infer geographic origin. They connect this information to Mayan ritual and spiritual practices using (3) textual evidence from the colonial period (that is, reports from missionaries, etc.), (4) ethnographic accounts of oral narratives, (5) architectural features of Mayan ruins, and (6) iconographic remains. Notice the use of regularities of varying fragility and locality in reconstruction: It is not merely that historical scientists care about fragile, but counterfactual-supporting regularities. Additionally, the integration of evidence from such regularities is essential for successfully reconstructing particular episodes from the past.

Carbon dating (a technique whose surprising fragility is a lovely example of how archaeologists adapt techniques to local conditions; see Chapman & Wylie 2016, chapter 4, 142–202) estimates the remains to between 430 and 600, a date corroborated in reference to surrounding pottery. Mazariegos et al. identified two human specimens—an adult (1B) and an adolescent (1A). They analyzed strontium isotopic values in both, and carbon and oxygen isotopes in 1A. These values can be matched to the Mesoamerican isotopic background: Strontium isotopes follow a distinct pattern across the Mayan region; oxygen tracks areas that are dominated by rainfall from different regions (i.e., the Atlantic or the Gulf of Mexico); carbon reflects the consumption of maize, which leave distinctive patterns depending on regional soil types. Each of these allows a match between the traces in the individuals and their geographic origin. Here, wide-ranging regularities (how oxygen isotopes are fixed in individuals based on their environment, for instance) are geared to local conditions—matched to the paleochemical record of the region and to the profiles of various samples of varying ages. Individual 1B didn't hail from Tikal (although it is ambiguous where he was from), whereas 1A was a local boy.

Taphonomic studies link the morphology and chemical properties of the bones to their combustion. For instance, 1A's remains were front down and bent backward with a hyperextended spine, a readily explained posture: "Muscle contraction during combustion produced a characteristic 'pugilistic' pose, with the spine and head bent back and the extremities in a semi-flexed state" ([Mazariegos et al. 2015, 193).

Here, regularities about the relationship between muscles, anatomy, and combustion are employed to explain the remains' appearance. Mazariegos et al. age individuals using tooth morphology—we have heard a fair bit about the information-preserving wonders of mammalian teeth! Both are male; 1A is fourteen, and 1B is around thirty-five to forty years old.

Further details of the combustion effects provide a clearer picture of the individual's positions and articulation, where the fire damage is concentrated, signs of green-bone combustion are indicative of long exposure to heat, and so forth: "the combined combustion attributes speak of a sustained but partial cremation of two fleshed bodies in one single process" Mazariegos et al. 2015, 196). These anatomical, pathological, and taphonomic features are highly suggestive of a ritualistic death—and this is backed up by comparisons with other sites across Mesoamerica and beyond. Reconstructing the specific circumstances of death rely upon fairly general regularities about how fire effects bodies, and the link to ritual is made via comparison with other sites. How about the details of the ritual, then?

The site fits into a common pattern across Mesoamerica, specifically the distinctive "E-Group" architecture: "E-Groups consist of an open plaza with a square pyramid on the west side, and a long rectangular platform on the east side" Mazariegos et al. 2015, 200). This building pattern is remarkably robust across the Mayan classical period. Their design typically tracks the solar year: "the compound's pervasive solar connotations provide a basis to think that they include commemorations related to the solar cycle and solar deities" (Mazariegos et al. 2015, 202).

This association with the solar is important because of the central role of the sun in Mayan cosmology, myth, and ritual. The birth and death of the sun and moon are important mythic themes, recorded both in ethnographies dating from European contact, contemporary ethnographic reports from indigenous people, and in the remains of Mayan iconography. According to the sixteenth-century *Popol Vuh* narrative, the twin heroes of the piece became the sun and moon "after they died by throwing themselves in a pit oven provided with heated stones and burning coal" (Mazariegos et al. 2015, 203). This basic story is retold and reiterated with variation throughout the region—this is taken to signal the myth's deep history. In many versions, the hero associated with the sun, and that associated with the moon, have different characteristics. The solar hero is strong, but suffers (he is often destitute, has damaged skin, and is in pain); the lunar hero is wealthy, good-looking, and often associated with beautiful women. The solar hero takes the pyre's full force (either to shield the lunar hero or because he does not hesitate in making the sacrifice) and thus is hotter and brighter as the sun when compared to his dimmer lunar twin.

These stories are connected to Classical Mayan art, which often depict "a pair of heroes with contrasting characters that approximate those described in colonial and modern narratives" (Mazariegos et al. 2015, 203). Mazariegos et al. identify two heroes who are often represented similarly to the solar and lunar heroes of the *Popol Vuh* narrative (figure 7.3). God S is depicted as a simple hunter, with ritualistically damaged skin. The Maize God, by contrast, is rich, receives the attention of women, and so forth. The thought is that S and the Maize god are forerunners of the sun and moon in the *Popol Vuh* narrative. The similarities between the art and the descriptions in the later narratives "provide indications that ancient Maya concepts about the origin of the luminaries were not radically different from the versions documented throughout Mesoamerica" (204).

Figure 7.3
Two Late Classical Lowland Mayan gods. On the left, God S, on the right, the Maize God. Drawings taken from a Late Classical vase by Oswaldo Chinchilla Mazariegos. (Detail from figure 15, Mazariegos et al. © Cambridge University Press.)

Mazariegos et al., then, link the myth of the sun and moon's birth to the two partially cremated unfortunates. There are many gaps in the hypothesis' support. First, the burial's association with E-Groups is somewhat tenuous, since it is not technically located within the structures where most of the ritual activities appear to have taken place. Second, there are several discrepancies in the match between myth and burial: The differences in the victim's age and their hailing from different regions undermine their being youthful twins. Third, in Classical Mayan mythic genealogy the Corn God is depicted as the *father* of the sun and moon and is associated with the morning star—not as having lunar associations and being the brother of the sun. Fourth, Mayan myth, both in the Classical period and the Colonial period of the *Popol Vuh*, is extremely rich, while Mazariegos et al.'s presentation is selective and in places depends on debatable interpretations.[9]

Regardless, to my mind, the hypothesis that myth and burial are connected is not unreasonable. Moreover, the truth or otherwise of the hypothesis is not what matters for our purposes here. Regardless of the plausibility of each step in the argument, it should be clear that the hypothesis itself is defended via a range—a network—of very different regularities and patterns. As in the case of Martian rivers (and as opposed to Cope's rule) in this circumstance we are primarily interested in uncovering a particular event, and yet in doing so Mazariegos et al. use, develop, and construct localized models geared toward uncovering particular information. In my discussions of methodological omnivory (2015a, last chapter), I highlight that historical science does not simply involve utilizing differing lines of evidence but also constructing localized epistemic tools and models in order to probe the past. This practice is very much on display in Mazariegos et al.'s paper:

We have learned in the course of this and other, similar contextual studies that a conjoined approach, combining different perspectives and levels of approximation, involving archaeological and osteotaphonomic analyses in combination with studies of textual and artistic media, is well suited to analyse the links between myth, ritual and human sacrifice in Mesoamerica and elsewhere. (205)

This case allows me to reiterate a few themes of the book so far. Methodological omnivory is an investigative strategy that successfully increases our inferential reach into the past. This does not involve simply the cooption of various techniques and lines of evidence but also the actual construction of localized models that allow these techniques to operate in that local context. Moreover, although such reconstructions are often aimed at token events, they do so by conceiving of these events as tokens of types—the burial is unique, but aspects of it may be inferred via comparison with other events (I will make this point in much more detail in the next section). The study is significantly richer and more plausible than a pessimist would expect—and moreover it achieves this by refusing to play the game of methodological conservatism that is sometimes associated with pessimism. Such historical reconstructions are anything but parochial.

7.2 Analogy

We have just seen that historical science is not parochial, that is, geologists, paleontologists, and archaeologists are not only interested in explaining actual patterns, but in situating their targets in possibility space. That is to say, their scope is not limited to what has actually happened—they are interested in various ways in which the world could be, and how what actually happens relates to those possibilities. Of course, they are also not interested in something so wide-scoped as the logically possible. On my view, historical scientists are interested in a variety of different modal spaces, depending on both the kind of question they are asking and the resources they have at their command to answer it. Moreover, they are interested in two related questions. First, understanding the conditions required for the phenomena to be instantiated and persist, second, understanding what things would be like under rather different conditions than those which in fact hold. They are intimately interested in counterfactuals, and not just those concerning actual difference-makers.

But, you might ask, how do we get an empirical grip on potential difference-makers? Especially considering that we are interested in unlucky circumstances, where there just aren't many traces about? You may have anticipated my answer: Traces are not the only empirical route into the prehistoric world. When scientists examine Cope's rule by looking at placental mammals, Australian marsupials, *Obdurodon tharalkooschild*, and sauropod dinosaurs, they are not interested primarily in the dependency relations between these animals. Rather, their evidential relevance lies in their being shaped by similar processes. When NASA scientists compare conglomerate rocks on earth and on Mars, they are not interested in whether these rocks have the same origin or source or whether the rocks have influenced each other—they are interested in whether they were formed by the same types of processes. That is, the rocks are unified as token results of the same type of process, rather than common products of the same token of a process. The anatomical features of 1A and 1B are unified with other potentially combusted remains, not because they are unified via ancestry, but because each has been partially cremated. They have undergone similar processes.

I have, then, provided examples of historical scientists gaining knowledge of fragile systems for their own sake and employing that knowledge in reconstructing token events. But what is required for them to gain this knowledge? They need empirical examples of the instantiation of such systems to use as data points. I'll call these "analogies," and it is my aim in the final part of this chapter to provide an abstract characterization of this notion. Let's get an example on the table.

Humans, like all mammals, have inevitably lived with, in fact evolved in concert with, ectoparasites such as fleas, lice, and their brethren. It is generally accepted that true fleas evolved originally to parasitize on mammals before adapting to feathered hosts in the Cenozoic (Lehane 2005).

Fleas are tiny wingless insects (8 mm at their largest, with lengths of < 3mm being much more common), with legs specialized for jumping, and sucking mouth parts adapted to, well, sucking. True fleas have been around since relatively early in the Cenozoic, but none have been found in the Mesozoic (they are thought to have evolved from scorpionflies). However, the early evolution of these critters is still mysterious, owing to the record's gappiness: "the origin, morphology, and early evolution of parasites and their associations with hosts are poorly known due to sparse records of putative ectoparasites with uncertain classification in the Mesozoic, most lacking mouthpart information and other critical details of the head morphology" (Gao et al. 2012, 1). Gao et al.'s description of two new mid-Mesozoic ectoparasites, then, could potentially tell us a lot. They describe two specimens, one from the mid-Jurassic and another from the Cretaceous. They are both, in some ways, remarkably flealike: "*Pseudopulex jarassicus* and *P. magnus* possess many ecto-parasitic characteristics such as wingless body covered with stiff, posteriorly directly setae and bristles, ocelli absent, eyes reduced, and legs with stout setae, and robust and elongate blood-feeding mouthparts modified for puncturing" (3). However, they are in other ways quite different from fleas. Neither *P. jurassicus* nor *P. magnus* are adapted for jumping. Their legs, rather than being built for power, are rather long and spindly. Poinar (2012) suggests that this fact indicates that they parasitized scaled dinosaurs, since such legs would be a hindrance in furred or feathered environs. Their proboscises are over-developed and serrated, rather than short as in true fleas. Taxonomically speaking, they are not true fleas at all; rather, they are a separate lineage of insect, adapted to the flea lifestyle.

Oh, and they are freaking enormous. ... Where your standard modern flea stands at something like 3mm, the Mesozoic's monster fleas are 17 to 22mm long.

Now, imagine we were committed to the kinds of trace-centric views I discussed in chapter 6—let's say we considered common cause explanation to be the modus operandi of historical science. Why, we might think, is such a fuss being made of these new finds, in regard to their relationship with respect to modern fleas? The three lineages' traits are not relevantly historically unified. Although of course they will share a common ancestor in the deep past—some ur-insect—their flealike morphologies evolved independently. They should not, then, be unified by common causes. So what is their evidential relevance? The answer is that they are not evidentially relevant *as traces*, but rather as *analogues*. The ripple model of evidence, it turns out, does not capture all we need to know about the past's accessibility. Even if the ancestors of true fleas—the missing links between Mesozoic scorpionflies and Cenozoic ectoparasites—have left no trace, this is not the end of the story. Because there might be causal structures, representable by causal models, which invertebrate ectoparasites generally follow. If so, other invertebrate ectoparasites can help us test and build such causal models. Even in unlucky circumstances, then, with rich analogues to draw on, much can be learned about the past.

7.2.1 Two Kinds of Evidence

You may remember that when I provided an account of traces (way back in chapter 3), I said that traces must be "downstream" of the relevant events and made a promissory note to explain this later on. That occasion is now. Briefly, the trace must be downstream insofar as it is ancestrally related to the past event.[10] By "ancestrally related," I mean that one can follow a line of causal descent from the trace to the relevant past event. To see the difference between "trace" evidence and "analogous" evidence, it will be helpful to distinguish between two different roles which theory can play in reconstructing the past.

Theory mediates our access to the past in several ways. Midrange theory, as we know, explains how processes in the past could lead to contemporary traces. *Models of causal systems*, by contrast, capture the system dynamics that explain events in the past themselves. In other words, midrange theory explains the creation and decay of traces; models of causal systems tell us how target systems behave. The distinction is not arbitrary or linguistic. Rather, theories play different epistemic roles (indeed, the same theory could in principle play both roles). It is useful to characterize them abstractly.

Imagine we are interested in some event in the past, *E*. *E* is related to (at least) two other kinds of events. First, some events are unified with *E* through joint history. This is to say that some events are causally upstream, some downstream, and others unified by common causes. Ancestral relations are like this. My mother is causally downstream of my grandmother.

She is upstream of me, and my sister and I are united via a common cause—our parents. These causal and informational continuities reveal facts about E. Midrange theory explains the formation of those causal lines, and so justifies that epistemic route.

Second, events are unified as types. Two events can fail to (interestingly)[11] share history, but nonetheless be instances of the same kind of process. E is treated as an event type, lumped with a bunch of other events (E*, E** and so on), and unified via a model that represents the essential dynamics common to both. The clearest examples of this are from physics. Carbon-14 is a rare isotope used for dating archaeological, paleontological, and geological traces. It has a half-life of 5,730 years, and so measuring its decay provides a relatively fine-grained measure of age. Although any two carbon isotopes might be independent of one another causally, they obey the same probabilistic laws. The invariance of these laws allows us to learn something about all ^{14}C by examining a small number of occurrences. Regularities are typically not so invariant in the historical sciences, but nonetheless, as we have seen, they exist and are utilized. Indeed, although the basics of Carbon-14 dating are robust, their application to the local conditions of particular historical investigations required arduous conditioning (Chapman & Wylie 2016). The fleas of the Mesozoic are not connected to modern fleas through ancestry—but they do converge on the flea lifestyle, thus providing inroads to the evolution of flealike parasitism.

Figure 7.4 illustrates the two ways some event E can be connected with other events. They can connect via causal pathways, or via event type. On the basis of this distinction, we can discern two roles that theory and evidence can play. Some (midrange) theories connect historically related events; others (models of causal systems) capture the dynamics of (often fragile) causal systems. Some observations (traces) are evidence for events in the past by virtue of our knowledge of these historical relations; other observations (analogues) are evidence by virtue of supporting models of causal structures.

Recall the "fleas" of the Mesozoic. These are related to true fleas insofar as both are insects. However, their evidential connection is largely based on analogy. Scientists are interested in how flealike ectoparasitism evolves in insects generally. By examining a collection of different flea-analogues, they can test models of causal systems pertaining to the evolution of

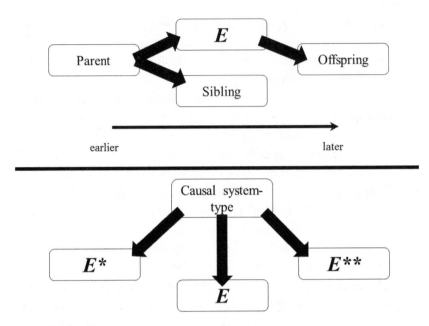

Figure 7.4
Two ways past event E is unified with other events.

ectoparasitism in insects. Briefly, the different lineages are data points, ways of testing and informing the causal models.

There has been philosophical discussion of analogous reasoning before, of course (see Currie 2016a and Wylie 1985 for archaeological discussion; Currie 2013, 2014b for discussion in biology; and Bartha 2013, 2010, and Hesse 1966 for more general discussion). And analogues play an enormous number of roles in science, from framing and grounding scientific debates, to possibility proofs, to more justificatory roles. In a justificatory mood, most discussion has focused on the formal features of the inference (Bartha 2013). Mill (1904), for instance, took the inference to work on the basis of two objects, a and b, which resemble one another in some known respect. By virtue of that resemblance, a proposition that is true of a will also be true of b. Here, analogous reasoning is a kind of Aristotelian syllogism. More reasonably, it is an induction: Given some relevant similarity between two objects, it is likely that some known fact about b will be also true of a.

I have the same worries about such formal approaches as I expressed about common causes in chapter 6. There, I argued that attempts to understand the epistemic warrant of common cause explanations in formal, abstract and general terms were problematic. Primarily this is because such approaches obscure the local warrants of such inferences. If we want to understand actual scientific practice, the license of inference is often provided on a more or less case-by-case basis. Here, too, I suspect that trying to account for the license of analogous reasoning in formal terms at best misses where the action is. The license of shifting from an analogue to a target or, as I have expressed it here, using an analogue to inform a model of a causal system that pertains to the target, turns on facts on the ground. What we can learn about the evolution of ectoparasitism in insects from analogues depends very much upon background facts about insect developmental systems, their evolutionary trajectories, and so forth. The action is local, not abstract (Norton, manuscript, chapter 4).

I do not see analogous inferences as *direct*, as in Mill's reasoning: One does not move from analogue-features to the target having features without mediation. The mediation in historical science is often via some process type that is taken to have been active in both analogue and target (see figure 7.2). I don't doubt that this could be captured in more general, formal terms, but I think that this way of understanding them both makes the difference between analogue and trace reasoning explicit and is sufficient for my purposes.

Paul Bartha's "articulation" model of analogous reasoning is, I think, about as close to an abstract or formal account that suits my purposes we could have (see Bartha 2010). He understands analogies as requiring a *prior association* and a *potential for generalization* for validity. In regard the former, there must be some connection (some prior association) between the known similarity of target and analogue and the unknown feature of the analogue to be inferred. In the convergence between modern and Mesozoic fleas, there is a prior association between the fleas having certain morphological features and them having an ectoparasitic lifestyle. For instance, their elongated, puncturing mouthparts are strongly associated with parasitism. The potential for generalization ensures that there are no disanalogies between the analogue and target that could undermine the subsequent extrapolation. If, for instance, we have reason to think the mouthparts of Mesozoic "fleas" could not have pierced scales, we would be very cautious to infer that they were ectoparasites from that morphology.

Analogous evidence is not trace evidence. The latter is evidence by virtue of being relevantly downstream of the target—that is, being at least minimally dependent on one another in a fashion picked out by justified midrange theory. The former is quite different: Here, object and target are unified by virtue of being tokens of the same system type, instantiations of the same causal system. These systems are represented by models. Analogues are evidentially relevant to one another by virtue of informing these models.

Why is this important? Recall one of chapter 5's reasons for pessimism about unlucky circumstances:

Our available evidence about the past is limited to traces.

This claim is false. Historical scientists can draw on both analogous evidence and traces in reconstruction. As we will see in chapter 11, this gives us reason to be more optimistic about our capacity to know about the deep past. In the remainder of this chapter, and the next, we will delve a little further into the nature of analogous reasoning. First, by considering an objection. And then, in the next chapter, examining the causes of the restricted causal systems which proliferate in the historical sciences— historical contingency—and its relationship to analogous reasoning.

7.2.2 An Objection

Turner (2005, 2007) is suspicious of the evidential value of analogues, especially appeals to constructed and living analogues in paleobiology.[12] If Turner's argument is a good one, then it undermines my appeal to analogous evidence.

Turner collects cases where reliance on analogy has led to long-term errors in paleobiology. For instance, a mistaken analogy between ducks and Hadrosaurs misled early researchers. Based upon structural similarities between their feet, it was thought Hadrosaur feet were webbed, and thus an aquatic lifestyle was inferred. This turned out to be a mistake (Hadrosaur feet are now thought to be padded like those of camels). And indeed, analogies have frequently led paleobiologists astray. Turner points out that philosophers of science need to explain the failures of scientific theories as well as the successes. In this case, he claims, the blame for failure falls on the use of analogues:

In these cases from the history of paleontology, the observable analogues available to scientists—the living ducks, shrimp, lizards and so on—help explain the mistakes. In each case, scientists drew false conclusions precisely because they judged that some long extinct creature must have been rather similar to some living organism. (2005b, 179)

Turner constructs a pessimistic metainduction from these failures. Pessimistic metainductions (see Laudan 1981) note scientific failures, explain these failures, and then (assuming the causes the explanation cites still hold) take past failure as reason to expect contemporary and future failure. Because past paleobiological mistakes are explained by the use of analogues, we have reason to be suspicious of the use of analogues generally.

I will respond to Turner's worry in three ways. First, the examples Turner cites were not problematic due to the use of analogues tout court, but due to the uncareful use of analogues. Modern paleobiologists attempt to control for such mistakes by testing the analogues, and incorporating them with other evidence streams. Second, he thinks of analogues in terms of simple similarities, whereas I think of them as data points for causal models. Third, his argument overgeneralizes. If I can draw on examples of scientific failures and missteps in order to undermine a particular scientific method via this kind of metainduction, then almost any scientific method can be undermined. This is because failures and missteps are so common in science. Let's look at one example.

One common scientific error is "external validity" failure in experiments. That is, the isolation and control afforded by experimental setups can themselves create artifacts that undermine the relevance of the experimental result to its "wild" target (Currie & Levy, under review). Let's pause to examine an infamous case.[13] Experimental study of addiction often utilizes rats as model organisms. Early experiments used solitary, caged rats with access to heroin, food, and water via levers. Some rats became so addicted to heroin that they ignored the food and water levers, eventually dying of dehydration. A disturbing result, and one from which conclusions were drawn about the effects of drug addiction on human behavior. Alexander et al. (1981), in their "rat park" experiments, challenged this. Using morphine instead of heroin, they allowed rats to socialize with one another during the experiment. Rats that interacted with conspecifics were significantly less likely to choose morphine over food and water. It seems that the rats' solitary confinement was to blame for the original studies' results. And yet, a central aspect of experimental design is controlling variables. Isolating the rats was, in a sense, the right thing to do, as it (or so the scientists thought) controlled for unknown variables that could stymie their attempts at uncovering the causal relationships between addiction and mammal behavior.

In cases of external validity failure, experiments do, in a sense, what they are supposed to do: isolate variables under experimental conditions. But sometimes the subsequent behavior is due to those experimental conditions. If such cases are common (and I'm guessing they are), it seems as if an analogous argument to Turner's could be constructed against controlled experiments. Part of the explanation for the mistake is an attempt to control and isolate variables. This is still an essential part of scientific practice, and so we can pessimistically infer that we ought to be suspicious of experimental design. Turner's argument, then, overgenerates. Scientists make mistakes often, and for all sorts of reasons, but it does not follow that we should be pessimistic of scientific methodology. What matters is that such missteps can be identified and corrected.

Clearly, the early rat addiction studies were operating under a mistaken assumption—a bad background theory. This (I assume implicit) theory held that rat behavior was not overly affected by social interaction, or lack thereof. The "rat park" experiments showed that this was mistaken, thus correcting the background theory. This point carries over to analogues. Analogous reasoning does not simply rely on similarities between analogues and targets in the simple way that Turner supposes (or, at least they ought not to). Rather, they are evidence for fragile systems—they provide empirical support for causal models that capture the common dynamics between analogue and target. I have emphasized that historical hypotheses are supported by an array of resources, and leaning too heavily on any particular resource would be misleading. The capacity to draw on analogues such as (putatively) homoplastic adaptations, and surrogates more generally, is an important part of the story about how historical hypotheses are developed and supported. But such use must be understood in the context of the total available evidence. The features bought together to establish an analogy between true and Mesozoic fleas is informed by both background theories about what matters in insect evolution and careful examination of the relevant specimen.

And so Turner's pessimistic metainduction fails. Analogies can be misleading, but so can any scientific method! When properly managed—balanced against the other evidence we have—analogy is a critical stream of evidence for reconstructing the past.

I have argued that historicity—contingency—does not entail parochialism, and that by avoiding parochialism historical scientists gain access to a range of non-trace evidence. In the next chapter, we will look more closely at the nature of history, and the epistemic upshots of that nature.

8 Exquisite Corpse: Historicity and Analogy

Hey farmer, how do you get to Little Rock?
Listen stranger, you can't get there from here.
—"Arkansas Traveler," traditional

History bends possibility: Once inviting routes are closed off, once remote paths become neighborly. In this chapter I discuss "historicity" or "contingency," how the unfurling of events and processes through time effect a trajectory's possible paths, and its subsequent influence on historical reconstruction. Historicity has a negative and a positive aspect. The farmer of the folksong "Arkansas Traveler" captures the negative aspect of historicity: Although at some earlier stage you could have reached Little Rock, it is not accessible from where you are now. Negative historicity expresses itself in Nature's conservativeness. For all of the wild and wonderful post-Mesozoic mammalian diversity, many conceivable morphologies are not realized. For hundreds of millions of years, neither the basic vertebrate body form, nor eye and digit number, nor the use of homeobox *Pax 6* genes in eye development regulation have changed.[1] These features have become *entrenched*. Why aren't there any six-limbed, nine-eyed, exoskeletoned critters in the vertebrate family tree? Because you can't get there from here.

The farmer in "Arkansas Traveler" doesn't say where you *can* get to from there, which is historicity's "positive" side. As we saw in chapter 2, sauropods were gigantic because they were the right lineage, in the right place, at the right time. To travel the road to gigantism, their distinctive body shape and egg-laying habits were required. Then, increasing predatory pressure during the Jurassic selected for larger sizes. Such contingencies, such as those that produced the sauropod phenotype, are *generative*. It was only from a platform of their already distinctive phenotype that sauropods were able to occupy the remarkable areas of morphospace that they did.

Bill Wimsatt synthesizes negative and positive aspects of historicity in his concept of "generative entrenchment," which he harnesses to understand the nature of evolved systems (and much more besides). For Wimsatt, because evolutionary structures are built cumulatively, such structures will accumulate dependencies: "With accumulating dependencies, seemingly arbitrary contingencies can become profoundly necessary, acting as generative structural elements for other contingencies added later" (2007 p. 135). He also states, "the product of runaway positive feedback processes, results in a contingent—but once started, cumulatively unavoidable—freezing in of things essential to their production" (136–137).

This explains why evolutionary systems are both conservative and creative. They are conservative because removing or altering their contingent foundations—the generatively entrenched features—would have too many knock-on effects for the system to function. This occurs because as time passes the system as a whole begins to depend more and more on the stable production of those features. Applying this idea to theory building in science, Wimsatt says, "The more fundamental a proposed scientific change (or change of any other sort!), the broader its effects would be, and consequently, the less chance that it would work" (138).

We do not see nine-eyed vertebrates because the dual-eyed vertebrate phenotype, and the developmental systems involved in its production and replication, have become entrenched. Changing them would likely ensure an unviable organism.

Such features are creative because they are often required for new structures. Evolution provides what Wimsatt calls "generative foundations" from which new areas of possibility can be explored. The distinctive sauropod phenotype is surely one example. Their unique level of gigantism was only achievable once that phenotype (oviparity, long necks, nonmastication, etc.) was in place: "Once each dynamical structure is established, it becomes part of the supporting repertoire or "scaffolding" for further innovations, and as such becomes reinforced for increased reliability and productivity" (136).

What does all this have to do with our purposes here? The historical processes Wimsatt identifies have prima facie upshots about our evidential access to the past. As he says, "Something that is deeply generatively entrenched is in effect a foundational element, principle or assumption" (140). If that is right, then last chapter's parochialism rears its head once more. If our foundations are contingent—generatively entrenched—then it is likely that analogues will rely on different foundations. And this might undermine their evidential relevance to one another.

Some scientists—most notably Stephen J. Gould—separate "historical" sciences such as evolutionary biology from more traditional "experimental" sciences on the basis of the contingent nature of biological and other historical processes (see Beatty 2006; Gould 1989; Lewontin 1967). Philosophers thus far have focused on what it takes to be contingent, and how these notions play into Gould's arguments in favor of the autonomy of macroevolutionary forces from natural selection (Beatty & Desjardins 2009; McConwell & Currie 2016; McIntyre 1997; Powell 2012; Turner 2011b). In contrast, I care about contingency and historicity for their epistemic upshots. In chapter 7, I provided three counterexamples to the idea that historical investigations are "parochial." The first was macro-level investigation of size increases over evolutionary time. The second was the extension of Earth-based geomorphology into a Martian context. The third was the production of confluence in the reconstruction of Classical Mayan ritual. We saw that it would be a mistake to restrict historical interests to the production of narratives. Moreover, we are not limited to trace-based reasoning but can draw on "analogous" reasoning: Similarity relations between events and entities can be used to reconstruct the past, and to bolster midrange theory. If historical contingency generates unique systems that lack analogues, then this response to parochialism is undermined.

My focus in this chapter, then, is on the epistemic and methodological upshots of contingency. Given that many historical targets are contingent, what does this mean for our potential access to them? I argue that the power of trace-based, "ancestral" reasoning rests on the historicity of these processes: Contingency is information-preserving. However, the contingency of historical processes generates apparent "uniqueness"; as we will see, contingent paths appear unlikely to have analogues. However, there is good news and bad news on this point. The good news is that contingent—even unique—events are still amenable to analogous treatment. However, the analogues will be imperfect. I call the use of imperfect analogues the "exquisite corpse" method. Basically, different analogues perform different roles, compensating for each other's failings. I have bad news for accounts of historical evidence that emphasize consilience. Consilience depends on distinct evidence streams doing the same work—supporting the same conclusion—and their results converging. In the cases I discuss, there is often epistemic division of labor, that is, the evidence streams support different aspects of the reconstruction, thus diminishing consilience's role.

I will start by explaining what I take contingency to be, and its evidential upshots. I then use an example—reconstructions of a mysterious South American saber-toothed mammal—to illustrate the exquisite corpse methodology. Lessons are then drawn.

7.1   Sensitivity to Initial Conditions and Path Dependence

"Contingency" and "historicity" are ambiguous and amorphous terms that have bred a plethora of analyses. Although it is a common notion that history matters, and that history's mattering makes a difference to historical and ahistorical enquiry, when and why history matters is difficult to nail down.[2] This is in part because philosophers have different takes on both the relevant sense of "history" and the relevant sense of "mattering" (compare, for instance, Ben-Menachem 1997; Desjardins 2011; Ereshefsky 2014; McConwell & Currie 2016; Turner 2011b). My task, however, is relatively straightforward: I am interested in how some entity or events' past history matters *in terms of our epistemic access to it*. That is to say, how do properties of the processes that shape events affect our epistemic access to those events? One way, already well covered, concerns the production and maintenance of dependencies between an event's occurrence and its traces. Some events are dispersed, will breed many disparate traces, while others are faint and gappy: Their causal descendants will quickly decay. History, then, matters for trace based evidence. Another way in which history matters is contingency's breeding of "unique" events. This will be my focus. And happily, Eric Desjardins (2011) has two related notions that serve my purposes well. He distinguishes between "sensitivity to initial conditions" and "path dependence." I think a good way to understand the importance of this distinction is by first considering the difference between "chance" and "contingency."[3]

Desjardins splits mere chance from contingency. In the scheme of things, it is true that both 60-ton reptiles and glaciers at the equator are improbable. But there is an important sense in which the explanations I have described are not "chancy": That is, they do not appeal to the event's chanciness. Let's examine a vintage example to distinguish between what I will call a "chancy" explanation and a "contingent" explanation. Say I win the lottery. This could happen by my getting lucky, or by the lottery being rigged in my favor. In most circumstances, the right explanation of my win is that I just happened to buy the winning ticket. But perhaps there are circumstances where my win isn't lucky after all: The lottery was set up to guarantee my win. And so, if someone asks, "Why did you win the lottery?" he or she may be enquiring about whether the game was rigged or not. In rigged lotteries, there is good reason why I, and not someone else, wins. Something unsporting has occurred: Perhaps every number in the draw is my number, for instance. However, in other circumstances I simply get lucky. In this context, the explanation is that someone had to win the lottery, and I met the necessary condition for being that someone (I bought a ticket), and the lucky someone happened to be me. The explanation incorporates the event's improbability. Naturally, one could *also* explain the event by providing a highly detailed narrative consisting of all the events that culminated in my lotto win, but that is not the point. The point is that in some contexts, a good explanation will *cite the improbability of the event* (see Feltzer 1981 and Railton 1978 for discussion of improbability explanations).

Contingency and chance explanations come apart analogously to how explanations of rigged and fair lotteries come apart. That is, an event that is highly improbable, unconditionally speaking, can have a high conditional probability due to either (1) an unusual set of initial conditions or (2) an unlikely cascade of events. Recall the Snowball Earth explanation. Here, an improbable event—glaciers in the tropics—was explained as the result of a particular set of initial conditions—continental clustering. Although the end-state is highly improbable, a set of initial conditions "rig the lottery"; guarantee the outcome. That is, the end-state is likely *given those conditions*.[4] We can contrast "contingency explanations" and "chance explanations," then, in the following way. A "chance" explanation appeals to the unconditional probability of the event, while a "contingency" explanation appeals to the event's likelihood, that is, its conditional probability. In the snowball case, continental arrangement receives a chance explanation. Any particular continental arrangement is improbable, but they have to be arranged in *some* fashion. And during the Neoproterozoic this just happened to be an arrangement involving continents clustered around the equator. The snowballing, however, receives a contingent explanation: Given the continental arrangement, snowball conditions are likely.

Now consider explanations of the Cambrian explosion that appeal to snowball events (Runnegar 2000). In chapter 6 we discussed whether the Cambrian radiation occurred as it did because snowball events capped metazoan evolution. During snowballs, metazoan lineages diverged phylogenetically, building up different stocks of evolutionary potential. Balmier conditions arrived, allowing these potentials to be realized in the dramatic radiation. In this case, the explosion's occurrence does not only depend on initial conditions, but on events happening *in the right order*. For the Cambrian explosion to occur, metazoan life had to emerge before various snowball events. Had the snowballs occurred prior to the emergence of metazoa, the radiation as we know it would not have occurred. This is an example of *path dependence*; the right *sequence* of events is required to reach the end-state. In a sense, this has the same conditional form as the last contingency explanation we saw, although in this case we do not care only about initial conditions but also about the timing and order of events.

Both of Desjardins's notions, sensitivity to initial conditions and path dependence, can be understood as the probability of reaching some end-state. Let's take Snowball Earth and the Cambrian explosion as two exemplar end-states. As I described them, snowball events are *sensitive to initial conditions*, that is, their occurrence depends on the right circumstances holding. In particular, snowballing is contingent on continental arrangement. Once those conditions are met snowballs are highly likely to occur. Events are sensitive to initial conditions, then, when the probability of an end-state depends on initial states. On worlds without continental clustering, snowballs are highly unlikely. On worlds *with* continental clustering, they are to be expected. Events that are sensitive to initial conditions (but not path-dependent; see the following), then, have the following modal profile. First, there are multiple possible beginning states. Second, there are multiple possible end-states. Third, there is a dependency between beginning and end-states. Events that are not sensitive to initial conditions, then, will be convergent: No matter where you start, you will end up at the same end-state. Although you can't get to Little Rock from the farm in Arkansas, according Chaucer and others, you can always get to Rome.[5] Many systems in developmental biology have this quality: Developmental systems have mechanisms that ensure the stable generation of the relevant phenotype. Redundancy is built into their design so that different initial conditions will reach the same end-states (a viable organism) regardless (see Wagner 2013; Wimsatt 2007).

The Cambrian radiation, by contrast, was not merely sensitive to initial conditions, but also depended on the timing and ordering of events—it was *path-dependent*. Here, the initial setup mattered, but so too did events downstream. The explosion's occurrence required a particular sequence of events. First, metazoans needed to evolve. Second, Earth must snowball for long enough to act as a filter. Third, balmier conditions must arrive. If events are reordered, no explosion occurs. Either life would stay "larval" or would have evolved differently. For an event to be path-dependent, then, there must be multiple possible paths (i.e., a path with snowballing first, or a path with metazoans first) and the probability of reaching an end-state (a dramatic radiation) changes depending on which path is taken.

And so, we can distinguish between two kinds of explanation: chancy and contingent. The former focuses on an event's probability, the latter on its likelihood (conditional probability). That is, chancy explanations understand events in terms of how unlikely they are unconditionally, while contingent explanations cite the prior happenstances that make the event likely. As Desjardin shows, the form that contingency explanations take depends on whether they are sensitive to initial conditions or path-dependent. This underwrites my discussion of evidence and historicity.

8.2   An Advantage and a Cost

The contingency or historicity of past events increases the power of a certain kind of reasoning and undermines another. Contingent events, as opposed to those that are not path-dependent or sensitive to initial conditions, *retain history*. As Desjardins has put it,

both convergence and chance can erase history. When different populations adapt similarly to a given environment, history is erased because past differences in the value of a state or variable cease to exist. Chance on the other hand can make derived populations more scattered and thus create a situation where it is impossible to see a relationship between changes in the initial states and the probability of different evolutionary outcomes. When this happens, having different evolutionary histories will not affect distinctively the probability of reaching one or another (set of) outcome(s). (347)

Contingent sequences are a "sweet spot" between randomness and total convergence. If the distribution and decay of an entity's traces are random, the signal from the past is lost. Any particular beginning state could lead to any particular end-state. If the traces converge, if the same end-state is reached irrespective of past states, then no signal is recoverable. We saw this in chapter 4 when I argued that because entropy guarantees convergence, Carol Cleland cannot appeal to it when grounding the overdetermination of the past by the present. The end-state is always one of maximal entropy, so initial states are irrelevant. It is by virtue of their path dependence and sensitivity to initial conditions that past events are recoverable from their remains.

Moreover, contingency generates embeddedness. Recall from chapter 3 that two processes, entities, or events are "embedded" insofar as they are connected through multiple dependency relations. Embeddedness is an epistemic boon insofar as increasing our knowledge of one dependency relation is likely to have positive upshots for our knowledge throughout the interdependent network. Contingency—at least sometimes—produces such networks. Deeply generatively entrenched properties will accumulate dependencies as time goes by—they will become increasingly embedded—thus increasing the benefits of trace-based reasoning.

And so, the contingent nature of historical processes is a boon for historical scientists insofar as the past is recoverable on the basis of traces, and that such processes will often promote embeddedness. However, I have argued that overemphasizing traces leads to an impoverished view about our access to the past. We must also consider *analogous* reasoning. Recall the difference: On a trace-based account, events are unified *causally* as ancestors, descendants, siblings, and cousins. On an analogous account, events are unified as *tokens of a type of process*. The alluvial processes on earth and Mars are not (so far as I know) causally connected, but they are tokens of the same processes. Sauropods and *Obdurodon tharalkooschild* are both possible examples of Cope's rule, but their (extremely distant) ancestral relations are probably irrelevant to their conforming to that pattern. Analogous reasoning connects Earthling and Martian alluvial formations, and gigantism in reptiles and monotremes, as instances of the same general processes, and thus informative of each other.

However, contingency appears to undermine the use of analogues. In brief, if an event's occurrence is path-dependent and sensitive to initial conditions, we should not expect many events of that nature to occur. After all, part of the mystery of sauropod gigantism is their uniqueness: No other terrestrial animal comes close to this size. How can we then find analogues of sauropods?

Sticking with biology, we can illustrate the problem by drawing on Paul Griffiths's arguments for the priority of relations of ancestry (homology) over functional similarities (analogy) in taxonomy and character delineation (Griffiths 1994, 2006, 2007). Griffiths claims that homology provides a better ground for taxonomic decisions on the basis that ancestry is a stronger "force" in shaping and maintaining phenotype than natural selection (or at least, ancestry combined with maintenance selection). You know more about a critter from its phylogeny than its niche. Here is an example from Griffiths (1994). Consider the ecological category "Top-Rank Fresh Water Predator." Crocodile and anaconda are two occupants of that niche. Which of these animals' similarities are due to their occupancy of a common niche? Other than large size, very few. There are, of course, many similarities between these animals—but these are due to their common ancestry, not their common way of making a living. To some extent, this lesson generalizes: "It is a truism in comparative biology that similarities due to analogy (shared adaptive function) are 'shallow.' The deeper you dig the more things diverge. Bat wings and bird wings have similar aerodynamic properties but their structure diverges radically, despite their deep homology as tetrapod limbs" (Griffiths 2007b, 216).

An organism's phenotype is more dependent on lines of ancestry—on path-dependent, "frozen accidents"—than on natural selection, (at least insofar as it is involved in reshaping organisms as opposed to maintaining stasis). If this is right then, prima facie, the utility of analogous reasoning fades. I will not attempt to ascertain this claim's scope—although I find it plausible in many biological contexts, I am not sure how to translate the primacy of ancestry over analogy into other historical sciences. Geological processes are often less historically contingent than biological processes (indeed, this is probably why NASA had so little difficulty translating earthling geomorphology into a Martian context). And archaeological contexts are potentially even worse (see Currie 2016a). Regardless, my aim here is to see what analogy can do *even when* similarities are "shallow." This is in line with my approach thus far: In the hard cases, what good is analogy?

Contingent processes, then, seem to breed *unique* or *quasi-unique* events. Readers with impressive memories might recall my mentioning "uniqueness" in chapter 4's discussion of underdetermination. There, I touched on Aviezer Tucker's account of unique events, and how they generated a kind of underdetermination. I also promised to return to the notion once we had sufficient machinery to understand it. Now is that time.

The term "unique" is ambiguous, and unsurprisingly has a rather complicated philosophical history. Happily, Tucker does the conceptual work for us, zeroing in on the relevant notion. The claim is this:

there are no acceptable scientific theories that can explain unique events because unique events:

Occur once and only once. Their significant properties or parameters, specified in the topic description of the why-question, are either:

(a) Not shared by any other event, apart from spatiotemporal location and self-identity (or it is unknowable whether they are shared by other events). Or

(b) Too complex or chaotic for effective comparison with other events. (Tucker 1998, 65)

Tucker's sense of uniqueness leans on van Fraassen's pragmatic account of explanation, which needs a little unpacking. Van Fraassen understands explanation in terms of the pragmatics of "why" questions. These specify a "topic description," the target of the explanation, and a relevant "contrast class," a set of possibilities. A good explanation will distinguish the target from the contrast class. Take sauropod gigantism, for instance. Often, the topic description is specified in terms of sauropod terrestrial size, say, upward of 25 meters long. The relevant contrast is typically mammalian giants (in this case, the contrast is actual rather than merely possible). We can understand the question as asking: What distinguishes sauropods from mammals, by virtue of which the former and not the latter are terrestrially viable at such impressive sizes? A successful explanation will pick out sauropod properties that do this work. For instance, laying eggs allows sauropods fast population recovery, thus mitigating the cost of small populations engendered by large individuals.

A "unique event," then, will have no other events that match it in the salient properties picked out by relevant why questions. Why is this problematic? Tucker's point is that without similar events, the specific background theories pertaining to the target cannot be confirmed. Without other instances we cannot distinguish empirically between competing theories that cohere with the case in question. Consider the discussion in Sellers et al. (2013) of the problems involved with reconstructing sauropod gait: "whilst we can get a great deal of useful information from studies of locomotion in the largest living terrestrial vertebrates, we should expect the locomotor kinematics of the largest sauropods to differ from those seen in modern animals since they are potentially an order of magnitude larger, and have their own unique musculoskeletal adaptations" (1). Did sauropods have unique musculoskeletal adaptions and, if so, which ones did they have? Tucker might say, well, even if you *find evidence* of such adaptations, if they have no analogues, we have extremely limited potential to confirm our hypotheses regarding them.

by definition [unique] events exclude the possibility of other events that can be relevant for confirming the theoretical backgrounds that can participate in their explanation because: Unique events are not recurrent; their significant properties are either not shared by any other event or it is impossible to know whether other events share them; or they are too chaotic and complex for effective comparison with other events. Explanations of unique events are radically underdetermined because any theoretical background that is relevant for their explanation is underdetermined. (Tucker 1998, 66)

It is worth contrasting Turner's and Tucker's conceptions of underdetermination in terms of my discussion in chapter 7. There, I distinguished between *ancestral*, or trace-based reasoning, and *analogous*, or surrogative reasoning. The former relies on tracing causal histories, the latter on unifying independent events as members of the same event class. Turner's underdetermination is a problem for ancestral reasoning. There will not be sufficient traces for us to follow the lines of history back to their targets. Tucker's is a problem for analogous reasoning: There will not be enough analogues to unify the target event with. We cannot construct a well-confirmed model using analogues if there are no analogues.

Historical contingency generates uniqueness. Some historical processes— those that cumulatively build "generatively entrenched" features—will often generate targets that apparently lack analogues. This is because historically contingent end-states require highly specific initial conditions or sequences of events to occur, and so shall be rare if those conditions or sequences are. This undermines the use of analogous reasoning. Such reasoning requires a set of events that have been formed by the same processes. If such processes are historically contingent, there will be few such events.

In chapters 9 and 10, we will examine one sort of response to this problem: The construction of simulations, or "proxy experiments." I will now focus on a different strategy: Instead of looking for perfect analogues, or constructing them, scientists move between several imperfect analogies. I will call this the "exquisite corpse" method. To illustrate, I move to another example of paleobiological reconstruction: that of saber-toothed "cats."

8.3   All the Better to Eat You With: *Thylacosmilus atrox*

South America was isolated for much of the Cenozoic, and this allowed many odd critters to flourish. Sloths and armadillos are representative remainders of this menagerie, and the *Phorusrhacids*, enormous predatory "terror birds," deserve special mention. *Thylacosmilus atrox* was another weird animal. She is a member of *Sparassodonta*, a clade of marsupial-like[6] carnivorous mammals who thrived throughout the South American Cenozoic, presumably competing with terror birds for the position of apex predator. *T. atrox* was roughly leopard-sized, weighing between 80 and 120 kilograms. She had strong forelimbs and neck, and a truly remarkable skull (see figure 8.1). Most strikingly, she sported enormously elongated upper canine teeth. She was, in short, a pouched, marsupial-like, saber-toothed "lion."

T. atrox is a remarkable example of a nonplacental saber-tooth (although, as we shall see, there were others), but what makes her mysterious is her bite force. Wroe et al. (2005) calculated the "bite-force quotient" (BFQ) of a range of mammalian carnivores. An animal's BFQ is its raw bite force mapped against its body size. In the figure below (figure 8.2), bite force is represented on the y-axis, and body mass on the x-axis. The further an organism falls from the central line on the graph, above or below, the more extreme its BFQ in the relevant direction. *Thylacoleo carnifex*, the Australian marsupial predator, for instance, is represented by the dark square marked "*Th. c.*" Its BFQ is extremely high (see Currie 2015a). By contrast, *Thylacoleo atrox* is circled. It falls far, far below the line. Its bite force proportional to size is, apparently, tiny.

Figure 8.1
Thylacoleo atrox skull. (Wikimedia Commons.)

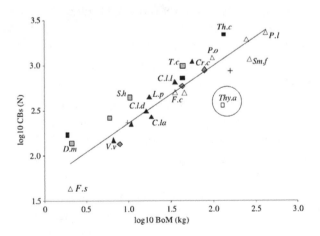

Figure 8.2
BFQ for various carnivorous mammals, *T. atrox* is circled. (From Wroe et al. 2005; © Royal Society.)

This is surprising. *T. atrox*'s extreme dentistry was presumably utilized in killing large, dangerous prey, and there is strong correlation between bite force and prey size. Yet her bite force is pathetic. Our question, then, is how did *T. atrox* manage to kill?

This has all the hallmarks of being an "unlucky" scenario. First, *T. atrox* remains are few and far between—reconstructions are based upon an insignificant number of incomplete specimens. Second, phylogenetic traces are nonexistent.[7] That is, poor *T. atrox* has no surviving descendants or close relatives. Third, as for sauropod gigantism, *T. atrox*'s outlier status appears to undermine comparisons with analogues. The deep contingency of *T. atrox*'s evolutionary history undermines the utility of analogous reasoning. Moreover, possible analogues, such as placental *felid* saber-tooths, are themselves long extinct.

However, as we shall see, this third point is far too quick.

The saber-toothed morphology crops up in carnivorous animals relatively often. At least two examples are found in the "mammal-like reptiles," and five examples are found in Cenozoic mammals. I will summarize some recent comparative work on these lineages, culminating in Wroe et al.'s (2013) reconstruction, which is based on comparisons between the true felid saber-tooth *Smilodon fatalis* and *T. atrox*. As we shall see, the lack of perfect analogues does not undermine analogous reasoning. Rather, scientists navigate between *several* analogues, at different levels of analysis, to construct a model relevant to the target.

Let's begin with a short evolutionary history of synapsid predators.[8] Synapsids are an ancient clade that includes both mammals and their mammal-like reptilian ancestors. A major difference between synapsids and their diapsid cousins (the reptilian lineage including dinosaurs and birds) is their versatile jaw, which allows a significantly higher range of motion. Where birds and dinosaurs are restricted to fore-and-aft chewing, we and other synapsids can chew side-to-side. This, as well as other innovations, led to an extraordinary diversity of jaw and dentition systems.

The synapsids had two golden ages. The first, and longest, covered 80–100 million years from the late Carboniferous to the early Triassic. Here, synapsids boasted the first land-based mega carnivore (*Dimetrodon*) and, important for our story, two separate lineages of saber-toothed predators. Notably, the general trend in synapsids throughout the period is away from a *kinetic inertial* bite system to a *static pressure* system. In the former method,

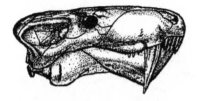 

Figure 8.3
Two nonmammalian synapsid saber-toothed skulls. Left, the therocephalian *Lycosuchus*. Right, the gorgonopsid *Leontocephalus*. (Detail from Van Valkenburgh & Jenkins 2007, 270. © Cambridge University Press.)

prey is killed by momentum—the jaw "snaps shut." Crocodiles are a fine exemplar. In the latter, prey are killed by steady pressure. The big cat killing strategy—strangulation—is an example. As static-pressure systems evolved, so too did saber-toothed forms in the therocephalian and goronopsid clades (see figure 8.3).

As the Mesozoic got into full swing, synapsid predators were replaced with dinosaurs and faded into insignificance—until the K-Pg extinction gave another synapsid lineage a second chance. From 55 million years ago until the present day, mammalian predators have often dominated ecosystems (with the notable exceptions of the terror birds' reign in South Africa, Australia's enormous goanna-like reptiles, and the leopard-like giant eagle of New Zealand). This second golden age lacks the progressive character of the first: Mammalian carnivores did not display an overall trend from a kinetic-inertial bite to one using static pressure (unsurprisingly, as the latter method was already the mammalian modus operandi). However, both share a pattern: A lineage of carnivores arises and increases in specialization (including the occasional evolution of saber-teeth), until they are replaced by another lineage. As van Valkenburgh and Jenkins put it, "Both radiations of predators show a pattern of dynasty replacement. That is, one or a few clades evolve large size and seem to dominate the carnivore guild for several million years, but then decline and are replaced by new taxa" 2002 p. 284).

Four of these mammalian synapsid groups evolved a saber-tooth morphology, including *T. atrox* (see figure 8.4). There appears, then, to be bountiful opportunity for analogous reasoning. Given their frequent evolution, saber-toothed canines must be good for *something*; again quoting van

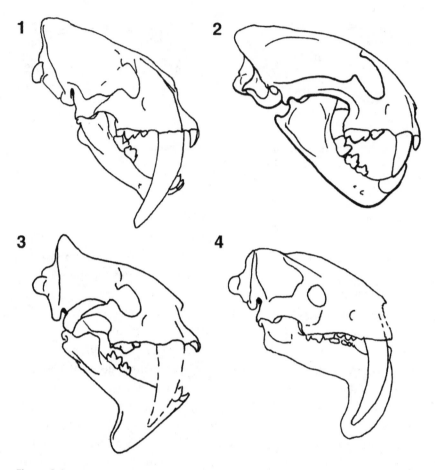

Figure 8.4

Extinct mammalian saber-tooths (1, *Smilodon fatalis*; 3, *Barbourofelis fricki*; 4, *Thylacosmilus atrox*) with an extant conical-toothed felid (2). (From Van Valkenburgh & Jenkins 2002, 280. © Cambridge University Press.)

Valkenburgh and Jenkins, "Without a doubt, saber-like canines are advantageous, but their absence today frustrates our ability to explain their obvious success" (287). But good for what? And how were they used?

Regarding that question, consider the following. First, note there are two basic saber-tooth morphs. "Scimitar-toothed" cats have shorter canines and are built for speed, while "dirk-toothed" cats have longer, pointed canines and have strong limbs built for power; Martin et al. (2000) argue that there is a third type that combines the power of the dirk-tooth with the dental morphology of the scimitar-tooth. *T. atrox* falls firmly into the latter camp. It is often suggested that the strong limbs indicate a stalk-and-surprise hunting strategy. Second, recall that saber-tooth morphology has only emerged once the static-pressure bite system evolved. This tells us something about how the teeth are used: Sabers are not for quick stabs, but for pressing deeply into prey. Third, reconstructions of the more derived saber-tooth bite show that the required gape exceeds 90°, angles that would require significant differences in jaw musculature than normal cats (Christiansen 2011). Given these points, it may be possible to unify the outlier *T. atrox* with dirk-toothed cats, and thus draw on analogues.

Paleobiologists have reached something like a consensus on how felid dirk-tooths such as *Smilodon fatalis* delivered their killing bite. In standard cats, bite force is largely a function of the musculature involved in jaw-adduction. *S. fatalis*, by contrast, had relatively weak jaw muscles, but built-up "head-depressing" neck muscles. Roughly, the start of the bite is something like a downward head butt, with the neck muscles taking most of the strain, jaw musculature taking over in the latter part of the bite. This "canine-shear bite" is, to an extent, an extreme version of the distinctive big cat bite (see Arkeston 1985). If an analogy can be drawn between *S. fatalis* and *T. atrox*, then we may have inroads to the latter's killing behavior.

Wroe et al. (2013) digitized the skulls and musculature of two cats and one "cat": *S. fatalis*, *T. atrox*, and the extant conical toothed *Panthera pardus* (figure 8.5). This involved the examination, scanning, and reconstruction (missing parts were inferred) of the skulls. Using simulations, they then proceeded to infer both the bite strength, and relative skull stresses, of two kinds of bites. The first, characteristic of *P. pardus*, used jaw abductor-driven bites. The second used neck muscle–driven bites.

One difficulty with the study is the extraordinary gapes of saber-toothed cats. Compared to the 65–70° of *P. pardus*, Wroe et al. estimate *S. fatalis* to manage 87.1°, and *T. atrox* was a whopping 105.8° (look at the difference in figure 8.5)! This matters because there is a relationship between the length of a muscle fiber and its ideal amount of tension, and so wider gapes are often achieved at the expense of bite force. Wrote et al. suppose this may have been mitigated, pointing to similar phenomena in marmosets, and circumstantial evidence in extant felids. Eng et al. (2009) compare the functional morphology of two tiny monkeys: cotton-top tamarins and

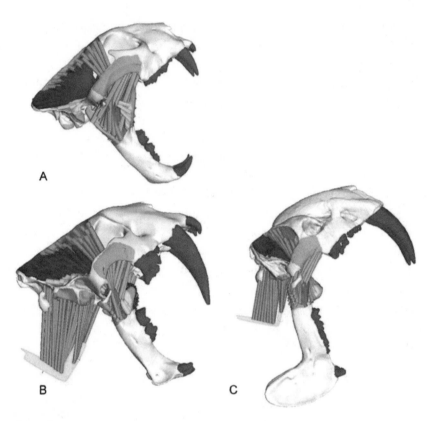

Figure 8.5
Digitized skulls and musculature for *Panthera pardus* (A), *Smilodon fatalis* (B) and
Thylacosmilus atrox (C). (From Wroe et al. 2013, 4.)

marmosets. Both have a similar diet, except marmosets actively use their
jaws to gauge trees. Eng et al.'s findings suggest that marmosets have muscular features (increased sarcomere number, leading to longer muscle fibers)
that maximize their bite force at wider gapes. On this basis, Wroe et al.'s
"modeling effectively accepts that muscle tensions do not decrease with
increasing gape" (2013, 4). The upshot of this is that their results might
overestimate the bite force at wide gapes—this is surprising, since the jaw-
driven bite forces are extremely weak.

How weak? The 68 kg *P. pardus* has a bite reaction force of 484 Newtons
(N), while the enormous 259 kg *S. fatalis* managed a proportionately embar-
rassing 519 N, and the 82 kg *T. atrox* weighed in with a truly pathetic 38 N.
Additionally, both saber-tooth skulls showed significantly more strain on

their skull musculature using jaw-driven bites. To put things in perspective, "At maximum gape, *Thylacosmilus atrox* would need to generate 14.5 times the jaw adductor muscle force of *P. pardus* in order to achieve a bite force consistent with its body mass (Wroe et al. 2013, 6).

However, once neck muscles are bought into play, *T. atrox*'s performance significantly improves: "When neck muscles forces only were applied ... mean 'brick' element stresses were comparable between *S. fatalis* and *P. pardus* and relatively low in *T. atrox*" (6). Wroe et al. take this to suggest that *T. atrox* did not bite as *S. fatalis* did. Rather,

The extremely low jaw-adductor-driven bite forces predicted at all gape angles for *T. atrox* suggest that the jaw muscles played an insignificant role in the dispatch of prey by the metatherian. Moreover, our findings suggest that in order to minimize stress on the canine teeth and resistance as the canines were inserted, *T. atrox* needed to move its head considerably further forward and downward relative to the positions of the jaw-joint than would *S. fatalis*. (6)

And so, in terms of bite, it seems that *S. fatalis* sits somewhere between a standard feline and *T. atrox*. Where a standard feline's bite utilizes jaw muscles in the main, and *S. fatalis* uses both neck and jaw, *T. atrox* foregoes the jaw and focuses on neck strength. This explains why its bite-force quotient is so low when calculated based on its jaw musculature. Wroe et al. do insist that this is a case of convergence: "Our simulations provide further evidence for convergence in these two highly derived mammalian predators with respect to the mechanics of the killing bite" (6); however, *T. atrox* is a *more extreme* version of adaptation to the "saber-tooth" killing morphology.

What about killing behavior? Wroe et al. suppose that the morphology of such long "dirk" saber-tooths could not withstand the stress of holding unconstrained prey, and suggest that "whether the metatherian ambushed or ran down its prey, we consider it likely that it was immobilized and secured first because the particularly long and laterally compressed canines would have been especially vulnerable to breakage" (7). This is supported by two appeals to analogy. First, the behavior has been observed in extant bears. Second, there is a strong correlation between forelimb robustness and upper canine length in other saber-tooths, "indicating that powerful forelimbs may be a prerequisite needed to immobilize prey in placental saber-tooths and that this becomes increasingly important as canines become longer and more fragile" (7).

Thus, we have shifted from *T. atrox's* incomplete remains to a picture of both its bite mechanism and its killing style. All this despite lacking a "perfect" analogue, either living or dead, of the organism. This was achieved by drawing on analogues both from other saber-tooths, and from other organisms. Our picture of *T. atrox,* as we will see, is an "exquisite corpse."

8.4   Good News, Bad News

Here are two things to notice about investigation of *T. atrox*. First, that lineage's apparent uniqueness was no block to the use of analogous reasoning in its reconstruction. Scientists drew on saber-toothed carnivores from throughout the synapsids. They also drew on non-saber-toothed animals—modern cats, marmosets, and bears. Second, notice that different analogues perform different evidential roles. Here is a breakdown:

• The tendency for nonmammalian synapsids to evolve saber-teeth *after* the shift toward a static pressure system suggests that such a system is a prerequisite for saber-tooth morphology.

• The correlation between robust forelimbs and canine length in placental saber-teeth suggests that the forelimbs function to stabilize prey (thus mitigating the costs of longer, more brittle upper canines).

• The similarities and differences in simulation performance between *S. fatalis* and *T. atrox* is taken to support the view that *T. atrox's* jaw-adducer muscles played very little role in its bite.

• The marmoset's physiological mitigations for loss of jaw strength at wide gapes is taken to suggest that wide-gaped saber-tooths may have had similar features.

• Ursid use of forelimbs to immobilize prey is taken to suggest the possibility of similar behavior in some saber-tooths.

Although they do not make it explicit, it is clear that Wroe et al.'s study is intended to be expanded to saber-tooth morphology generally. They say they "propose that the metatherian presents a more complete commitment to the already extreme saber-tooth 'lifestyle'" (2013, 1). There is an implicit "saber-toothedness" model, of which *T. atrox* is taken to be the most extreme example. At one end of the model, we have critters with a standard carnivorous bite: Bite force is a function of jaw musculature and size, and the static pressure system is employed. Prey animals are largely immobilized via a crushing bite, with some aid from forelimbs. *S. fatalis* is a rather extreme saber-tooth: Much of her bite force has been transferred from the jaw to the neck muscles and the forelimbs are increasingly important for steadying prey. *T. atrox* is the most extreme so far discovered: The bite force is almost entirely drawn from neck musculature.

We see, then, that a general model of "saber-toothedness" has been inferred from synapsid saber-tooths. The model is obtained by drawing from both saber-tooths and other relevantly similar animals. The uses of the analogues range from quantified comparisons (as between the skull shape, musculature, and bite mechanics of *S. fatalis*, *T. atrox*, and *P. pardus*) to qualitative possibility proofs (the appeal to bears' use of forelimbs in stabilizing prey). The "uniqueness" of *T. atrox*—and thus presumably the path dependence and sensitivity to initial conditions of its evolution—is no block to the use of analogues.

Analogues are used to construct a general model into which the target is fitted. Although no other saber-tooths were perfect analogues of *T. atrox*— no other had such a low bite-force quotient—the species can be understood as an *extreme example* of the dirk-toothed carnivorous morphotype.

Moreover, the use of imperfect analogues depends upon their specific matches with, and divergences from, the target. The nonhomologous similarities between marmosets and *T. atrox* are certainly "shallow" in Griffiths's sense. However, if both have evolved jaw and neck musculature that mitigate the trade-off between gape size and bite power, then, at a coarse level of analysis, they may inform one another. An important factor when considering the evidential relevance of an analogue is its "match" with the target (see Currie 2013; McConwell & Currie 2016). Both bats and birds independently evolved flight, it is true, but they diverge on morphological and biomechanical levels. Bird wings are stretched across their forelimbs and they fly by flapping their "arms"; bats' wings are stretched across their fingers and they fly by flapping their "hands." This does not mean that studies of bats and birds cannot inform one another. Rather, the lesson is that their similarities need to be considered at the relevant level of grain. Differences between their morphology and biomechanics, of course, can also be extremely informative.

This need for analogues to match at the relevant level of description is mitigated by the "exquisite corpse" methodology. This characteristically involves (1) using several analogues, at (2) different levels of grain, to (3) construct models relevant to the target (à la the implicit "saber-toothedness" model) and (4) support particular aspects of target reconstruction (à la the appeal to marmosets). Let's turn to this methodology now.

Exquisite Corpse is a game developed by surrealist artists (most notably Andre Breton) in 1920s France, and is descended from Victorian parlor games. In its original version, the game involves each player writing a word, or a sequence of words, either following a rule (for instance, the first player writes an adjective, the next a noun, and so on) or by hiding much of the previous writing from the new player. Its name comes from a sentence generated in this way "*Le cadavre exquis boira le vin nouveau*" ("The exquisite corpse shall drink the new wine.") (Brotchie & Godding 1991). A version of the game salient here involves drawing rather than writing. Together, players create an odd creature. The first player draws the feet, then folds the paper over so the next player can only see the tops of the ankles. The new player draws the legs, folding the paper similarly and passing to the next player, and so forth until the creature is drawn.

In essence, the exquisite corpse strategy works by navigating between a group of different analogues, each of which is similar to the target in some respects and dissimilar in others. As a completed critter generated from a game of exquisite corpse is a composite drawn from several sources, so is the target of a historical reconstruction conceived as a collection of traits informed by various analogues. By comparing and contrasting the various cases, a model is constructed drawing elements from the various lineages. This model can be used to both explain the features of the target organism, and to infer further facts about it. Let's present this schematically.

Target organism T has features f_1, f_2, f_3. For each of T's features, there is a corresponding hypothesis, h_1, h_2, h_3 that accounts for that feature. In order to confirm h_1–h_3, analogues A_1, A_2, A_3 are available. Each analogue possesses one of the relevant features (or a relevantly similar feature), and so may be able to test the corresponding hypothesis. I will indicate these analogous features using asterisks. For instance, A_1 possesses $f_1{}^*$, and so if h_1 is true of it, this (via analogous reasoning) provides reason to think that h_1 is true of f_1 in T. And so on for A_2 and A_3.

In this highly schematic version of the strategy, T is in effect taken to be an exquisite corpse: a combination of $f_1{}^*$ in A_1, $f_2{}^*$ in A_2 and $f_3{}^*$ in A_3. We have seen this in reconstructions of sauropods. The sauropod neck is analogous to that of a swan, or is mechanically similar to a beam.

Its digestive capacity and dietary requirements are reconstructed on the basis of comparisons with large endothermic mammals such as elephants. Their metabolic requirements are examined via large predatory reptiles. The role of its long neck is examined in comparison with vacuum cleaners. An aspect of each analogue is taken to inform the general model of the target.

It is worth pointing out that this is not a complete response to Tucker's worry about uniqueness. For I have simply shown that *T. atrox* is not unique: There *are* relevantly similar critters that can underwrite its reconstruction. I have shown that analogues need not be "perfect." Tucker anticipates the exquisite corpse methodology but appears to suspect that the interactions of the various parts the corpse may be too complex: "The French Revolution happened only once. Many of its historiographically relevant properties, perhaps all, can be found elsewhere, but their unique combination in 1789 may be so intertwined, complex and chaotic that it is impossible to combine *acceptable* theories to produce, or deduce from, a theory that can explain it" (68, italics in original).

Perhaps so. Assuming we cannot reproduce our target, we may be in a situation where such underdetermination simply will not be resolved. In short, the full-on uniqueness Tucker is concerned with may exist, and is certainly possible. However, the question is, how often will we find it? My bet (and, I'm guessing, Tucker's as well) is that cases of true uniqueness will be rare. This is because, as Tucker nicely illustrates, as scientific investigation proceeds, apparent uniqueness often dissolves. In the next chapter, we will see a stronger response to uniqueness. I will argue that through modeling we can in fact reproduce aspects of apparently unique targets.

In short (and here is the good news), in reconstruction, different, incomplete, analogues are *combined*, in addition with other sources of information like direct reconstruction from traces, modeling of the target, ancestral information, and so forth. The imperfect analogues are taken simply to be another line of evidence weaved together to support the reconstruction.

And the bad news follows from this.

Recall Wylie's distinction between *vertical* and *horizontal* independence. Evidence streams that are *vertically* independent do different epistemic jobs. Consider the roles of taphonomy and calculations of bite force in Wroe et al.'s study. By providing a mechanistic explanation of the process of fossilization, taphonomy justifies Wroe et al. taking the remains of *T. atrox* as being revelatory of the extinct animal. The bite-force calculations, by contrast, give Wroe et al. reason to think that *T. atrox's* bite was comparatively weak. The bite-force calculations required, among other things, taphonomy for validity. They are vertically independent.

Horizontally independent sources do *the same* epistemic job; these are the cases of consilience that Forber and Griffith emphasize. Their argument for the success of historical reconstruction relies on an epistemic effect that occurs in such cases. In consilience, two measures converge on the same result. Because the measures rely on different auxiliary assumptions, the chances of both assumptions failing, but *nonetheless* converging on the same result, are much less than the chances of the measurement being true.

However, in an exquisite corpse we saw that each feature (each f), had a corresponding hypothesis (h) that was corroborated by a feature (f^*) in a relevant analogue (A). For consilience to occur, the various analogues, say A_1 and A_2 would both have the same feature, and so provide independent support for h. This is not what we see. Instead, A_1, A_2, and A_3 each support a *different* hypothesis about an organism's features. Consider that the static pressure method of biting is a prerequisite for saber-teeth is supported by nonmammalian synapsids; that *T. atrox's* bite relied significantly more on its neck muscles was supported by contrast with *S. fatalis*; that *T. atrox's* forelimbs were used to immobilize prey was supported by the correlation between upper canine length and limb strength in placental saber-tooths.

The problem is this: If many historical targets exhibit contingency in the sense discussed in this chapter, and if contingency breeds "uniqueness," then many historical targets will lack perfect analogues. Given this lack, the exquisite corpse methodology can be employed. However, this decreases the importance of consilience, since consilience requires that evidence streams be horizontally independent, but in exquisite corpse cases they are often only vertically independent. That is, as opposed to h_1 being supported by *both* A_1 and A_2, A_1 and A_2 do separate, but complementary jobs: supporting h_1 and h_2 respectively.

This argument should not be taken too far. After all, the reconstruction of *T. atrox* does contain horizontal independence. For instance, at least two convergent cases of nonmammalian synapsid were drawn on to establish the connection between the static pressure bite method and saber-teeth, and four separate mammalian analogues were drawn on to construct the model of saber-teeth of which *T. atrox* is an extreme example. How much horizontal independence we can find is an open question. Moreover, I have argued that many kinds of dependence are utilized in historical reconstruction, particularly in "unlucky" circumstances. I take it the reduced emphasis on consilience can be mitigated by how our overall picture "hangs together"— the confluence of evidence matters. h_1 and h_2 may be interdependent; there may be dependency relationships between those aspects of the animal and they may bolster support. This seems likely in the *T. atrox* case.

Consider the hypothesis that *T. atrox* used its forelimbs to immobilize prey, and the hypothesis that her bite style used static pressure, driven by her extensive neck muscles, to drive the long upper canines through tough flesh. As we have seen, there is independent reason to believe both hypotheses. But the two are also related. Notice that one reason the forelimb's purpose is plausible is the fragility of the canines. An animal using that bite method would need some other way to stop prey from escaping. This is an example of using viability considerations to inform a reconstruction (see Chapter 2's discussion of Sauropods, Currie 2015a). And so, although the analogues are doing different work, our general picture of *T. atrox* can make for a robust reconstruction as it reveals dependencies like that between brittle teeth and strong forelimbs.

Nonetheless, if we are tempted to emphasize independent evidence streams, this example should give us pause. There could well be less horizontal independence than we first thought. However, this does not undermine the good news. The use of imperfect analogues to construct exquisite corpses shows that the "historicity" or "contingency" of targets in historical science does not do irreparable harm to our capacity to use analogous reasoning.

The use of imperfect analogues is especially powerful when it is coupled with the capacity of historical scientists to produce their own analogues. Thus far, I have assumed that we are restricted to what the world provides. It is time to see if we can do better.

9   How to Build Sea Urchins and Manufacture Smoking Guns

The capacity to conduct controlled manipulations is a scientific boon. Experimenters (as opposed to those relying on mere "passive" observation) can generate their own evidence. One central difference between observation and experiment is the experimenter's ability to *bring about* test cases, while observers can only "happen upon" correlations. On the face of it, this boon is denied to those seeking to uncover the deep past. Historical targets tend to be at large spatial and temporal scales—not easily admitting of experimental treatments—and besides, they are long gone. I mentioned in chapter 5 that this experimental impotency drives pessimism about "unlucky" historical science. When facing homogeneous, gappy, and degraded trace sets, historical scientists are unable to generate new evidence: They are stuck with what nature provides. If no analogous saber-tooths had been preserved, then tough luck. In this chapter, I argue that this view is mistaken.

As we have seen, Carol Cleland (especially in her 2002) distinguishes between two archetypical scientific methods—those of *experimental science* and *historical science*:

There are fundamental methodological differences between prototypical historical science and classical experimental science vis-à-vis the testing of hypotheses. These differences represent different patterns of evidential reasoning. ... Insofar as they are concerned with identifying particular past causes of current phenomena, historical researchers cannot directly test their hypotheses by means of controlled experiments. (Cleland 2002, 494)

Although Cleland argues that each method has distinctive advantages, when we consider the degradation of traces, it is difficult not to pity the historical scientist, forced as she is to make do with what she has. Derek Turner makes this suspicion overt. Recall that, according to Cleland, historical science proceeds by identifying "smoking guns," test cases that empirically discriminate between past hypotheses. Turner points out that, without experimentation, historical scientists must rely on luck to locate such evidence. As he says, "Although they can develop new technologies for identifying and studying potential smoking guns ... historical scientists can never manufacture a smoking gun. If, in fact, every single dinosaur heart was destroyed by the fossilization process, there is nothing anyone can do about it" (Turner 2007, 58). In Turner's parlance, studies of historical targets cannot play a "producing role," that is, create new phenomena; they can only "unify" observable evidence. By his lights, experimental scientists both postulate entities that unify their observations and go on to manipulate those entities in order to explore the systems concerning them. The path through a cloud chamber is explained by the presence of an electron, and the properties of electrons are revealed by repeated manipulation in controlled settings. By contrast, historical scientists only unify: They might postulate a worldwide freeze to explain Neoproterozoic glaciation, but they cannot perform repeated, controlled studies of the Neoproterozoic.

In this chapter, I will argue against such views.[1] Historical scientists *can* manufacture smoking guns. That is, they are able to generate evidence by conducting controlled manipulations. This undermines pessimism by showing that historical scientists' epistemic fate is in their own hands after all. The basic idea is that historical scientists can experimentally explore their targets *indirectly*. I will illustrate the in-principle evidential role of such indirect approaches via Zachos and Sprinkle's (2011) study of ancient echinoderm development. After explaining their work, I will show how it could count as a smoking gun by appealing to the law of likelihood and validation practices before considering other possible readings.

As we will see in chapter 10, where I generalize and build on our current discussion, many philosophers have recently pointed out functional and epistemic similarities between simulations and experiments (Parke 2014; Parker 2008; Turner 2009a; Winsberg 2003). To some extent, that work assumes that simulations and other surrogates can indeed play an evidential role. However, I intend to make an explicit argument to that effect.[2] Specifically, I want to show that the results of studies of constructed surrogates can count as smoking guns. Let's begin with the case study.

9.1   How to Build a Sea Urchin

Sea urchins and sand dollars are representative modern echinoderms. They are slow moving, somewhat alien critters that use their spines and adhesive tube feet to creep over hard surfaces, feeding mostly on algae and largely immobile invertebrates such as brittle stars, sponges, and shellfish. They are covered in a distinctive, spiny shell or "test." Although Echinoidea first emerged in the late Ordovician (say, 450 ma), and had high disparity (if relative rarity) throughout the Palaeozoic, the clade almost went extinct at the Palaeozoic's close (around 250 ma). Only a single body plan survived into the Triassic to give rise to modern echinoderms.

It is thought there are significant differences between the developmental systems of Palaeozoic echinoderms (the *stem group*) and Mesozoic and modern echinoderms (the *crown group*). The stem group is made up of the ancient, diverse echinoderms; the crown group is made up of their morphologically impoverished descendants. Differences in development are postulated to explain the striking differences in stem and crown group disparity. The echinoderm test is constructed from plates arranged into columns (other than the genital and ocular plates at the apex of the echinoderm, those associated with the mouth and anus). All post-Palaeozoic echinoderm tests are constructed from a mosaic of ten ambulacral (underside) plate columns and ten interambulacral (central) columns. Palaeozoic echinoderms, by contrast, had wide disparity in plate columns, from 15 to more than 150 and exhibit a variety of complex structures (see figure 9.1).

In light of the immense differences between stem and crown group morphology, it is theorized that stem group echinoderms developed largely by

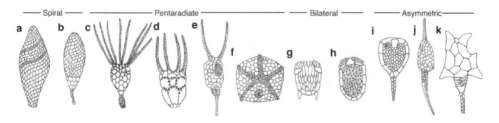

Figure 9.1
Stem-group echinoid disparity. (From Smith, Zamora, & Álvaro 2013; © Nature Publishing Group.)

plate *addition*, whereas crown group echinoderms develop largely through plate *accretion* and modification. Development occurs via "accretion" when the organ first grows, then divides. Human digits develop in this manner: Our hands begin as finlike paddles, before fingers are divided via cell pruning. Development via "addition" involves the separate growth of each part. The digits of some newts develop in this manner: Each "finger" grows separately (Hall 2003). So far as I can tell, it is assumed that the high disparity of the stem group compared with the crown group is explained by the more restrictive accretion and modification mode of development.

To test this hypothesis, we need to uncover the developmental systems of 250-million-year-old echinoids. How is this possible?

Developmental biologists study crown group development by directly examining living organisms, an option off the paleobiological table. Indeed, fossil-based reconstruction of complex developmental processes is extremely difficult—scientists must either infer development from (often incomplete) adult remains or, if they are extremely lucky, partly developed (juvenile) remains. As Zachos and Sprinkle put it: "The complex patterns of plates in Paleozoic echinoids make it very difficult to deduce the processes involved in their formation directly" (76).

This is, then, an unlucky circumstance: Historical scientists face a gappy, biased trace set and, on the face of it, cannot make their own luck. But wait—they go on: "However, it is exactly this geometric complexity that may hold the answers to these questions, and modelling is an approach that can evaluate hypotheses regarding the growth of these animals" (76).

As we will see, Zachos and Sprinkle use modeling to make up for a lack of traces: Models compensate for missing data. In the next section I argue that, in principle at least, their results could count as a "smoking gun."

Zachos (2009) constructed a simple model of crown group development—what is in effect a program for growing virtual sea urchins. Variables include the growth rate, addition rate, and maximum perimeter values (roughly, maximal size) for ambulacral and interambulacral plates. With these set, the program adds and grows various geometric shapes taken to represent the plates that make up the sea urchin test. That is, by adding a certain number of plates, of a certain size, at a certain speed, a geometric shape develops with striking similarities to a sea urchin.

The resulting models are sensitive to initial conditions, but in most cases converge on a relatively simple pattern of alternative plate addition and growth. Although the calculations are made over a spherical frame of reference, the resulting models can be deformed to display a typical sea urchin shape. (Zachos & Sprinkle 2011, 86)

Figure 9.2 shows one of Zachos's (2009) models of crown group echinoids.

Zachos and Sprinkle perform various modifications to the original model with a mind to producing stem group morphologies. If developmentally plausible transformations can be simulated, it is thought likely that the model represents stemgroup developmental sequences (see the following). Zachos and Sprinkle focus on the "Ocular Plate Rule" (OPR) in echinoid development. By this rule (valid across the crown group) all new interambulacral plates develop connected to the ocular plates, effectively restricting plate insertion to two loci during development. Zachos and Sprinkle found they could produce many stem group morphologies by relaxing this rule.[3] When the interambulacral plate insertion points are increased from two, they produced shapes similar to fossilized forms. Figure 9.3 reproduces Zachos and Sprinkle's model with six insertion points and compares it with a stem group echinoid fossil.

The thought is, then, that the developmental difference between crown and stem group echinoderms is the number of points where plates are inserted. In the crown group plates are inserted at only two locations—OPR holds—whereas in the stem group OPR was relaxed. According to the model, breaking OPR opens up significant areas of echinoderm morphospace. Palaeozoic, stem group echinoderms took advantage of this.

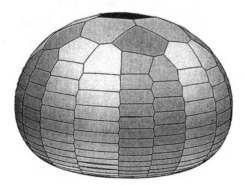

Figure 9.2
Model of crown group echinoid. (From Zachos & Sprinkle 2011, 79; © Springer.)

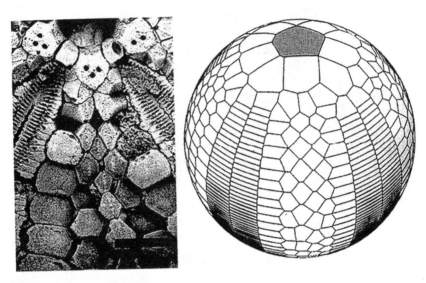

Figure 9.3
On the left, fossil of stem group echinoderm; on the right, model with six insertion points. (Detail from Zachos & Sprinkle 2011, 91; © Springer.)

Enforcement of OPR restricts modern sea urchins and sand dollars to their more conservative morphology. Just why, evolutionarily speaking, only conservative echinoids survived into modern times is a further question. Moreover, it isn't clear how this model could bear on that issue. However, progress has been made. It seems that Palaeozoic echinoderms are not as developmentally divergent from their modern cousins as we first thought: "Paleozoic echinoids appear odd because of their range of morphological disparity characterized by the number of plate columns (15 to over 150) but the same model for growth of individual plates can be applied to both modern and Paleozoic echinoids" (Zachos & Sprinkle 2011, 91)

Let's summarize. Stem group echinoid development is interesting because of the difference in disparity between stem and crown group echinoids. However, our trace set for echinoderm development is gappy. Fossil snapshots of echinoderm development are rare; that is, it is uncommon to find fossil echinoids at different developmental stages. Adult fossils, by contrast, are plentiful but often fragmentary. Moreover, such traces are faint: Inferring developmental sequences from adult remains is a long inferential shot. Models like Zachos and Sprinkle's go some way to compensate for this by providing a new evidence stream. By generating qualitatively similar morphologies to the stem group, Zachos and Sprinkle provide inroads to the evolutionary history of echinoids. Their model and others like it allow us to compensate for gappy, biased trace sets—or so I will argue. Such models give us information that we would have, were there a fuller trace set: Historical scientists can manufacture evidence.

9.2   How to Manufacture Smoking Guns

With the case study under our belt, I can argue that simulations can produce new evidence that discriminates between rival hypotheses. I will do this by providing a simple story about evidence using the law of likelihood and showing how models can be evidential on those grounds. To begin, it is worth getting clearer on what it takes to be a smoking gun.

Carol Cleland's concept is naturally (but mistakenly) thought of as a kind of "critical test": A smoking gun's discovery decides unambiguously between two past hypotheses. She has been read in these terms by Forber and Griffith (2011), who see smoking guns as "a naturally occurring *experimentum crucis*" (3). If this is the right reading of Cleland, we should be worried: True critical tests are few and far between—if they exist at all.[4] Fortunately, Cleland doesn't think this. Consider this 2013 statement of her view: "Considered in isolation, independently of the other lines of evidence, few traces would unambiguously count as a smoking gun for a hypothesis. A smoking gun for a hypothesis is a capstone piece of evidence; it can only be judged as a smoking gun when combined with the rest of the evidence available" (4).

Smoking guns need not be critical tests; they do not carry their epistemic burden alone. Rather, they are data that, in the context of discovery, shift investigations one way or another. Cleland is telling us how historical scientists go about generating evidence, and why it works, rather than committing to a view on the nature of evidence or the existence of "critical tests" (see Currie 2016b).

Cleland's methodological distinction between the historical and experimental sciences, then, is not that each provides critical tests in different ways (one via smoking guns, the other via controlled experiments). Rather, she claims that one generates evidence via intervention, while the other relies upon traces. That is, historical scientists search for evidence while experimentalists generate it. And so, where the experimentalist manipulates some system both to explore its properties and test hypotheses, the historical scientist generates a group of hypotheses accounting for current evidence, then hunts for new observations that could discriminate between them.

A smoking gun, then, is a *new* piece of evidence that has some evidential bearing on two hypotheses. It need not decide once and for all between them and, moreover, can play a relatively minor evidential role overall, as is suggested by Cleland's use of the term "capstone." I argue that, in principle at least, Zachos and Sprinkle's study has what it takes to be a smoking gun. Cleland does take her concept of a smoking gun to apply to traces only (personal communication), but this strikes me as stipulative and not of obvious epistemic import.

Let's distinguish more clearly between two hypotheses regarding echinoderm development. By the first, differences in disparity between stem and crown group echinoderms are due to their having radically different developmental regimes. Let's put this simply and say that stem group echinoderms develop via accretion, while crown group echinoderms develop via addition. Call this hypothesis "h_1":

H_1: Stem-group echinoderms develop via accretion.

By the second hypothesis, the two groups' developmental divergence is subtler. Both develop via addition; however, in the stem group the ocular plate rule is relaxed. Whereas for contemporary echinoderms plate addition is limited to two insertion points; their ancestors added plates at a wider variety of loci.

H_2: Stem group echinoderms develop via addition, with OPR relaxed.

Both hypotheses concern the developmental systems of organisms millions of years dead. What kinds of observational evidence could be bought to bear on them? Recall the *law of likelihood* (Hacking 1965), a simple way of understanding when some observation counts as evidence for one hypothesis over another:

Observation O is evidence for one hypothesis (H_1) over another (H_2) just when $P(O \mid H_1) > P(O \mid H_2)$

That is, an observation is evidentially relevant to two hypotheses when the conditional probability of the observation given one hypothesis is higher than its conditional probability given the other. The law of likelihood demands that there be a probabilistic connection between the occurrence of the observation in question, and the truth of the relevant hypothesis. According to the law, if the first hypothesis is true, I ought to find the occurrence of the observation *less surprising* than if the second is true. Note that I do not mean to imply that historical scientists can actually resolve or determine the likelihoods here. Many historical inferences, particularly in unlucky epistemic situations, are not amenable to strict statistical analyses.[5] Let's consider two observations that, prima facie, appear to count as evidence for H_1 or H_2.

First, consider the wide discrepancies in crown group and stem group fossils. On the face of it, this discrepancy would be quite surprising were H_2 true. After all, there being a wide variety of morphologies seems to suggest a corresponding variety of developmental systems. H_1 seems less surprising: If there were significant differences in development, this would suggest significant differences in morphology. That makes sense intuitively, and after all Zachos and Sprinkle present hypotheses like H1 as the received view. But how would the observation actually count as evidence? What is required for the observation—that there is wide disparity in stem group morphologies and low disparity in crown group morphologies—to have a higher conditional probability given H_1's truth than H_2's?

There must be some *linking principle* connecting the observation to the hypotheses. Such a principle has already been implied: If there is a robust link between disparity in development and disparity in morphology across echinoderms, then the observation of morphological disparity is less surprising if there is developmental disparity. That is, if we have a wide variety of developmental systems, we would expect them to generate a similar variety of subsequent morphologies. If that is right, then we have a linking principle that justifies taking morphological diversity as grounds for inferring developmental diversity. Thus, H_1 would gain ground at H_2's expense. The strength of the evidence, then, depends on the nature of the linking principle.

Second, consider observations of modern echinoderms. That sea urchins develop via addition is at least *some* reason to expect their ancient relatives to have also done so. This is because developmental systems tend to behave conservatively across ancestry. That is, over time they become increasingly entrenched. This is a form of *homologous* or *phylogenetic* inference, which we met in chapter 6's discussion of common cause explanations (Currie 2014b; Griffiths 1994, 1999; Levy & Currie 2015). Such inferences follow lines of ancestry: One organism's possession of a trait makes it more likely that its conspecifics, its ancestors, or its descendants do. This relies on *phylogenetic inertia*: That is, traits are more likely to be conserved across generations than not. Different traits are more or less inert: Highly labile traits are more likely to shift, while highly entrenched traits will remain stable. Inferring my natural hair color from my father's is probably safer than inferring my musical taste from my father's. This is because hair coloration is more heritable, relying as it does on a more robust process. Both my father and I have reddish hues in our beards, but my father's taste for progressive rock was only inherited incompletely (and my penchant for hip hop is a novel trait in our lineage). A trait's evolutionary lability, then, is an essential component of the strength of a phylogenetic inference.

Considering modern echinoderms' ancestral relations, the observation that sea urchins develop via addition is less surprising if H_2 is true (that is, if stem group echinoderms also develop via addition) than if H_1 is true, so long as the linking principle that echinoid developmental systems are phylogenetically inert is likely.

Thus, observations do not count as evidence alone, but require linking principles that ensure they are probabilistically related to the hypotheses in question. Midrange theories, for example, provide linking principles that license inferences from a trace to its causal ancestors. It is by virtue of such principles that observations count as evidence and thus potentially count as smoking guns. It is time to turn back to Zachos and Sprinkle's study.

Notice that the observation in question is of the output of a computer simulation, *not* an observation of the remains of our target, neither their fossils nor their homologues. Zachos and Sprinkle's simulation results are not, on the face of it, downstream descendants of the developmental systems of long-dead echinoderms. We are not dealing with trace evidence then. At a first pass, this could look problematic. Zachos and Sprinkle's results are surely determined by how they set up their model, not the properties of extinct echinoderms. To see the problem, consider how we might alter the world in order to alter their simulation results. Surely altering ancient echinoderm development would not change the outcome of Zachos and Sprinkle's model. If the echinoderms were different, the results would come out the same. This is because the behavior of a simulation depends on facts about geometry, about software and programming, and about the relevant hardware, not echinoderm developmental systems. How, then, might the two be linked?

To identify the linking principle, we need to examine briefly how simulations are *sanctioned*. As Parker (2009) has stressed, simulations are not confirmed tout court. That is, it doesn't make sense to say that a model is "true" of a system. Rather, a model is *suitable for some purpose*. So "confirming" a model involves confirming that it can fulfill some epistemic function. For instance, it might be able to predict the behavior of a system under a certain range of conditions (see chapter 10 for a fuller discussion of modeling). Simulationists themselves divide the sanctioning of models into two parts: *verification* and *validation* (see Winsberg 2010).

Verification concerns the internal properties of the model and its relationship with surrounding theory. In a sense, verification asks *did I create the model I wanted to*? Scientists test for verification by (1) checking for internal consistency, (2) confirming that model outputs, or the outputs of its components, approximate the results of relevant theory, and (3) "sensitivity testing"—that is, scientists check whether the simulation responds as expected given variations on initial conditions.

Validation concerns the *external* properties of the model, in particular, whether it relates to the world in the intended way. Validation tests are carried out by comparing the model to the world: either directly to the target system, testing the model's components separately, or by testing it against an analogue of the target. Recall Hyde et al.'s (2000) simulation from chapter 2, which they constructed in order to explore Snowball Earth scenarios. The simulation was tested by seeing if it could reproduce the last glacial maxima of the Pleistocene ice ages. Their capacity to retrodict the ice extent of that event was taken as reason to think that the model is also a good guide to the Neoproterozoic. This is an example of validation.

Let's focus on the validity of Zachos and Sprinkle's model. In 2009, Zachos built the model to mimic crown group echinoderms such as sea urchins. Because of their transparent embryos and simple developmental progression, sea urchins have been a favored model organism for developmental biologists since the invention of the microscope. Consequently, we know a lot about crown group echinoderms (for instance, the sea urchin genome was sequenced back in 2006; see Sondergren et al.). This knowledge gives us good reason to think that the model's aping of echinoid forms is not mere coincidence: It does so because it resembles, albeit in simplified terms, real sea urchin developmental systems.

To see why this matters, consider the following imagined case. Imagine that, instead of constructing a model, Zachos and Sprinkle were able to directly manipulate OPR in live echinoderms; perhaps it turns out that insertion point number is amenable to simple genetic manipulation. Imagine further that, upon making those manipulations, the resulting mutants looked like stem group echinoderms. On the face of it, this would lend very strong support to H_2. If the difference in morphology between two related groups of animals can be generated using relatively simple variations in developmental rules, then it would be very surprising if some other route were taken to generate that variation.

Given that we have good reason to think that the model in question correctly captures the developmental systems of extant sea urchins, it seems that the results of Zachos and Sprinkle's study should be read in similar terms. That is, their simulations count as *proxies* for manipulating OPR in live echinoderms. It is the ancestral relationship between crown and stem group echinoderms, coupled with the model's success with the crown group, which licenses the study.

So, the relevant linking principle is this: If H_1 is true, Zachos and Sprinkle's simulation would not have been sanctioned. Clearly, had they constructed the same model, it would provide the same results. However, these results would not have been treated in the same way, nor would the model have been constructed as it was in the first place. Perhaps Zachos's original 2009 model would not have captured crown group development so successfully.

Michael Weisberg (2006) distinguishes between two kinds of success conditions for models, which clarifies my reasoning. First, we care about our model's *dynamical fidelity*: That is, does the model produce qualitatively or quantitatively similar outputs to our target system? Second, we care about the *representational fidelity* of our model. That is, does the structure of the model resemble the structure of the target system? These can come apart. In climate modeling, for instance, many processes that make a difference to climate dynamics are too complex to represent in the model: We lack either the requisite understanding of the system or the computational power to include them in already overloaded simulations. In these cases, scientists *parameterize* components. In the extreme, rather than represent components with simplified causal models, they represent them with a single value intended to capture *well enough* the influence of that factor (Steele & Wendl 2013). In such cases, representational fidelity is sacrificed to improve dynamical fidelity.

If Zachos and Sprinkle's simulation provided the output that it did, but H_1 is true, then, (1) the model does a very nice job of mimicking crown group morphology (moreover, in a way that we think has dynamical fidelity) and (2) simple changes to that model mimics stem group morphology (in other words, it has dynamical fidelity). Yet, (3) the representational fidelity between the model and the stem group fails. This is quite a coincidence!

The simulation, then, at least in principle, counts as a smoking gun: It could be the "capstone" piece of evidence that leads us to decide in favor of H_2. Moreover, if it is a smoking gun, it has been manufactured. It outruns trace evidence, and is come to via the controlled manipulation of a system. It does not work independently of other evidence—but then *no evidence does*. Its linking principles are indirect and perhaps somewhat tortuous (and torturous). It relies on claims about the fidelity of the model in regards to crown group echinoderms, the relationship between the model's outputs and stem group fossils, and phylogenetic relationships within the clade. Nonetheless, there are grounds to take this to be legitimately evidential. In principle, historical scientists can manufacture smoking guns.

Two caveats before turning to objections. First, as mentioned earlier, Cleland restricts her notion of "smoking gun" to traces—this potentially undermines my conclusion by fiat. I'm not too worried about what we call the results generated from proxies, so long as we agree that they sometimes count as evidence. Moreover, it strikes me that the "trace only" feature of Cleland's view is easily jettisoned. Second, you might be worrying about the account of evidence underlying my discussion. After all, you might think that evidence must be empirical, while simulations are theoretical. In response, it is important to see that distinctions like that between empirical "observational" evidence and theoretical, perhaps a priori aspects of scientific work have a shelf life. That is, they become quite artificial and unhelpful in some contexts. A lot of simulation work, on my view, is properly understood as synthetic—they sit *in between* the empirical and the theoretical—and thus are not necessarily well captured by that distinction.

Further, by "evidence" in this context I simply mean some observation that counts as evidential by the law of likelihood. Zachos and Sprinkle generated observations or data using their simulation, and demonstrate those observations' evidential relevance via validity testing. Studies of simulations and other proxies can generate evidence by the law of likelihood, which I take to be an intuitive and plausible notion.[6]

9.3 Alternative Readings

You might want to deny that Zachos and Sprinkle's model is appropriately read in evidential terms. In this last section, then, I will consider three alternative readings of their work. First, models merely generate hypotheses. Second, models draw out the empirical consequences of theories. Third, even if you agree that in principle models can generate evidence, you might be skeptical of whether this is often realized in practice.

Let's begin with the first alternative reading. That is, the idea that models merely generate or represent hypotheses. That simulations generate evidence is not always taken for granted, and is sometimes denied, among scientists. To take one example, A. J. Underwood's (1990) view on the use of models in ecology emphasizes their creative role, but this is limited to suggesting hypotheses. Models provide *how* or *why* explanations—hypotheses—that are then tested against the world. They do not *themselves* provide evidence: "The most profitable method of research is to devise as many models as possible, and then to deduce from each the most different hypotheses (predictions), because this provides the maximum possibility of contrasting the various models" (Underwood 1999, 367).

Here, models represent or suggest hypotheses but do not play an evidential role: They do not support hypotheses. On this view, Zachos and Sprinkle's model generates a hypothesis, presumably H_2, but does not give us any reason to believe this hypothesis. This reading fails.

First, it does not line up with how Zachos and Sprinkle treat or discuss their model. It appears to provide a *surprising result*. If so, this contradicts the idea that we can draw an epistemic distinction between simulations and experiments on the basis of their capacity to surprise scientists. Discussing models in economics, Mary Morgan (2003, 2005; see also Harre 2003) puts this most clearly. She distinguishes between the capacities to *surprise* and *confound*. Whereas both simulations and experiments can provide unexpected results (i.e., be "surprising"), only experiments can "confound" investigation, that is, deliver a result that challenges investigators and drives research. The thought behind this is that:

In mathematical model construction, the economist knows the resources that went into the model. Using the model may reveal some surprising, and perhaps unexpected, aspects of the model behavior ... but in principle, the constraints on the model's behavior are set ... so that however unexpected the model outcomes, they can be traced back to, and re-explained in terms of, the model. (Morgan 2003, 325)

In other words, while I can explain a simulation's operation by referring to its setup, to explain anomalous experimental results, I must draw on and directly confront the theories I am testing. This position is mistaken (see Parke 2014 for a different tack). As we have already seen, the sanctioning of simulations grants them evidential relevance. Although I can explain how the simulation behaves in reference to software, the computer, and so on, this does not exhaust the relevant information that demands our attention. I must also explain the simulation's success, its sanctioning. There are two alternatives here. Either it is a surprising coincidence that Zachos and Sprinkle's model happens to generate both crown and stem group forms but only generates the stem group accidentally, or the model correctly reflects ancient echinoderm development. In the good case, a simulation result can indeed confound investigation (I develop this line of thought in much more detail in Currie, [under review]).

Second, it seems to me that, given my earlier analysis, the onus is on others to explain why simulations do not provide evidence. We shouldn't move from something's providing indirect or somewhat ambiguous evidence to its providing *no* evidence. Admittedly, the road from the simulation result to evidential relevance is a long one, but this does not mean the road cannot be traveled.

Now, consider the second interpretation. This is the popular view that models and simulations are *mediators* between our theories and the world. That is, their job is not to bring new empirical data to the table per se, but rather to draw out the empirical consequences of our theories. (See the papers collected in Morgan & Morrison 1999, although the "model as mediator" role expressed in these papers also emphasizes their autonomy. See also Frigg & Reisman 2009.) There is certainly something to this: Climate simulations, for instance, are frequently constructed from a base of fluid mechanics and other physical theories. These physical theories often do not have empirical consequences insofar as they are not resolvable computationally or in principle. Climate models, then, are *theoretically grounded*: Their construction begins from well-established but unresolvable and computationally intractable physical theories.[7] The theory is mathematically manipulated, typically by "discretization"—the replacing of differential equations by algebraic expressions (Winsberg 2003). This results in the model: a resolvable and tractable proxy of the theory. Components of varying levels of idealization and representational veracity are then added until the simulation is sufficiently complex to understand or predict some aspect of the climate (see Winsberg 1999 for a detailed discussion of the "hierarchies" involved in simulation construction). Although the simulations are constructed from a theoretical base, they are (semi) autonomous of theory (Morgan & Morrison 1999; Winsberg 2003, 2013). They have different a priori and a posteriori content because of how the originating theory is transformed in model construction, and due to the addition of subcomponents required for empirical relevance.

Although some models function by drawing out empirical consequences from theories, Zachos and Sprinkle's (at least) does not. Here, there is no *mathematical* theory that the model represents or is a proxy for. Rather, the theory of echinoid development, if there is one, is a causal or mechanistic model. How we get from a single cell to a differentiated, multicellular sea urchin is explained via a description of the relevant parts and their interaction. Perhaps Zachos's model *embodies* this explanation in some ways, but there is no explicit theory from which it helps to "draw the empirical consequences." Moreover, Zachos and Sprinkle's model, if it represents anything, represents the developmental systems of sea urchins. The simulation results counted as evidence on the basis of (1) the phylogenetic relationship between stem and crown group echinoderms, (2) the dynamical and representational fidelity of Zachos's (2009) model and crown group echinoids, (3) the dynamical fidelity of Zachos and Sprinkle's results vis-à-vis fossils from stem group echinoids. As such, it is not a *mediator* between theory and world, but rather a way of doing a proxy experiment. As I suggested, it is best read as a virtual way of manipulating a sea urchin's developmental system indirectly. I will clarify this notion in the next chapter.

Finally, the third alternative: One might agree that *in principle*, simulations play evidential roles, but argue that in practice, this barely ever happens. One could read some of Wendy Parker's work in these terms. Parker is much more skeptical of our capacity to solve validity than verification issues:[8] "in many cases it is difficult or impossible to directly test key assumptions relied on in constructing computer simulation models, because the target systems of interest are inaccessible in space and/or time" (2008, 177).

To establish the representational and dynamical fidelity of a model, we presumably need sufficient access to both model and target system. But in cases where the target is mind-bogglingly complex or inaccessible for some other reason (lost to the deep past, for instance), it seems that we lack such access and thus cannot validate.

Parker does not mention the use of analogue tests. We can test the external validity of a simulation by contrasting its results with analogues of the target. We see something like this in Hyde et al.'s appeal to their simulation's correct prediction of Pleistocene ice ages.[9] Although I concur with Parker that more work needs to be done to understand the justificatory relations vis-à-vis simulations and target systems, it seems to me that we can make the following preliminary points. First, as Winsberg stresses, validation and verification are not, in practice, separate activities, but a mishmash: "simulationists are rarely in the position of being able to establish that their results bear some mathematical relationship to an antecedently chosen and theoretically defensible model. And they are also rarely in a position to give grounds that are independent of the results of their "solving" methods for the models they eventually end up using" (2010, 20).

Winsberg argues that the "sanctioning" of models occurs by the "simultaneous confluence" of various tests for validation and verification: "Simulationists try to maximize fidelity to theory, to mathematical rigor, to physical intuition, and to known empirical results" (23). It is the combination of these factors that grants rational belief in the simulation's capacity to tell us about the world. Second, solving the model-world relationship is not a special problem for simulations (Frigg & Reiss 2009)—but rather, a more general problem about the relationship between representations and the world. Whatever story we might tell about that, is presumably applicable to simulations as well (see Weisberg 2013). Third, I emphasize a holism beyond Winsberg's. In deciding on the validity or otherwise of a simulation study, we do not simply rely on the study itself and direct tests, but the broader epistemic context that relates the study to our other theories, observations, and so on. Against this broad backdrop, there is much ground for optimism about establishing the external validity of simulations; even if our access to the target is indirect. The connection between Zachos's (2009) model and modern echinoderms, as well as the crown group's phylogenetic connection to the stem group, are long and indirect. Nonetheless, Zachos and Sprinkle's results do lean evidentially on hypotheses of stem group development.

The worry that simulation (and other proxy) evidence is generated only rarely might itself be an expression of the worry that I have not argued that simulations produce evidence that meets some general standard. I have not compared them against a general account of evidential sufficiency (the law of likelihood decides only what it takes to be evidence, not what it takes to be good evidence) or to some other method that we take to be successful—experiments for instance. As I said in the introduction, I am skeptical of the work that general accounts of evidential sufficiency can do. Epistemic and methodological success depends crucially on context, and I think should be taken on those terms. As we will see in the next chapter, I do think that non-proxy experiments provide better confirmation in many contexts (Currie & Levy, under review) but here that is a moot point. My task at present is to show that pessimists underestimate the epistemic resources historical scientists have at their command in unlucky circumstances. I have argued both that simulations can be evidentially relevant, and via a discussion of validation practices given reasons to think that such evidential relevance is often achieved. Moreover, even if the evidential power of proxies is somewhat limited, I have also argued in chapter 6 that support in historical reconstruction is often underwritten by the "confluence" of evidence—*many* small pieces of evidence playing a variety of roles. In this context, having more resources in your arsenal can make an enormous difference. The support a historical hypothesis has is not determined by the evidence in isolation. Rather, it is the combination of the pieces of evidence and their relationships that constitute support for a hypothesis. Even if, standing alone, Zachos and Sprinkle's evidence does not do much work, it might still *in combination with the rest of our knowledge* play an important role.

Chapter 5's third reason for skepticism—that historical scientists cannot generate evidence—is therefore undercut. By constructing and manipulating models and other surrogates, we are able to extend our epistemic reach beyond traces and actively generate evidence. In the next chapter, I will further my examination of the use of models and simulations in historical science by unifying it with more general philosophical discussion. In particular, I will suggest that such simulations are best understood as "surrogate experiments." This will lead me to a discussion of the role idealization plays in historical science—and to another reason to doubt pessimism.

10   Idealization and Historical Knowledge

Idealization is a prevalent and prima facie mysterious feature in science. Scientific representations are rife with omissions and outright distortions. Newton treated the moon and the Earth as a two-body system (ignoring every other body in the universe), Zachos and Sprinkle treat three-dimensional sea urchins as two-dimensional geometric shapes, population geneticists treat finite populations as infinite populations, and so on. These are examples of idealizations in models, theoretical constructs that are intended to be imperfect representations. However, Angela Potochnik has argued convincingly that idealization is "rampant" and "unchecked" in science. That is, idealizations are ubiquitous (and not restricted to models), and "de-idealization," the removal of distortions, is rare. In this chapter, we will meet two kinds of scientific tools that are highly idealized. First, simulations, such as Hyde et al.'s study of Snowball Earth, which I will suggest are sometimes best understood as "surrogate experiments." Second, "inference tools," which are less explicitly representational than models or simulations, and function to extend our epistemic grasp on the past. That is, they take the knowledge we have and use it to infer further past facts. Why is idealization puzzling? Well, if science aims for truth, explicit, outright falsehoods should not play such central, and apparently indispensable, roles. In this chapter I have three aims.

My first aim is this: In the last chapter, I argued that historical scientists can manufacture smoking guns. That is, they can use models like Zachos and Sprinkle's to generate evidence that empirically distinguishes between hypotheses. This capacity increases our epistemic access into the past, as it means the fate of historical investigation is not beholden to luck. They are not reliant on the traces and analogues they happen to find alone, but can make their own. I shall connect my discussion of surrogates in historical science to the more general literature on modeling and idealization. I will argue in the first section that models such as Zachos and Sprinkle's, and simulations such as Hyde et al.'s, are best understood as "surrogate experiments." As we will see, these are not true experiments because of their indirectness. However, they do have some of the advantages and functions of experimentation.

My second aim is to argue in sections 10.2 and 10.3 that the purpose of idealizations in historical science is often to facilitate testing—idealizations ensure information is provided at the right grain to be relevant to the hypotheses concerned. Falsehoods, omissions, and distortions, then, further our epistemic access to the past.

The third aim concerns what I have called *investigative scaffolding* (Currie 2015a). In order for us to discern some past facts, we must already know a fair bit about the past. Historical investigation proceeds in a piecemeal fashion. A body of knowledge must be sufficiently supported, must "form a scaffold," in order for further evidence to become relevant. I argue that in historical science, models often function and idealize as they do in order to construct, reach, and exceed these scaffolds. Scaffolding is a general feature of historical (and perhaps all) scientific investigation, and as we will see, plays an important role in my arguments against pessimism.

Here is a sketch of the chapter. I shall compare experiments and simulations in order to articulate the notion, and limitations, of surrogate experiments. I then contrast the role idealization plays in historical science with idealization in other contexts. And finally, I will turn to investigative scaffolding. To begin, however, we should look in more detail at the nature of simulations. I will delve deeper into Hyde et al.'s investigation of Snowball Earth. The point of doing this is to provide a more sophisticated example of a simulation that is more explicitly "experiment-like." There will be a little mathematics, but don't panic: My aim is to give a taste of the complexity of the simulation. Getting a sense of how the variables relate is all that's required for my purposes.

Parker (2009) defines a simulation as "a time-ordered sequence of states that serves as a representation of some other time-ordered sequence of states; at each point in the former sequence, the simulating system's having certain properties represents the target system's having certain properties" (486). By this account, an investigation is a simulation by virtue of there being a structural and temporal mapping between system and target. That is, particular states in the model system map onto states of the target in an ordered fashion. By this view, a simulation need not be a computer program; all that is required is the right kind of relation between the simulant and its target. Parker points to uses of wind tunnels to investigate stresses on car or aircraft parts. Orreries are another example of a nondigital simulation. I can imagine (although I know of no cases) geologists constructing a physical model of the earth's atmosphere, and using that to simulate possible snowball and slushball events. Realistically the cases that interest us are digital. It is not clear whether the kinds of state ordering we need in order to successfully simulate global climate events are achievable in a physical simulation. There is not merely the question of cost, but also of the possible scale-dependence of the behavior of such systems. That is, the increased size and duration of the target may undermine extrapolation. Moreover, the complexity that such models attempt to capture lends itself to a digital format: Controlling the number of variables in a physical realization would be a nightmare. Although I call on Parker's definition for ease of exposition, I take it that much of my coming argument will hold for nonsimulations as well. In the next section, for example, I will introduce the use of idealization in a different scientific product, "inference tools."

(Computerized) simulations are often constructed from analytic differential equations that themselves represent some theory (Winsberg 1999). For various reasons—in part a lack of data, in part to provide computational tractability, and sometimes for more principled motives such as identifying a theory's empirical consequences—simulations are constructed "from" theoretic models. Frequently, these are constructed because the original model is not solvable: "In the types of systems with which the simulation modeler is concerned, it is mathematically impossible to find an analytic solution to these equations ... it is impossible to write down a set of closed-form equations" (Winsberg 2010, 7).

In response to these challenges the simulation is constructed by "forcing" the differential (continuous, infinitesimal) equations of the model into discrete, finite equations. In doing so, the simulation both abstracts, ignoring parts of the equations in the original model, and idealizes, crudely representing complex, dynamically essential aspects of the model with simple equations: "These rough-and-ready, theoretically unprincipled model-building tools typically involve relatively simple mathematical relationships that are designed to approximately capture some physical effect in nature that may have been left out of the simulation for the sake of computational tractability" (Winsberg 2010, 12).

Let's look at some simulation-based "experiments." To keep things (relatively) simple, I will refer to some fairly old studies from the late 1990s and early 2000s, but the basic principles hold in more cutting-edge work. Recall that Hyde et al. (2000) are interested in testing whether or not early life could have survived the Neoproterozoic, specifically, whether or not "refugia," or open areas of water, are expected in snowball events. This is pressing because of the emergence of complex metazoan life relatively soon after the close of the Neoproterozoic. Given the Cambrian explosion, we have good reason to think that early (perhaps metazoan) life survived the Neoproterozoic. If so, how? Hyde et al.'s model has two parts: an ice-sheet model, which predicts ice extent, and an energy balance model (EBM), which outputs global and equatorial temperature. The model is extraordinarily complex, and I lack the space to give anything close to a full description, but will nonetheless provide a taste. In short, the ice-sheet model tells us about the relationship between ice pack, continents, and the sea, while the EBM tells us about the interaction of climate and temperature on a global scale.

Here are the basics of EBMs. EBMs attempt to predict the earth's temperature, and from this to infer the surface temperature at the equator and at other latitudes. T_e, or the earth's overall temperature, is calculated given the earth's radius, solar luminosity and albedo. That is, the planet's size, the amount of energy from the sun, and how much of that energy is subsequently returned to space. Once T_e is calculated, surface temperature is inferred via the following equation:

$$T_s = T_e + dT$$

Where T_s is the Earth's surface temperature, T_e is Earth's temperature (from space) and dT is the greenhouse gas increment. And so, Earth's temperature is taken to be a function of how much heat is radiated (T_e) and how much is captured (dT). As Hyde et al. point out, basic EBM models fail to predict snowball events, as the predicted temperatures are too high. That is, the atmosphere traps far too much energy unless atmospheric CO_2 is set to half current levels. In response, Hyde et al. couple an EBM model with an ice-sheet model that allows them to represent both the Neoproterozoic geography and continent-sea-ice dynamics. The ice sheet model is taken from

Tarasov and Peltier (1997). That model, which Tarasov and Peltier used to model the ice ages of the last 100,000 years, consists of four submodels, each predicting "ice flow, mass balance, temperature and bedrock sinking" (Hyde et al. 2000, 426). Ice sheets, like glaciers, depend on year-round global temperatures. If sea-ice becomes dense enough to resist the summer melt, the sheet will grow as thicker ice from the colder climes "flows" into warmer waters. Moreover, the height and position of landmasses influence the icepack by influencing temperatures and interfering with the advancing ice. To get a grip on ice sheet behavior, then, we need to know about geography, temperature, as well as the physics of sea and icepack. Tarasov and Peltier's ice sheet model looks like this:

$$\frac{\partial H}{\partial t} = \nabla_h \cdot \left[B(\rho_1 g)^m H^{m+1} (\nabla_h h \cdot \nabla_h h)^{\frac{m-1}{2}} \nabla_h h \right] + G(\vec{r}, T)$$

(from Tarasov & Peltier 1997, 21669)

Again, don't panic! In the equation, h is the surface elevation of the ice, p is ice density, g is gravitational acceleration and G is mass balance. B is the average flow parameter, determining how far sea-ice increases over time. H is overall ice thickness. Roughly, Hyde et al.'s simulation works by coupling some of the variables in the ice sheet model to variables in the EBM. For instance, the output of the ice sheet model (H) partly determines the albedo in the EBM, while the output of the EBM (T_s) determines precipitation in the ice sheet model. The interaction between the ice sheet model and the EBM is, then, complex. And, of course, idealized. Some data is simply unknown, and maintaining tractability requires simplification. For instance, the average flow parameter (that is, how quickly the ice advances) is treated independently of the temperature output of the EBM, and the rate of bedrock sinking (that is, how the height of the sea floor reacts to the advancing ice-sheet) is kept at a steady pace. Let's look at some of Hyde et al.'s "experiments."

Hyde et al. ran a series of simulations testing the sensitivity of ice-sheet extent to different levels of CO_2 atmospheric concentration. Their aim was to see how sensitive snowball events were to atmospheric CO_2.[1] Concentrations from half the Holocene value, to slightly over twice the Holocene value, were tested. Roughly, Hyde et al. were exploring the relationship between how much CO_2 we have, and how likely snowballing is, while holding other factors fixed at Neoproterozoic levels. They found snowballing occurs at the lower end of the scale, while at the higher end "ice volume approaches that of the Last Glacial Maximum" (426), the fullest extent of the more recent ice age (around 20,000 years ago). In each case the shift from cold to frozen solid occurs at a blistering geological pace (in fewer than 2,000 years). As experimenters often explore their systems by holding aspects of the system in question fixed, and establishing causal influence by intervening on further aspects, it's not surprising that Hyde et al. call their studies experiments.

Another test investigated the relationship between CO_2 and continental freeboard (that is, average sea-level relative to continents). Between them, these two variables determine ice volume. Because sea has a lower albedo than land, increases in freeboard (and thus decreases in exposed land), lead to lower overall albedo. And so, if freeboard is high, presumably for snowballs to occur CO_2 must be decreased. Hyde et al. found that for higher freeboard values, snowballs only occur under fairly dramatic decreases in CO_2 concentration. Finally, a series of experiments tested the occurrence of snowballs against different continental arrangements. A dispersed paleogeography versus a "megacontinent" model makes no difference to the qualitative results. Here, other variables are held at roughly Neoproterozoic levels, but continental position is manipulated.

We see, then, that Hyde et al. carry out three "experiments." They are interested in establishing and exploring various causal claims: If freeboard increases, snowballs are less likely, for example. Hyde et al. set CO_2, or freeboard, or paleogeography, to a specific value—create the antecedent circumstances—and then see which consequent arises. In doing so, they seek to establish various dependencies within their target system. This is strikingly similar to experimentalist reasoning.

Here are Hyde et al.'s conclusions from these experiments:

An ice-covered Neoproterozoic land mass is predicted by our coupled climate/ice-sheet model with only two significant changes in boundary conditions from present values—a solar luminosity decrease consistent with estimates from solar physics, and a CO_2 level within about 50% of the present value. ... These conclusions are relatively robust, in that they hold true for different land-sea configurations and continental freeboard. ... A major finding of the study is that an area of open water in the equatorial oceans—which could have allowed for the survival of metazoans—is consistent with the evidence for equatorial glaciation at sea level. (2000, 428)

In short, these simulations are taken to support the hypothesis that the ancestors of the Cambrian explosions could have survived in the chill (but not frozen) seas of the Neoproterozoic and further inform us about the conditions required for snowball events. The simulation shares many features with experiments—let's look at this a little closer.

10.1   Surrogate "Experiments"

Arnon Levy and I have been developing a view about the epistemic nature of experimentation (Levy & Currie 2015; Currie & Levy under review). In this section, I will draw on aspects of that view to articulate the notion of "surrogate experiments," which I argue captures Hyde et al.'s and Zachos and Sprinkle's investigations. The point of this is to build on the work in chapter 9. There, I argued that historical scientists draw on evidence from surrogates in order to learn about the past. Here, I suggest that the concept of a "surrogate experiment" captures that practice.[2] I will start by comparing two kinds of definitions of experiment.

Compare a *procedural* and a *substantive* account of experiment. A procedural account of experiment defines an experiment as a kind of activity, specifically an intervention on some system. An example of such a definition is Wendy Parker's (also see Woodward 2001, 2003): "An *experiment* can be characterized as an investigative activity that involves intervening on a system in order to see how properties of interest of the system change, if at all, in light of that intervention" (Parker 2009, 487).

On such accounts, an activity is an experiment just in case it involves an intervention on some *object* in order to learn about some *target*. Hyde et al.'s object is a computer simulation and their target is the Neoproterozoic climate (the object/target language is due to Parke 2014 and Winsberg 2009). They intervene on their simulation in order to discover whether, for instance, continental arrangement or freeboard range effect snowball events. On procedural accounts, then, Hyde et al. perform an experiment. That is, they manipulate a system to learn about its properties. Although manipulation is a critical part of experimentation, on my view such accounts are largely stipulative: They make minimal claims about the epistemic power of experimentation.

Levy and I prefer a *substantive* account of experiments. Here, an account of experimentation ought to pick out what is *epistemically noteworthy* about experiments. We focus on experiment's role in confirmation—suitable for inferring past states of affairs. By our account, a surrogative experiment is not a true experiment. Let me explain.

Paradigm experiments have two features. They are *controlled* investigations of *specimens*. I will start with control. "Control" targets the interventional nature of experiments. By isolating and repeatedly manipulating various features of their objects, scientists are able to produce robust effects and confirm causal relationships. Samir Okasha (2011) explains this latter power, which suffices for my purposes here. Many causal claims can be thought of as conditionals, $P \to Q$. Say, if freeboard increases, the amount of CO_2 required for snowballing also increases. Most "passive" natural observations (or worse still, trace-based examinations!) are not of this logical form. They are rather conjunctions (or correlations), $P \& Q$. The layer of shocked quartz and increased iridium are correlated, or "conjoined," at the K-Pg boundary. The strength of the inference to causal dependence from co-occurrence is relatively weak, logically speaking. Control, in Okasha's terms, allows us to *bring about* the conditional's antecedent, and see if the consequent results. Hyde et al. are able to set freeboard to various levels, and then see which amounts of CO_2 are required for snowballing. Without control we must infer from the conjunction to the conditional, a weaker position. Naturally, this logical weakness can be mitigated in various ways— most clearly when there is a relatively obvious or well-understood causal reading of the conjunction.

Already we see some advantages of surrogate experiments over trace-based methods. Not only is hunting for smoking guns beholden to luck— the traces must have been preserved, and then you must find them—they are also only correlative. Hyde et al., by contrast, are able to exercise control: By bringing about the relevant antecedent conditions, they are able to see whether the consequences proposed in the relevant hypotheses (a snowball versus a slushball) are the result.

Surrogative experiments such as Hyde et al.'s, then, are experiment-like insofar as they involve control (à la Parker's definition). However, they are not investigations of *specimens*. In brief, a specimen is a typical instance of the thing scientists are ultimately interested in studying. Here is one way of getting at the notion. In an investigation of a specimen, the object is a subset of the target. That is, they are one and the same kind of thing.[3] A specimen is analogous to a statistical sample. It is drawn from the natural world.

This means that to infer from the object to the target, we must extrapolate on the basis of the object's typicality. If I want to know how long banjo strings will keep good tone without being played, I might conduct a study of a group of banjo strings. I could leave them under Paul Griffiths's desk for months on end, and see how long it takes for them to become dull. If the sample size is sufficient to warrant extrapolation, and there is nothing odd about the strings in my study, the results can then be extrapolated to the general class of banjo strings. This is not, of course, the situation facing Hyde et al., or Zachos and Sprinkle. They investigate surrogates; that is, their object and target are very different kinds of things (see figure 10.1). Computerized software and hardware are neither climates nor echinoderm developmental systems. In such cases, the relevance between object and target is made on the basis of the object *representing* the target in some way. There is philosophical discussion about the nature of the relationship between model and world, but I will remain silent on it[4] other than taking models to be, in some sense, intentional representations. Regardless of their nature, my argument from chapter 9 established that such representational relationships can be evidentially relevant.

And so, surrogate experiments are controlled investigations of surrogates, as opposed to true experiments, which are controlled investigations of specimens. Although this does make the epistemic relevance of such tests comparatively limited, this doesn't undermine their value (Currie, under review), particularly in unlucky circumstances. What matters for present purposes is this: Historical scientists, like experimental scientists, are to at least some extent masters of their epistemic fates. Even in circumstances where our access to past events and processes is indirect—relies on surrogates—we have the capacity to actively generate new knowledge about the past. Moreover, as I argued in the last chapter, when evidential confluence is the name of the game, indirect, unstable evidence can still be important, as warrant is provided by a large number of quite different evidential sources.

I have discussed two kinds of surrogate. In chapters 7 and 8 I discussed analogues such as the use of the placental sabre-tooth cat *Smilodon fatalis* to investigate the South American *Thylacosmilus atrox*. Here, contemporary and past objects are unified with past targets as tokens of process types. That is, they are explained by virtue of being products of the same kind of process, and are thus evidentially related insofar as they grant epistemic

Specimen   Surrogate

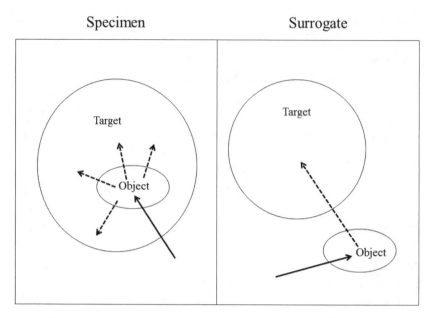

Figure 10.1

In a *specimen*-based investigation (left) an intervention (solid arrow) is made on a subset of the target and then extrapolated to the broader class (dashed arrow). In a *surrogate*-based intervention (right) the object is not a subset of the target; rather, it represents it.

access to the process in question. The reconstruction of the functional morphology and evolutionary history of *T. atrox*'s magnificent sabre-teeth relied on other lineages, both saber-toothed and non-saber-toothed. These were understood in terms of a model that represented the range of variation in dental morphology, head musculature, and bite mechanics across mammalian carnivores. In the last chapter and expanded upon here, we have seen that historical scientists also construct and manipulate surrogates. I have characterized some models as "surrogate experiments." In these cases, new knowledge is actively generated by intervening on a representation of the target. This suggests that at least some of the epistemic gifts of experimentation are available to historical scientists. By conducting controlled investigations of their simulations, Hyde et al. seek to understand the causal dependencies between Earth's climate, geography, and so forth.

It is time to switch our focus to a property that is typically associated with modeling but is in fact a common feature of much scientific work:

idealization. As we will see, considering the role idealization plays in historical reconstruction reveals hitherto unnoticed features of the practice.

10.2 The Purpose of Idealization

I have said idealization is puzzling. If science aims for truth, why do scientists knowingly and explicitly introduce falsehoods into their work? As is so often the case with science, the answer turns out to be: *for heaps of reasons*. Sometimes we might be just wrong about science's aims (Angela Potochnik argues forcefully for this), sometimes idealization mitigates our cognitive or technological failings, and sometimes idealization aids us in uncovering the truth. I am going to explore the latter purpose. A closely related issue concerns de-idealization. When scientists abandon idealizations (if they do), what governs such decisions? I am going to suggest a new function for idealization, and thus a novel reason for de-idealization. In brief, I will argue that in some cases idealizations ensure that a hypothesis is articulated at the right "grain" in order for the data available to be evidentially relevant. De-idealization occurs when either insufficient detail has been included for the empirical situation at hand, or because what I will call an "investigative scaffold" has been reached. Michael Weisberg (2007) has a nice framework for understanding idealization's purposes, which will help set my claim apart. In this section, I will discuss his view and introduce a case study. In the next, I turn to investigative scaffolding. Weisberg identifies three functions.

First, *tractability*. Many of the targets of scientific investigation are complex and only partially understood. This limits our capacity to represent them accurately. Because of this, various aspects of our representation will be based on guesswork. Moreover, the complexity of the target system often outruns our cognitive and computational capacities. In order for human understanding to interact with such vast, interwoven intricacies we require frequent black-boxing, distortions, and so forth. Weisberg claims that such "idealization takes place with the expectation of future de-idealization and more accurate representation" (641). That is, given that our purpose is accurate representation, we should view such idealizations as performing a heuristic role in support of accuracy. And so, as our understanding, knowledge, and power increase, we should expect such idealizations to be abandoned. As Bill Wimsatt has put it, false, idealized models can help us build truer theories (Wimsatt 1987, 2007). Moreover, idealization can aid in pedagogi-cal goals. Distorting and simplifying the facts can increase uptake of infor-mation and skills (Walsh & Currie 2015).

However, we should be cautious of thinking that de-idealization is always on the cards. Angela Potochnik (forthcoming) argues that the aim of much of science is not a complete or veridical representation of the world, but rather a kind of *humanized* representation of the world. That is, science aims for what she calls "understanding"—a combination of a representation's approximate truth, and its capacity to produce appropriate psychological states in human agents. If so, then we shouldn't expect de-idealization to be a priority (and she argues that it is not). More generally, if we accept pluralism about epistemic goods—if we think science is in the business of generating a wide range of different kinds of outputs for different kinds of purposes—then the veridical sacrifices that models make could very well be in order to achieve other kinds of goals. I think reflection on historical science should motivate this kind of view (I will discuss this more fully in chapter 13), and thus should underwrite these kinds of expectations about idealization. Note that the next two roles for idealization also reflect this kind of pluralism.

The second role for idealization that Weisberg identifies is establishing *explanatory salience*. I have previously discussed the idea that explanation sometimes requires omission. A good explanation of a phenomenon is not necessarily its complete description. Rather, a good explanation distin-guishes the target from the relevant contrasts, or identifies the salient dif-ference makers, or otherwise represents the explanatorily relevant aspects of the case at hand.[5] Idealization can aid us in this. Weisberg identifies what he calls *minimal models*. "A minimal model contains only those factors that make a difference to the occurrence and essential character of the phenom-enon in question" (641).

While the advance of computational and theoretical power can lead to the redundancy of idealizations made for tractability, minimal mod-els are insulated from this by virtue of isolating explanatorily privileged factors. Assuming the model gets those factors right, then no matter how fine-grained our information gathering, and how powerful our comput-ers, it will still remain a good explanation of the target. It is worth point-ing out that idealizations for the purpose of explanation are not restricted to models. This is a fairly general feature of explanation. Insofar as an explanation isolates the causal dynamics of a system's behavior in a way salient to relevant context, the explanation succeeds and added accuracy—de-idealization—is unnecessary.

Third, *conflicting desiderata*. You can't always get what you want. In some circumstances it might be impossible to construct a model that is both minimal, in the sense discussed above, and precise (see Matthewson & Weisberg 2009 for more on trade-offs in model building). A single representation may be an inadequate tool for meeting several conflicting desiderata. In such circumstances, "multiple models," with different features tracking different representational goals, might be constructed in order to navigate that tension. The use of multiple models also aids in predictive purposes when dealing with complex systems. When various models with divergent mechanics converge on similar results about the same system, a kind of robustness emerges, a sort of theoretical version of consilience (Calcott 2011; Elliott-Graves 2016; Parker 2011; Weisberg 2006).

There is a common theme to Weisberg's categories, and discussion of modeling and idealization in general. They are understood as reactions to the messiness and complexity of the world, and the plurality of scientific goals. Introducing distortions to scientific representations is a way of rendering our theories accessible to computation and human understanding, targeting the relevant aspects of a system for explanatory purposes, navigating conflicting desiderata, and better predicting the behavior of labyrinthine phenomena. Although historical scientists also examine complex targets, toward a multitude of goals, reflection on the use of surrogate experiments reveals another purpose, which deepens our understanding of historical investigations. Let's turn to this now.

Some idealizations in historical science have a different character from those Weisberg identifies. These differ from minimal models. Although they typically aim to represent minimal causal detail—to capture the essential causal dynamics of the system—their function is not explaining phenomena via isolating the most salient causal factors of the target in question. Rather, they are interested in inferring new facts about the past. There is an important similarity, however: Many simulations in historical science attempt to isolate and represent the relevant causal details to some level of grain. They come apart by virtue of what determines the appropriate grain. Instead of explanatory salience, as we will see, grain is set by the requirements of investigative scaffolds. The cases I am interested in depart from idealizations for tractability in that they are not heuristic responses to complexity, but rather epistemic tools that compensate for missing data. Historical scientists, then, sometimes idealize in response to missing data, for the purposes of inferring facts about the past.[6]

As I have said, idealization is not restricted to modeling. I will now discuss another class of scientific objects that I have called "inferential tools" (see Currie 2015a). Although these are highly idealized, this is not primarily for representational purposes, but rather in order to facilitate inference. I will draw on these to argue that the point of idealizations in historical science is often to achieve the *right grain* to test between hypotheses.

Wilkinson et al. (2012) blame the elevated temperatures of the Mesozoic on prodigious increases in atmospheric methane gas due to sauropod digestive processes. That is, they hypothesize that high temperatures during that time were fuelled by sauropod greenhouse gas emissions. To do this, they estimated both the amount of methane per tonne of sauropod biomass, and global sauropod biomass. These, in combination, provide an estimate of global sauropod greenhouse gas production. Their results are comparable to current anthropogenic gas output, and they claim that the production could account for raised temperatures during the Jurassic and Cretaceous.

Van Loon (2012) turns a skeptical eye on Wilkinson et al.'s argument. He questions both the methods for determining the individual methane output of sauropods, and their global biomass. To work out this latter value, Wilkinson et al. need to estimate both the mass of an individual sauropod, and total sauropod population. Let's focus on the latter estimate: How many sauropods were there?

Wilkinson et al. base their calculations on an average of ten 20,000 kg sauropods per km^2. This estimate of population density is from Farlow et al.'s (2010) investigation of the Morrison Formation in the western United States. Farlow et al. estimate population density by considering (1) the carrying capacity of the landscape, (2) average sauropod body mass, and (3) their metabolic energy needs. It is surely false that those three factors alone determine population density in sauropods. Moreover, Wilkinson et al.'s co-optation of Farlow et al.'s tool involves idealization. The conditions at the Morrison Formation are taken to reflect worldwide conditions. As we saw in chapter two estimating the energy produced by Mesozoic flora (and thus the carrying capacity of the landscape), the body mass of sauropods, and their metabolism, is exceedingly difficult.

Wilkinson et al. make their argument using something that isn't quite a model, but is certainly idealized. Let's call it an "inference tool." These depart from models insofar as their purpose is not so much to *represent* populations, but rather to *infer* their properties. This certainly is not a clean distinction. Many highly model-like tools, such as Hyde et al.'s simulation, are also used to make inferences about the past. However, they do so slightly differently. Hyde et al. construct a model system that is intended to bear some representational relationship to their target system. It is by virtue of the former representing the latter that results may be transferred from one to another. In contrast, inference tools are not so explicitly representational. They are more directly concerned with inferring past facts. Their success turns on whether they output approximately trustworthy inferences given empirical input. That is, a good inference tool isn't such because it is a good representation. Instead, to some degree of approximation, it generates corrects inferences.

For present purposes, however, this contrast is not too important, since we are concerned with the role that idealization plays in the historical sciences (and it is similar for both simulations and inference tools). I am going to argue that the role that idealization plays is ensuring *relevance*. As we will see, historical hypotheses are at a certain grain, a certain level of description, and idealizations in inference tools and models are justified by virtue of producing information at the relevant grain. This will become clear if we consider van Loon's objections to Williams et al.

Van Loon's objection is, in essence, that the model is too simple and (even with plausible values for carrying capacity, body mass, and energy) will overestimate sauropod numbers. However, the objection is not about precision per se, but about providing the right level of grain required to test Wilkinson et al.'s claim that sauropod-fueled elevations in greenhouse gas would rival current anthropogenic levels. I will summarize these objections, then use them to show that historical scientists sometimes use idealizations to aid inference.

Wilkinson et al.'s model doesn't account for predation pressure. Van Loon points to (in my eyes) implausible evidence of velociraptor flocks hunting sauropods.[7] More persuasively, he highlights the role of parasite-borne infection and disease, which lower actual populations from their theoretical maxima.[8] As we have seen, Gao et al. (2012) and Poinar (2012) discuss fossil finds of remarkable flealike insects dating to both the Jurassic and Cretaceous. In contrast with true fleas, they were significantly larger (17–22 mm compared to 8mm at absolute maximum for modern fleas), had larger mouths and (fortunately, for my capacity to sleep at night) were not adapted to jumping. It is plausible that individual sauropods supported thriving ecosystems, and that infection and disease from these parasites depressed sauropod numbers.

Van Loon also questions the inference from the population density of the Morrison Formation to global population density. The Morrison Formation included shallow marine areas and fluvial plains, zones that often support high population densities. This undermines Wilkinson et al.'s extrapolation from the Morrison Formation to global km² averages. Moreover, it is plausible that (at least some) sauropods were migratory (although see McNab 2009) and lived in herds. If so, populations would be patchily distributed, and extrapolation from density at a single location would be highly unstable. A further factor is food availability. Sauropods needed prodigious fuel: Their environmental impact almost certainly constrained their population size. Ten average sized sauropods per km², according to van Loon, "would then have consumed the complete available lush vegetation in 16 years" (144). This is too fast for flora to recover, even if plant growth is accelerated by higher atmospheric CO_2. Finally, vegetation itself is patchily distributed, and so inferring vegetative productivity from one location to global averages is unstable.

As we have seen, van Loon offers two kinds of objections. The first are problems with data collection, and inferences from it, such as inferring from the carrying capacity of the Morrison Formation to the globe. The second, which concerns us here, is the failure to include variables in the model that make a relevant difference to population size. This second objection does not target the fact that the inferential tool is idealized, but that it is *badly* idealized, given the precision required for Wilkinson et al.'s hypothesis. Global population size is not a function of carrying capacity, body mass and energy requirements as (1) predation and parasitism take their toll, (2) environments are patchy, (3) animals may be migratory, and (4) external disruptions (floods, storms, bad seasons, and so on) cause fluctuations in population densities. These objections rely on the idea that extra variables, not represented in the model, make a difference to actual population size.

And, moreover, these variables make a difference to such an extent that the inference tool is simply uninformative of Wilkinson et al.'s hypothesis. It could be that for other (presumably coarser-grained) purposes the inflation would not be so problematic. This departs from the objections we might see in meteorological estimations, where real-world feedback allows models to be tweaked. The combination of this feedback and the use of multiple models make predictive simulations quite different from retrodictive ones. In many cases, meteorologists have access to feedback that allows them to tweak their models, and thus may take a more instrumentalist stance to the representational fidelity of their work. A meteorologist predicting the weather need not be so concerned with whether her model represents the causal dynamics of weather systems. Rather, she can focus on whether it provides true predictions. In historical science we seek inferential tools that include only those causal details influencing the target system to the level of approximation required to support or undermine the hypotheses that concern us.

So, although the cases I am interested in are not minimal models, since their idealizations and omissions are not primarily the provision of explanations, they are similar to such models insofar as both aim to represent the central difference makers of the target system. Moreover, the difference makers matter relative to certain variances. Historical hypotheses rarely require (or get) high-precision tests (Cleland 2013): Wilkinson et al.'s hypothesis doesn't require a close match between sauropod-produced greenhouse gas and contemporary human output. They rather need to be within a relevant range. That is, Wilkinson et al. want to compare a sauropod-driven greenhouse effect to current anthropogenic production. To do this, they needn't know sauropod output particularly precisely. By using an idealized inference tool, they attempt to achieve the level of grain appropriate to that task, given the information we have. Part of van Loon's objection just is that they fail to do so. The approximate nature of such claims helps identify which difference makers are relevant. Van Loon's objection is not quite that Wilkinson et al.'s model is insufficiently precise. Rather, it fails to include the relevant causal details required to be relevant to the hypothesis in question.

So, in addition to the various features that idealization plays in science, we should add another: producing evidence at the right grain for evidential relevance. That might be all well and good, but what is so interesting about adding yet another role to an already long list? We know very well that idealization is (in Potochnik's terms) rampant in science, so cataloguing these differences isn't all that exciting. Happily, this role matters for present purposes for two further reasons. First, it allows me to illustrate and explain a feature of historical inquiry that will matter for pessimism about historical science: investigative scaffolding. Second, it will provide insight into de-idealization.

10.3 Investigative Scaffolding

If I have seen further it is by standing on ye sholders of giants.
—Isaac Newton, in a letter to Robert Hooke, February 1676

A scaffold, in engineering, is a structure that exists for the purpose of allowing other structures to be built. In constructing a fort using chairs and blankets, some chairs may be initially load bearing, but they can be removed once other chairs are added and take up the strain. Although the completed structure is self-sustaining (assuming this is a well-constructed fort), the temporarily load-bearing chairs are necessary to get from a pile of blankets to the finished product. Scaffolding can often be understood in terms of path dependence, which we met in chapter 8. There is a limited number of ways to successfully construct a pillow fort (particularly when there are constraints in terms of time, material, or funds!) and success often requires a particular sequence of actions. The scaffolding must be in place *before* we can erect the blankets. Continuing our reference to chapter 8, scaffolds are a way of talking about historicity's positive aspect. Some parts of possibility space are only accessible through a very particular set of actions. So far as I can ascertain, the world's longest blanket fort at time of writing was constructed by Lincoln Military Housing. It is 475 feet long and constructed of PVC pipe and, well, blankets. There are only so many ways to build such a structure, and the scaffolds are not only physical—you need a willing community, the available funds, and the engineering know-how.

"Scaffolding" is becoming a common notion in philosophy of science. Kim Sterelny has applied notions of scaffolding to hominid evolution (2003, 2012), and a recent volume (Caporael et al. 2014) continues this, considering human cognition,[9] culture, and evolution. Wimsatt's work has long applied notions of scaffolding to diverse phenomena (2007).

Recently, Chapman and Wylie apply it to archaeological evidence (2016), and Kirsten Walsh to Newtonian theorizing (under review). I shall apply a similar notion to the generation of scientific knowledge. Although I shall focus on historical science, I am fairly confident that similar claims could be made across the board. After all, what are the shoulders of giants if not scaffolds?

I have argued that idealizations in historical science sometimes serve to ensure that tests are at the relevant grain to distinguish between, or support, the hypotheses under consideration. If I am right that idealization sometimes functions to facilitate scaffolded historical inference, then, in some cases we should expect them to be abandoned—for more detail to be added—once finer grained hypotheses have been articulated. This is another way in which historical idealizations come apart from the examples Weisberg is interested in. More importantly, this provides insight into the scaffolded nature of scientific investigation. I am sure that scaffolded investigations are a ubiquitous feature of science,[10] but scaffolding is particularly relevant for this discussion, because of its role in pessimistic judgments of future scientific success, and because the piecemeal nature of historical inquiry makes scaffolding more blatant.

In my (2015a) I argue that investigation in paleobiology is scaffolded, that is, progress is made by first establishing a scaffold of relatively coarse-grained facts. In that context I was defending the tendency for paleobiologists to conflate various senses of function. For instance, paleobiologists at times do not clearly distinguish between claims about etiological function (i.e., trait x evolved for purpose y in organism z) and, say, activity function (i.e., trait x played role y in organism z). Paleontologists investigating the role of sauropod necks in increasing browsing efficiency rarely distinguish between different kinds of functional ascriptions. Compare, for instance, two hypotheses: (1) the necks were selected for browsing efficiency, that is, sauropods with longer necks outperformed their conspecifics, and thus left more long-necked offspring in the next generation; (2) the necks were in fact required for the viability of the specimens the paleobiologists examine. These functional hypotheses are empirically distinct as they cohere with different stories about the origin of the trait in question. The long necks could have been selected for browsing efficiency early in sauropod evolution but then were co-opted for other purposes; they could have been selected for a different reason (mate-selection perhaps) but then been co-opted for a further use in the actual specimen. My diagnosis of such conflation is that in order to discriminate between functional hypotheses, paleontologists must often reach an "investigative scaffold," that is, a rich enough knowledge base is required: "in order to discriminate between some functional hypotheses, a set of coarse hypotheses must already be on the table. Once these are established (or near enough) further tests become relevant ... not all ultimately relevant evidence is discernible prior to reaching a scaffold" (Currie, 2015a).

As I have mentioned, a pressing question about scientific idealization concerns when and why idealizations are abandoned. As we have seen, philosophers often argue that some idealizations are permanent.[11] For instance, if the point of idealizations is to navigate tensions between different, conflicting, desiderata, or if they pick out the relevant contrasts required for explanation, then idealizations are, in a sense, eternal. They are not a stopgap on the way to a better theory but are an end goal in themselves.

As we saw in the last section, there are similarities between the kinds of idealizations we see in historical science and the "minimal" and "multiple" (those that compensate for conflicting desiderata) models that Weisberg discusses. All aim for a particular "level," let's call it, or grain of representation. They differ in terms of what governs the correct level. For minimal models success is measured in terms of explanatory adequacy. That is, a model is idealized correctly when it succeeds at generating the explanation we are after. For multiple models success is measured in terms of whatever desiderata are in play, say, precision and generality. A successful model finds the right balance with respect to the aims of inquiry. How about the idealizations I am concerned with? I will argue their success conditions turn on the requirements of scaffolding, and that this tells us something about de-idealization.

Both Wilkinson et al. and Hyde's tools predominately aim to test hypotheses. Hypotheses are never so specific as to pick out a single way that things could be, rather, a range of possibilities cohere with a hypothesis. The truth conditions of a hypothesis, then, pick out a set of possible worlds, not a single world. Let me explain. Consider the Snowball Earth hypothesis, as originally presented in chapter 2. The hypothesis claimed that the cause of Neoproterozoic glaciation was a feedback loop triggered by the clustering of landmasses in the tropics. Having more land in the tropics increases the absorption of CO_2 (thus thinning the atmosphere), and also global albedo, because land reflects more sunlight away from the earth than water. The effect of equatorial landmasses, then, is a colder globe. Decreases in temperature would be met by increases in icepack. Because icepack also has higher albedo than water, as total icepack cover increases, so too does the amount of sunlight deflected from the Earth. Further decreases in temperature due to plunging albedo will further increase ice cover, until even the tropics are left glaciated. This scenario coheres with a wide range of possibilities. Most strikingly, both snowball and slushball scenarios count, it seems to me, as variations on Kirschvink's original theory. And so, when one is testing a hypothesis, one is very rarely narrowing possibility to a small number of scenarios. Especially in historical cases, the modal range of hypotheses remains quite wide. As we will see, this aspect is important for understanding investigative scaffolding and de-idealization.

We could think about the content of such hypotheses as disjunctions of the various possibilia that cohere with them. For instance, we might say that the original Snowball Earth proposal implicitly consisted in (at least) both snowball and slushball scenarios. It may be that Kirschvink himself did not state his hypothesis in these terms: Indeed, he may have specified the scenario more finely. The point, however, is that much of the evidence for snowballing cannot distinguish between those finer-grained hypotheses. For instance, the dolostone caps are important evidence for the theory, since their presence tells us that the turnaround from an icehouse to more normal conditions was rather sudden. And this is just what Kirschvink's theory predicts. However, dolostone cap evidence speaks to both snowball and slushball scenarios. If we wish to distinguish between these hypotheses, the dolostone caps alone only provide local underdetermination.

However, one person's underdetermination is another's scaffold. Once the Snowball Earth hypothesis gained sufficient traction—sufficient support and understanding—scientists began looking for ways to further distinguish between variations of the hypothesis. As we have seen, this was done in multiple ways. Let's look at two examples.

One way in which the Snowball Earth hypothesis has become more complex since the 1990s is a reconceptualization of the explananda due to a finer-grained understanding of the traces. The dolostone caps are marked by ^{13}C depleted rocks directly before the event, and ^{13}C rich rocks directly after. It is recognized that the increase in ^{13}C after the event is explained by the Snowball Earth hypothesis. ^{13}C is an isotope of carbon whose proportion tracks the amount of atmospheric carbon during the rock's formation. As we have seen, the snowball contains the seeds of its own demise. Because Earth's carbon output (volcanoes and such) would be unaffected by the snowball, but Earth's carbon absorption (most obviously, biotic activity) would be cut off, we should expect a massive increase in atmospheric carbon. This increase in atmospheric greenhouse gas would cut off the feedback loop between increasing icepack and temperature, and thus ensure a fast warming. However, landmass clustering in the tropics shouldn't lead us to expect such a sharp drop-off in atmospheric greenhouse gas. What could explain the suddenness of the biotic collapse? To answer this, it has been suggested that the Earth's atmosphere had a different composition than it does now—specifically, a "methane greenhouse," wherein methane replaces carbon dioxide as the main greenhouse gas. Methane is significantly less stable than carbon dioxide, and so it is much more liable to collapse. On this view, the increases in CO_2 absorption caused by continental clustering was enough to upset the delicate, methane-based, atmosphere, causing a much faster collapse than would be expected based on today's more resilient atmosphere.

And so, reevaluation of trace evidence allows us to distinguish between variations in the more general hypothesis, and search for new tests. This reevaluation occurred against the backdrop of that more general hypothesis (snowballing or slushballing) *already being relatively well supported*. That is, it provided a scaffold for further investigation. In a sense, the previous coarse-grained hypothesis played a role analogous to that played by the temporarily load-bearing chair in our blanket fort. It played an essential role, but could be discarded once that role was discharged.

Here is my second example. As we saw in chapter 6, consideration of the relationship between Snowball Earth and the Cambrian explosion led to a reevaluation of the nature of the former event. Roughly, because early life survived the Neoproterozoic, the snowballs can't have been so extreme so as to wipe out these early beginnings. This thought motivated Hyde et al.'s simulations—they were interested in testing whether less extreme snowballs were likely. Their simulation gave us reason to expect slushballs rather than snowballs. Later simulations are more complex still, adding finer-grained details about the specifics of Rodinian geographical breakup and basalt flood emplacement (see Donnadieu et al. 2004, for example). Again, we see investigation carried out from a scaffold. We are now in a position to characterize this notion abstractly:

investigation is *scaffolded* when the following conditions are met. Consider two fine-grained hypotheses, H_1 and H_2. In order for the information required to empirically distinguish between H_1 and H_2 to be relevant, the coarser-grained disjunction of those hypotheses, H_1 or H_2, must be relatively well understood and established. That is, only once the disjunction of H_1 and H_2 is relatively established can the individual hypotheses be evidentially disambiguated.[12]

I understand "fine" and "coarse" grains in terms of modal breadth—how many possible circumstances the hypothesis in question coheres with. The basic Snowball Earth hypothesis is quite coarse-grained: It coheres with snowball and slushball, as well as CO_2-greenhouse and methane-greenhouse scenarios. In order to distinguish further between these, scientists first require sufficient knowledge and understanding of the dynamics common to the set. In scaffolded investigations, then, the evidential relevance of some information can only be discerned once enough information about the coarser hypothesis has been gathered. I will discuss how it explains both idealization and de-idealization in some historical investigations before drawing a moral about pessimism.

Consider our epistemic situation prior to reaching a scaffold. To a greater or lesser extent, scientists will be unaware of the modal range of the hypothesis they are testing, or what information will be available to distinguish between variants of the hypothesis once it is better understood. Extremely fine-grained studies, then, are simply not helpful. Going with the abstract characterization above, when we are attempting to test between the disjunction of H_1 or H_2 and *some other* (mutually exclusive) hypothesis (call it H_3), we want to be able to empirically distinguish between the sets of possibilia indicated by H_1 and H_2 and the set of possibilia picked out by H_3. Roughly, how much detail we want in a test depends upon the hypotheses we are attempting to distinguish between. In scaffolding cases, once a coarse hypothesis is established (say, we have good reason to prefer the disjunction of H_1 and H_2 over H_3), and our grasp on the possibilia cohering with that hypothesis is sufficient, then we can begin to disambiguate them. Indeed, in some situations new evidence may lead us to carve up the modal space differently. Although continental clustering has been a steady presence in discussion of Snowball Earth, it is possible that at some point in the future that aspect could be abandoned, while other aspects of the explanation (say, the methane-based atmosphere) could be retained. New empirical tests, then, sometimes do not simply pick between disjuncts included in a coarse-grained hypothesis but can lead us to alter coarse-grained hypotheses, as some aspects are retained, some added, and others abandoned.

Idealization, I take it, aids in getting the grain right. Consider van Loon's objections to Wilkinson et al.'s proposal. Wilkinson et al. argued that the methane output of sauropods would be sufficient to increase greenhouse gas levels to something comparable to modern-day levels, and moreover account for the increased temperatures in the mid- to late Mesozoic. Some of van Loon's objections—for instance, those that accuse the inference tool of missing depressed sauropod numbers due to parasitism—do not target the imprecision of the tool per se. Rather, van Loon targets the tool's inferences in terms of getting the estimate to within a reasonable range, given the kind of hypothesis they are trying to test. The demand for de-idealization is driven by the requirements of the hypotheses. We should expect, then, de-idealization to occur once scaffolds have been reached. And indeed we see this in the increasing sophistication of the simulations used to examine Snowball Earth hypotheses. As scaffolds are reached and, for instance, we are able to discern further subtleties within hypotheses, a demand for more fine-grained evidence, and fine-grained tests, is the result. In the face of such demands, where possible, idealizations are abandoned.

We see, then, a surprising role for idealization. Its point is to ensure that studies hit the right level of grain to test between the relevant hypotheses. It actively helps us gain empirical access to the world. Although this is a surprise, why idealization has such a capacity is clear: Given the scaffolded nature of investigation, idealization is necessary to generate information targeted at the right level to gain a scaffold.

We have seen that historical scientists use surrogate experiments in order to further their reach into the past, and that this reveals new insights into the purpose and nature of idealizations and modeling in science. I suspect this lesson carries over to other sciences as well. It is time, however, to return to my main purpose: understanding the attitude we ought to take toward historical investigation in unlucky circumstances. Recall one of the propositions underlying pessimism:

Historical scientists are unlikely to uncover further evidence.

This is an empirical bet about the future success of science, something we will discuss in more detail in the next chapter. Now, note that in making a bet, it is wise to take one's own epistemic position into account. As we have seen, prior to reaching scaffolds, scientists often cannot tell what evidence will be available. From that perspective, we typically don't know which new hypotheses can be disambiguated, and what kinds of evidence might be generated to distinguish between them. Scaffolding generates epistemic opacity about the future.

More than generating opacity, scaffolding also generates a bias toward pessimism. If we judge the ultimate evidence we might have regarding some past event, process, or entity, from our current standpoint, we are likely to underestimate. This is because, from a scaffold, evidence is far more likely to increase or stay the same than it is to decrease.[13] If that's right, then the above claim looks doubly foolhardy. Not only are we in a bad epistemic situation to judge future success, but we are also likely to be biased toward pessimistic judgments.

I have argued that historical scientists use highly idealized objects, such as surrogate experiments and inference tools, to extend their reach into the past. In such cases, idealization plays an active role in the generation of knowledge. It ensures that the information we draw is relevant—specifically, at the right level of grain—to the hypotheses we are testing. Idealization aids in establishing investigative scaffolds. These are well-supported bodies of knowledge that form a basis of further, often more detailed, investigations. In such cases, de-idealization occurs when scaffolds are reached, or if scientists have erred in regards to the appropriate level of grain, or got the causal facts wrong. Moreover, idealization in historical science is not a reaction to complexity but to gappy and faint data sets. Their role is evidential and compensatory. Finally, the scaffolded nature of historical investigation has something to tell us about pessimism. We are not in a good position to tell whether historical scientists will uncover further evidence, and indeed may be biased toward pessimism. It is time to carry these lessons, and those from earlier chapters, to chapter 11, where we can draw our main conclusions.

11 Optimism, Speculation, and the Future of the Past

Absent a time machine, we are trapped in the present and must rely on present traces to learn about the past.
—Elliott Sober and Mike Steel, "Time and Knowability in Evolutionary Processes"

Over the last five chapters, I have argued, by developing what I take to be an expansive view of the epistemic resources available to historical scientists, that we do not need a time machine to go beyond present traces. I began by discussing trace-based evidence, that is, evidence derived from contemporary observations linked to the past via midrange theory, the window to the past provided by causal ancestry. In addition to traces, I have identified other sources of knowledge. In chapter 6 we saw that historical scientists also rely on dependency relationships between noncontemporary events and entities: They expand their reach into the past by conducting "coherency" tests. In chapters 7 and 8 I discussed the role of analogous evidence. Here, evidential relevance to a past target is not generated by causal ancestry but by virtue of being an instance of, or having been formed by, the same process or system type as the target. I argued that historical scientists often explain past events in terms of fragile systems: exception-ridden but nonetheless explanatory regularities. Analogues grant epistemic traction to our investigations of fragile systems. And, as we saw in chapters 9 and 10, we are not restricted to the analogues nature provides; we can also construct them. In particular, surrogate experiments and inference tools are used to generate further evidence about the past. The historical sciences have rich epistemic resources.

The next three chapters draw implications and lessons from my view of historical reconstruction. Here, I will turn to optimism and pessimism—effectively tying together my main argument. In the next chapter, we will shift to a more applied key and ask after the instrumental value of historical science, and the kinds of funding strategies that should promote it. Finally, in the postscript, I will take a step back and consider to what extent what we have learned should affect some of the questions that have traditionally diverted philosophers of science.

You may note that this book lacks a proper concluding chapter. The next three chapters involve significant summarizing and packaging of the book's view and main argument, and in light of this a chapter focused on yet more of this seemed unnecessary. Metaphorically speaking, as well, there might be good reason to lack a conclusion: I am much more interested in this book's motivating further research rather than wanting to have anything like the last say. In light of this, tying it off with a concluding bow seemed simply too final, and insufficiently open-ended.

In this chapter, I have three tasks. First, I will begin by drawing together and reiterating how these points undermine the considerations of chapter 5 in favor of pessimism. Second, I shall discuss various optimistic theses and to what extent I have supported them. Finally, I shall draw a methodological lesson for historical science: Progress is driven by speculation. That is, the best approach to increasing our knowledge of the past is to put forward and test hypotheses that appear to outrun our available evidence.

11.1 Pessimism?

I am largely interested in establishing a positive thesis: We should be optimistic about our capacity to uncover the past. Part of establishing this thesis involves undermining pessimism, the view that much of what we would like to know about the past will remain unknown. Here, I want to run over my argument against pessimism. To what extent have I undermined reasons to adopt an attitude of doubt toward our capacity to uncover many, interesting, facts about the past? As we saw in the introduction, whether or not we adopt a pessimistic attitude matters. First, it underlies arguments for antirealism about the past such as Turner's. Second, it is related to certain conservative methodological themes in the historical sciences. To take one example, part of the explanation of paleontology's lack of theoretical input into biological theory before the 1970s was the perceived poor quality of the fossil record (see Sepkoski & Ruse 2009). Third, as we will see in the postscript, the attitude we adopt toward historical science has upshots for the instrumental value it has, and how it should proceed.

I have not been particularly explicit about what a pessimist thinks. Rather, I have highlighted a set of epistemic situations—unlucky circumstances—and considered which epistemic resources might be available in these circumstances. In essence, unlucky circumstances are epistemic situations that look inauspicious by the ripple model. In such circumstances, the past event in question is the following:

Of *low dispersal*: Its expected causal descendants are few and homogenous;
Gappy: Potential trace sets are incomplete;
Faint: Both locating traces and linking them to the past are difficult tasks.

Prima facie, If we are going to be pessimistic about any set of circumstances in historical reconstruction, it is these. We should expect the resulting trace sets to be homogenous. If so, we have limited lines of evidence, and thus are denied the advantage of consilience. We should expect the trace sets to be incomplete. If so, this limits the amount of evidence we have: Inferences from present traces to the past will be shaky. We should expect our capacity to interpret traces, to draw on midrange theory to connect the past and present, to be limited. If so, our theories will be highly unstable.

In chapter 5, I articulated three claims that, if true, would drive pessimism about unlucky circumstances. These were as follows:

Our available evidence about the past is limited to traces.
Historical scientists cannot manufacture evidence.
We are unlikely to uncover further traces.

If these claims were true, pessimism about unlucky circumstances would be unavoidable. This is simply because under those circumstances trace evidence is poor, and the claims restrict our capacity to mitigate such scarcity.

Happily, I have shown that each claim is unjustified. Our available evidence is not limited to traces. We can utilize analogies, dependencies between past variables, and surrogate experiments. Historical scientists can manufacture evidence. By constructing simulations and models, they build smoking guns and utilize them to discriminate between past hypotheses. We are likely to find more traces. As midrange theory develops we should expect further traces to be uncovered via refinement, and to learn surprising new things about the conditions under which the past's footprint is preserved. Moreover, historical investigation is scaffolded, and evidential quality and quantity often cannot be ascertained prior to reaching a scaffold. As our evidential pool is likely to either increase or stay the same, it is plausible that we will understate the total evidence ultimately available, and so will be biased toward pessimism.

Even in unlucky circumstances, it seems, we do not have a positive argument for pessimism. Presumably this has consequences for optimism. After all, if there is reason for hope when traces are not pulling their weight, we should be confident in an inquiry's epistemic fate when they are. That is, if unlucky circumstances are the best case for pessimism, and it fails there, then this provides at least some reason to adopt an optimistic attitude about historical reconstruction more generally. However, why should we barrack for either option? Refusing to bet on pessimism doesn't mean I should put my money on optimism. Moreover, surely there is a whole range of different ways of being an optimist or a pessimist. I haven't yet met my philosophical duty of rigor in this respect. In the next section, I will shift to optimism and discharge said duty. The various distinctions I am about to make may well have pessimistic analogues, but given my sunny disposition, I will focus on the half-full glass.

11.2 Betting on the Future of the Past

I have left the exact theses that "optimists" and "pessimists" might commit to purposefully ambiguous. In part this was because my main argumentative focus has been establishing an expansive picture of the epistemic resources available to historical science. It is now time to be a bit more precise. In this context, "pessimism" and "optimism" denote *predictive epistemic attitudes* about the occurrence or otherwise of some event. We could think about these attitudes in terms of credence: If I am optimistic that my banjo playing will improve with practice, then I have a relatively high credence in that proposition. Let's use the term "be willing to bet" as a shorthand for "having a sufficiently high credence."[1] As we saw in the introduction, this departs somewhat from the folk meaning of the terms "optimism" and "pessimism," which turn more on whether we emphasize the good or bad aspects of a situation, which is not an epistemic question (those who see the cup as half full and those who see the cup as half empty surely do not have an empirical disagreement). Moreover, I am going to stay silent on what is precisely required for "sufficiency." There are several reasons for this. It is extremely tricky to work out what an account of "sufficiency" would look like in this context, in part because I suspect that our interests might play a large role in determining such questions, and because I am suspicious that the highly disparate circumstances, methods, and evidence streams across historical science would be amenable to a precise notion of "sufficiency." If I am right about confluence, articulating a general standard of evidence would be misleading. Indeed, skepticism about general, context-insensitive standards has been a major theme of this book.

My strategy is to disambiguate six versions of "optimism," and consider the extent to which my earlier considerations support, or at least clarify, what we should think of them. These are summarized in table 11.1, at the end of this section. But let's begin by clarifying the nature of my argument. I am not making an induction from cases of scientific success to general scientific success. As I will claim regarding arguments in favor of realism or antirealism in chapter 13, such a method is unlikely to provide useful results. This is in part because it is difficult to work out both how many cases of success would be required, and how they should be weighed against apparent failures. As Turner says, "If a few examples of underdetermination do not support a generally pessimistic view of the epistemology of historical science, then neither do examples of success support a more sanguine view" (2016, 11).

Rather, I argue that historical scientists have a wider set of epistemic resources than we hitherto thought. This claim is at base *contrastive*, and this matters for identifying fruitful research paths. It is this realization that grounds a set of optimistic claims. That is, insofar as I support an optimistic view of historical science, it is by virtue of the epistemic resources I think they possess (as opposed to a history of success). Let's turn to notions of optimism now.

First, optimism could be *relative* to sets of epistemic resources. That is, I could be more optimistic given one set of epistemic resources than another. In other words, my credence in the future success of one enquiry could be higher than my credence in a different enquiry. Let's call this "optimism$_R$."

Optimism$_R$: I'd be more willing to bet on scientists with epistemic resources R than scientists with epistemic resources R^*

This is, I think, a notion of optimism clearly supported by my position. As we have seen, most methodological reflections about the historical sciences have adopted some kind of trace-centrism (call this R^*). On such views, the epistemic situations of historical science are captured by the ripple model of evidence. I have argued that historical scientists have a much wider range of epistemic resources: In addition to traces, they utilize dependency relations between variables in the past, as well as use and construct surrogative evidence (call this set of epistemic resources R). Given that R^* is a subset of R—I haven't argued that traces *are not* evidence—then surely optimism$_R$ has been established in favor of my view in contrast to more conservative accounts. If my inflationary picture of historical science's epistemic resources convinces you, then you should feel *more* optimistic. However, a pessimist could grant my extended picture of available epistemic resources, and agree that optimism$_R$ is established in this sense, but still complain that this is insufficient to grant optimism of a more general sort. For example, they might still insist that there is reason to bet on the future failure of historical science.

Let's remind ourselves why it is important to make such bets. Shouldn't we just be quietists about the future success or otherwise of science? As Turner (2016) points out, this is a mistake. The most obvious reason is that scientists themselves must make such judgments: In deciding which direction to take their research, scientists make bets on which directions will yield dividends. In the introduction, I briefly discussed the enormous unprepared specimen room in the Tyrrell Museum. Paleontologists must decide which fossils should have priority in preparation—there are precious few skilled preparators and an embarrassment of riches waiting to be prepared. So, in the unprepared fossil room, bets must be made about which research avenues will be the most interesting and fruitful. This point holds more generally: Historical scientists must decide which research to conduct. And moreover, funding bodies must decide which research to encourage. Scientists are making decisions about the direction of research, and this turns on comparing options: Where should our limited epistemic resources be spent? This is an extremely complex question, but I think my discussion thus far has given us grounds for some basic considerations, which I will discuss in the next chapter. For now, I will focus on further disambiguating and critically assessing explicitly noncomparative notions of optimism. Note that I will cash this out in terms of "epistemic goods."

I have earlier committed to a kind of pluralism about the outputs of scientific investigation: They generate truth, understanding, predictions, prognostications, and so forth. By "epistemic goods," I simply mean some subset of those.

And so, on to my second kind of optimism. We could be optimistic about a particular token good, a *local* optimism. Let's call this "optimism$_L$."

Optimism$_L$: For some token past good, g, I'd be willing to bet that historical scientists will learn g.

Given the arguments thus far, it is plausible that optimism$_L$ is established for each particular good on a case-by-case basis. What did *Obdurodon tharalkooschild* eat? How did *Thylacoleo atrox* bite? Is Cope's rule more often a driven or passive trend? To establish optimism$_L$ for some good, we must examine the epistemic situation and resources at hand. I have given us a fair amount of guidance in this regard. In addition to the ripple model, we should consider other factors, such as the availability of analogues. It will be helpful to list these.

In chapter 5, I provided a set of considerations that together determine the *accessibility* of some past good via the relevant traces. Suitably updated by my expanded epistemic picture, this provides a schema for establishing (or failing to establish) optimism$_L$ about some past good. The ripple model of evidence tells us about accessibility in terms of traces. This turns on g's dispersal, that is, the number and heterogeneity of its causal ancestors; g's faintness, that is, the ease of its traces' discovery and the stability of relevant midrange theory; and g's gappiness, or how "complete" we should expect its subsequent trace set to be. We should also take into account features of g's dependence with traces. Specifically, we should consider the number of g's variables we expect to be minimally dependent on the modern world, and with past events we know well (embeddedness); we should consider the strength of dependencies; and consider how informative they should be. In addition to this picture, we should also consider our capacity to discover or construct analogues and the dependency relations holding between g and other past events. Even when these considerations lead us toward pessimism about some fact, there is still reason for caution. If, having considered the dependencies, the traces, and the surrogates, our learning of g still draws long odds, we must keep in mind that our knowledge of these factors is limited. Recall from the last chapter that the scaffolded nature of historical investigation makes epistemic fortunes opaque. That is, prior to reaching a scaffold, it is often unclear what evidence is ultimately available. Moreover, there is reason to think that, by merely considering our *current* epistemic situation, we will underestimate g's accessibility. Regardless, these considerations show that establishing pessimism about a particular fact is tricky. There are many positive considerations—routes to empirical traction on a past fact—and the optimist only needs some of these to be open.

Third, we could be optimistic about some *epistemic context*—that is, some combination of epistemic situation and resources. Instead of adopting an attitude about a particular fact, or all past facts, we might restrict our claim to a context. Let's call this "optimism$_C$."

Optimism$_C$: For some set of goods falling under some epistemic context $\{g_1, g_2, g_3 ..., g_n\}$, I'd be willing to bet that we will learn most or all of $\{g_1, g_2, g_3 ..., g_n\}$.

I take it that the ripple model highlights an epistemic circumstance that breeds optimism: past events that have non-gappy, non-faint, and dispersed causal descendants. If we find ourselves in such lucky circumstances, we should be reasonably optimistic about our chances of success. But I have argued that pessimism often fails in unlucky circumstances. Have I established optimism$_C$ in these cases as well? This depends. Not all unlucky circumstances are created equal. Some have bountiful analogues, or are amenable to surrogate experiments, or are highly embedded with past events that we know a lot about. In at least some cases, however, we should not expect much extra-trace evidence with which to mitigate the situation. Call these *unluckier circumstances* (I will discuss this in more detail below). So, I see a partial victory for optimism here—and certainly, the considerations herein have shed valuable light on how to approach such questions.

Fourth, we could articulate a general optimism. Let's call this "optimism$_G$."

Optimism$_G$: I'd be willing to bet that we will learn every past good.

Optimism$_G$ is an extremely risky proposition. Surely there will be many things about the past that we will never know; such blind optimism is a bit much! However, if I am right that there are grounds for optimism *even in unlucky circumstances*, I am inclined to think that I've gone some way to making this view look at least *less* reckless. As formulated, though, it should surely be rejected.

Why reject optimism$_G$? Here are two related problems. First, as Derek Turner has pointed out to me, there are a *vast number* of things we might want to know about the past. Surely among that vastness there will be goods deserving of pessimism. In light of such vastness, if we make no distinctions between various past goods, we are basically guaranteed to lose. Pick an inconsequential morning in Napoleon Bonaparte's childhood and ask what he had for breakfast. Or ask what your maternal great-great-grandmother's first words were. Surely the option space is just too wide, and the records just too inconsequential, for us to hope to know the answers to such questions. Second, this becomes particularly pressing when we consider what I will call the "answers breed questions" problem. Basically, any progress on some historical question breeds further, more difficult, questions. As Turner has put it, "whenever scientists settle one question in dramatic fashion, new, often finer-grained questions crop up, and the answers to these new questions may well be underdetermined" (2016, 11). If I am right about scaffolding, this should not be surprising: As relatively coarse hypotheses are established, scientists begin seeking ways of differentiating between more finely grained variants. For instance, say we discover good reason to prefer a slushball to a snowball in accounting for Neoproterozoic climate, why not then ask for further detail? Which areas of land and ocean were left ice-free? How big were the glaciers? And indeed, if it turns out that Napoleon had eggs for breakfast, how many did he have? How were they prepared? Did France's future emperor prefer sunny-side up? What was the calorie content? And so on. There is always, it seems, a finer grained question to ask, and surely at some point our epistemic well will run dry. In light of both the vastness of historical facts, and the idea that answers breed questions, optimism$_G$ is surely just *too* optimistic. However, if properly restricted, a general optimism can be saved.

We can save a general optimism by restricting it in one (or both) of two ways. We can restrict our range to "important" goods, and we can make it defeasible. As these are not mutually exclusive, I will present them together. I'm more confident about the latter, so let's call this "optimism$_{GD}$":

Optimism$_{GD}$: For any (important) past good, g, I'd be willing to bet that historical science will learn g—unless g has particular features that make its recovery unlikely.

Let's start with defeasibility. What kinds of features might g have in order to defuse optimism$_{GD}$? We already have the resources required to guide this question. For instance, imagine a case involving what I have called "unluckier" circumstances: Both trace evidence and non-trace evidence are unlikely to help. Perhaps Napoleon has left traces of his breakfasting habits: He may have written reflections on his childhood in which such details are mentioned. Or we might be able to zero in somewhat on Napoleon's breakfast using analogues: Presumably there are records of what the typical diet of a child in Corsica in the 1770s was like. If it turns out that the diet is fairly homogeneous, then we could have a fair basis for inference. If not, however—if the diet is heterogeneous enough, or if records were not kept, only minimal traction could be gained. It is also difficult to ascertain how simulation studies might help, although this could be a failure of imagination on my part. It looks as if Napoleon's breakfast could be "unluckier," and thus not fall within the set specified by optimism$_{GD}$.[2] When such conditions do not hold, however, it seems that we have good grounds for optimism.

Further, and I am less sure on this point, we could restrict ourselves to "important" facts. As we have seen, much historical science is concerned with large-scale, big-picture questions about the processes and events that shape the past. These targets are important partly by virtue of their causal impact on history's unfolding. These qualities, to some extent at least, are just the features that can ground optimism. That is, important facts will tend to be retrievable facts. Moreover, investigative scaffolding highlights the enormous effort required to uncover the past. It could be that Napoleon's breakfast would be retrievable, if only we turned enough of our epistemic capital to the question. Consider the immense effort required to scaffold. We must construct surrogate experiments and inference tools, develop midrange theory, hunt for traces, and so forth. In light of this, pragmatic rather than epistemic constraints could be doing the work: If much of the past simply isn't that interesting to us, why go to all of that trouble?[3]

At any rate, it seems as if we have some reason to prefer even quite general versions of optimism.

Finally, we might support optimism about general progress. Call this "optimism$_P$."

Optimism$_P$: Historical scientists will generally increase their pool of knowledge about the past, discovering many important facts.

Optimism$_P$ is, surely, established. The opportunistic methodology, and varied epistemic resources, of historical science has allowed us to make significant progress over the last century, and we should expect this to continue in a similar fashion (assuming we don't, as Carl Sagan was fond of caveating, destroy ourselves). Of course this optimism is not particularly ambitious, and not so satisfying for those of us wanting a dramatic upshot (optimism$_{GD}$ is a bit more exciting in this regard!). Happily, my epistemic picture provides a lesson about how to ensure optimism$_P$, that is, a lesson about how historical science progresses. Let's look at that now.

11.3   Empirically Grounded Speculation

But, as for us ... who can only arrive at the truth by crawling from probability to probability, it is not for us to pronounce so boldly against hypotheses.
—Emilie Du Châtelet

In this section I want to argue in favor of a particular kind of speculation. As we will see in the next chapter, this has consequences for how we should fund and support historical investigations. Consider this quote from archaeologist Holly Hayter:

The archaeological record is a very biased sample of material remains which have survived selective natural processes of preservation, erosion, disturbance and other human forces of destruction and removal ... These material traces of past behaviour are often ambiguous as to what they signify or give evidence of ... We look to models, analogies and other aids to help explain patterning in archaeological materials in an attempt to derive some sort of information about the cultural, social and political aspects of the people living on a site ... It is now emphasized that one, when making inferences, should go beyond the visible data and attempt to make speculations and cultural reconstructions. People would rather read a story book than a laundry list. (1994, 40–41)

Somewhat ironically, Hayter has deep reservations about going beyond visible data and is not speaking of the prospect approvingly in this quote. However, I think we should be approving of the practices she describes, and I think we should take it even further. A storybook is not only more satisfying than a laundry list, it is also *better science*.[4] Why? Because, for

Table 11.1
Various forms of optimism

Form of Optimism	Definition	Considerations
Optimism$_R$	I'd be more willing to bet on scientists with epistemic resources R than scientists with epistemic resources R*.	On my view, trace-based evidences are supplemented with other evidential sources (analogues, surrogate experiments, coherency tests). The increased epistemic resources give us reason to be more optimistic than on "trace-centric" views.
Optimism$_L$	For some token past good, g, I'd be willing to bet the historical scientists will learn g.	Establishing local optimism requires examining the particular context, I have argued for these relevant considerations: • g's dispersal, faintness and gappiness; • available g-analogues; • our capacity to construct surrogates of g; • how embedded, strong, and informative g's dependencies with traces and other past events are.
Optimism$_C$	For some set of facts falling under some epistemic contexts {g$_1$, g$_2$, g$_3$... g$_n$}, I'd be willing to bet that we will learn {g$_1$, g$_2$, g$_3$... g$_n$}.	The same set of criteria as in optimism$_L$ can be applied to epistemic contexts rather than particular facts.
Optimism$_G$	I'd be willing to bet that we will learn every past good.	This kind of "blind optimism" is surely too strong, especially in light of the "answers breed questions" problem.
Optimism$_{GD}$	For any (important) past fact, g, I'd be willing to bet that historical science will learn g—unless g has particular features that make its recovery unlikely.	Plausibility turns on how specific we can be about the "particular features" that are required. Scoring low on the four criteria from optimism$_L$ is a good option. We might also restrict g to "important" facts (see text). Overall, established.
Optimism$_P$	Historical scientists will generally increase their pool of knowledge about the past, discovering many important facts.	Established.

scientists in unlucky circumstances, progress is driven by speculation. To show this, I will need to first get clear on what I mean by "speculation," and then show how my previous discussions support it. In brief, I argue that going beyond simple description into the bold and the speculative is more fruitful, generates more opportunities for testing, unlocks new avenues of investigation, links our little pockets of knowledge to each other and is (let's face it) more satisfying (and fun).[5] This is in stark contrast to the kind of pessimism-driven conservatism about method in historical science we saw in the introduction, and often expressed by historical scientists in conversation.

And so, what is "empirically grounded speculation?" Let's start with speculation and then move to empirical grounding. Speculation, as I understand it, is the practice of making claims that go beyond available evidence by some relevant margin. Scientists often talk as if there is a clear dividing line between what Hayter calls "visible data" and "speculation." This must be an idealization: The support and reliability mustered by different hypotheses and observation reports differ both qualitatively and quantitatively. Some claims are extremely firm: The specimen of *Dreadnoughtus schrani* that Lacovara et al. examined was 26 meters long. Others are less so: Lacovara et al.'s specimen was osteologically immature—that is, it was still growing—at death. Others even less so: *D. schrani* was endothermic, say. There is no clear dividing line between "speculative" and "secure." However, we could think of a speculation "scale": The less certain a hypothesis, the more speculative.

A hypothesis, then, is *speculative* insofar as its support diverges from certainty, and is treated speculatively insofar as we take it seriously despite such divergence. This is naturally understood in broadly Bayesian terms. Let's think of support in terms of credence, falling between 0 and 1, where "1" is complete certainty, and "0" is complete uncertainty (that is, certainty that the proposition in question is false). One way of understanding speculation is as follows. A hypothesis' level of speculation is proportional to (what I'll call) a "conservatively rational" agent's attitude toward it. Such an agent will believe in, and act on, only what rationality forces her to. In a sense, she will minimize her belief commitments and, more important, what she will act upon on the basis of those beliefs. That is, she will minimize her epistemic risk. By comparing her credence in a hypothesis to how seriously it is taken in investigation, we can get a sense of how speculative the hypothesis is. For instance, if, given the evidence available, such an agent's credence is 0.5, but the hypothesis in question is treated as garnering some higher value, say 0.8, then the hypothesis is speculative to 0.3. Don't take this too seriously: I don't think, of course, that this can actually be *determined* for many actual cases—evidence is often not quantifiable, particularly in the contexts concerning us here. Further, credence attaches to individual agents, but we are concerned with scientific knowledge, which is often a social phenomenon. It is not obvious whether communities have "credence" or make assertions. Also, depending on the breed of Bayesian you are, and how much evidence has been collected, the agent's initial credence might be both arational and partly determine later credences. Moreover, I'm sure many philosophers will balk at claiming that a "conservatively rational" agent is in fact rational. Given that I will soon suggest that we would ultimately do better—generate more knowledge— than such an agent, calling her "rational" could be problematic. But again, I do not mean this particularly seriously. It is just a helpful way of thinking about speculation. The level of speculation of a hypothesis is, metaphorically speaking, the amount to which it is taken seriously in an investigation compared to how much a "conservatively rational" agent would. A hypothesis that has little support, but nonetheless gathers significant investigative attention, that is, is used as a base for further hypothesizing and a target for various tests, is highly speculative in these terms. Naturally, "safe" hypotheses—those which the rational agent would consent to—can still be fruitful, both warranting and generating further empirical tests. Because of the opacity of epistemic fate discussed in chapter 10, however, often fruitfulness and credential confidence will not correlate.

So, we could distinguish between two (non–mutually exclusive) forms of hypothesizing. Some hypotheses are justified on the grounds of their support—that is, whether we think they are true. Other hypotheses, speculative ones, are justified on their fruits, which I will discuss shortly.[6] Of course, there will be a spectrum here: Some hypotheses will be more or less speculative. It seems as if this latter kind of hypothesis can be formulated, and tested, and largely treated as being "on the table" without us necessarily adopting an attitude of belief toward it. This practice, I think, drives progress in historical science (and surely other sciences as well)—especially in the non-ideal "unlucky" circumstances where methodological omnivory rules.

Now, what about "empirical grounding"? By empirical grounding, I refer to the potential tests, and tools geared toward tests, generated by the hypothesis. This is fruitfulness of a particular sort, and to understand it we should return to the notion of investigative scaffolding. By investigative scaffolding, in order to extend our empirical reach, a set of hypotheses must already be relatively well established. The initial formulation of hypotheses, then, is an crucial part of investigation. It is only by this process that certain kinds of epistemic or empirical goods can be generated: For example, identifying potential smoking guns, articulating potential dependencies between the event we are concerned with and other events in the past, discovering and constructing analogues and other surrogates, and so forth. So, the "empirical grounding" of a speculative hypothesis should be understood as the extent to which it potentially generates these routes to knowledge.

To combine these notions, then, empirically grounded speculation involves formulating a hypothesis that (1) significantly outruns the available evidence, and (2) generates epistemic or empirical goods that increase epistemic traction. It is important to see that the two conjuncts are related. The kinds of epistemic goods we are interested in can be (or perhaps are best) generated only by hypotheses that outrun available evidence. Hunting for empirical traction—the scaffolds—is motivated by the speculative nature of the hypothesis. Moreover, as we saw in chapter 10, even the failure of such hypotheses can play an important role both in directly informing questions about the past and developing midrange theory, models, and other inferential tools applicable in other areas.

In historical science, empirically grounded speculation encourages the reaching and exceeding of scaffolds, the development of new techniques and theories, and generally drives our access to the past. We see this pattern again and again. Williams's speculation about Earth's obliquity and its climate during the Neoproterozoic motivated new modeling of Earth's rotation, its effect on climate, and the distribution of subsequent geological deposits. The idea that *Thylacosmilus atrox* is an extreme saber-tooth motivated Wroe et al.'s exquisite-corpse modeling and quantitative comparison with the placental *Smilodon*. Wilkinson et al.'s suggestion that sauropod methane output drove up Mesozoic temperatures led to van Loon's more careful considerations of the factors effecting worldwide sauropod populations (and moreover made reconstructions of the parasites preying on

such giants evidentially relevant). Kirschvink's basic, and highly speculative, story of Snowball Earth led to an extraordinary burst of research that
has ultimately provided a more sophisticated picture of the Neoproterozoic
climate, the causes of the Cambrian explosion, and some of the tools necessary for understanding climate at a general scale.

Let's go back to the very beginning. Pian et al. (2013) did not restrict
themselves to formulating well-supported hypotheses about *Obdurodon
tharalkooschild*. That their specimen was a certain size, that she was a member of the platypus clan, and that she was a new species were certainly
well established. However, they also suggested that she preyed upon vertebrates such as soft-shelled turtles and that the pattern she represents
might be unified with general marsupial size increases during the Australian Mesozoic. These two claims surely outrun Pian et al.'s evidence. If they
were conservative and interested in only asserting, describing, and taking
seriously hypotheses with strong evidence, then no such claims would be
made. Why do they make them, then? The answer, I take it, should be
clear. Hypotheses are justified not only on the basis of their support but
on their empirical grounding—the new empirical and epistemic fruits that
they bear. This is not a new idea. But perhaps this is an idea worth repeating, and certainly in the context of historical science where pessimism
often drives methodological conservatism. The "unjustified" anticipations
of Pian et al. may very well bear important epistemic and empirical fruit;
only time can tell.

The long and short of all of this is that the scaffolded, opportunistic, and omnivorous character of historical investigation means that
empirically grounded speculation is the way forward. If we must judge a
historical hypothesis, we should do so by balancing its support and its empirical grounding. That is, we should compare the epistemic risk we expose
ourselves to—the hypothesis could, of course, turn out to false—with the
potential benefits, the empirical grounding, that it brings to the table.

Faced with the gappiness and faintness of traces, historical scientists are
often tempted to hide their heads in the sands of naïve empiricism: Adopt a
careful, stark, and methodical approach; be cautious in application, projec-
tion, and assertion. This is a mistake. Especially when the going gets tough,
when circumstances are unlucky, historical science should be wild, messy,
and creative. Such a method does not eschew scientific testing or scientific

rigor, nor does it lead to a subjectivist free-for-all. Rather, it leads to richer, more robust—and often *true*—pictures of the past.

And so, although my arguments about the nature and method of historical reconstruction do not establish a kind of naïve, free-for-all optimism about our access to the deep past, it does support a wide range of optimistic theses. In the next two chapters, I will draw further lessons from this tempered optimism, and the reasons underlying it.

12 Promoting Success in Historical Science: The Why and the How

The majority of this book was developed and written in two places: Canberra, Australia's capital city, and Calgary, the largest city in the Canadian province of Alberta. Both Alberta and Australia are economically marked by mineral wealth, and this brings with it an attached ambiguity about the value of historical science. Paleontology's and geology's understandings of the processes that form Earth's minerals are essential for efficiently locating and extracting that bounty. Alberta's paleontological resources are almost second to none, as is its economic dependence on both conventional oil and gas, and "oil sands" (as they call them). This means that there is often a close connection between historical science and mining. In Canberra, my chats with earth scientists often touched on a tension between those trained geologists who went on to high-paying jobs for mining companies, those who remained behind to work in government (regulating said companies!), and those who managed to stay in the academy (competing for funding from said companies!). Under such conditions, it is hard not to reflect on the value of historical science. What, exactly, is knowledge of the deep past supposed to get us? Considering the economic importance of some aspects of historical science—those that are critical for unlocking Earth's Pandora's box of mineral wealth—what reasons are there to spend our resources on other aspects that, let's face it, are hardly likely to help us to ride out an economic downturn?

My aim is not to answer this question in any kind of satisfying way here. Tackling the why and how of research funding, to say nothing about the point of science overall, is outside of my remit.[1] But I think the position thus far developed allows me to say a few things about the value of historical science. My points will be partly instrumental, reasons why we should care about, fund, and pursue knowledge, which do not depend on the value said knowledge holds in and of itself—that is to say, factors beyond the new knowledge it produces, the pleasure that knowledge engenders in us and (speaking for myself at any rate) the creativity, wonder, and fascination from understanding the practice itself. When we come to divvy up our epistemic resources, what arguments are there to spend some of these on investigations of the deep past?

Several common features of historical investigation are unfortunate from this perspective. First, historical investigation is often hit-and-miss. For instance, fieldwork can be a high-risk venture. Recall Ward's ammonite hunt from chapter 5. He was interested in determining whether ammonite biodiversity was reducing before the End-Cretaceous extinction. In the original investigation, he found very few ammonites. However, a later investigation in a nearby location found bountiful fossils. The often gappy, faint nature of historical evidence means that fieldwork often provides empty hands. Second, the results of such investigations are often ambiguous and imprecise. It is difficult to know what to make of Ward's results—which of the two studies provides a clearer biological signal? In many contexts (particularly unlucky ones), drawing together a large amount of different kinds of evidence is required, but this means that results are often not amenable to quantified representation. I have discussed several simulations that attempt to get traction on the extent of snowball events. It isn't obvious whether there are nonarbitrary ways of setting error bars for such studies. When the support garnered by a hypothesis draws on many different streams of evidence, it is unclear how this might be represented in a quantitative way. Finally, much historical work is costly: Fieldwork is expensive; preparing, storing, and preserving specimens is expensive; the technology required to analyze specimens is expensive.

Thus, historical science is often risky, expensive, and provides ambiguous results. This does not play nice with processes of dividing our epistemic resources, which prize efficiency, likely return, and clear outputs. The public, policymakers, and so forth want clear epistemic dividends on their investment. This problem is intensified for paleontology, which (see Sepkoski & Ruse 2009) typically does not have the institutional support that comes from having dedicated university departments.

I shall make three points about the value of historical science. First, I have argued that we should expect investigations of the past to be successful, and thus likely to pay epistemic dividends. An increase in expected dividends, on the face of it, would increase the value of pursuing that science. Second, my argument that analogy can inform reconstructions of the past is also applicable to our understanding of the contemporary and future world. The past matters if we care about understanding the future. Third, I have emphasized the capacity of historical study to produce epistemic tools. There is no reason to think such tools are inapplicable to investigation of contemporary and future matters. That is, examining the past can produce epistemic goods with wide applicability. These points about value will also underlie a point about the kinds of research that would encourage success in unlucky circumstances. I will argue that we should balance traditional—focused and centralized—funding methods with more diverse "exploratory" approaches.

I should start by making some caveats.

We are in the business of dividing up "epistemic resources": We have only so much energy, people, and cash to spend on knowledge-producing pursuits. Thus, in effect, I will treat the value of historical science in terms of resource allocation. That is, we have some finite amount of resource (funding allocated for epistemic pursuits, for instance), and so need to understand how such resources should be divided. We should be very careful here. Often, framing an issue in terms of resource allocation can obscure other perspectives. First, it matters which constraints we choose. Contrasting funding for biomedical research with funding for further investigation of *Obdurodon tharalkooschild* is a very different proposition from, say, contrasting biomedical research, extinct platypus studies, and defense funding. Second, as Colyvan and Steel (2011) have pointed out in the context of conservation, approaching issues in terms of resource allocation can lead to conservatism. Such strategies involve making do with the pot you have, that is, maximizing the efficient use of available resources given your aims. This can obscure other strategies, for instance, in some circumstances one might hold out for a bigger pot. That is, it may be that we should barrack for diverting more resources to epistemic tasks overall, rather than making do.

I am also going to restrict my discussion to only two factors relevant when considering the value of a scientific investigation. The first of these is *our chance of success*. Rather obviously, whether I want to devote resources to a project will depend in part on the probability of the project bearing fruit. If scientists want to go and search for more *O. tharalkooschild* specimens, how likely are they to find any? The second is the possibility of *indirect epistemic benefits*. That is, in addition to what I will call the "direct" fruits of the investigation, how likely is it that what we learn will help us in other epistemic arenas? Will knowledge be gained and epistemic tools be developed that will be applicable across other scientific contexts? Do I have any reason to think that investigation of extinct platypuses will yield epistemic dividends for nonhistorical scientists? These concerns are relatively narrow and tailored toward the points I want to make. Regardless, chance of success and indirect benefits are an important part of the story—and are topics that we have the resources to at least say something about!

12.1 Chance of Success

In deciding how to distribute our epistemic resources, an important consideration is how likely investigations are to succeed. Of course, the nature of success is a vexing question, about which I will say more. Note that we can compare sciences in more or less localized ways. For instance, I might rank whole fields: Ought paleontology get more of the pie than developmental biology or psychology? Or I could consider various questions within fields: Should I care more about investigations of sauropod gigantism or of *O. tharalkooschild*? Further, I could consider the merit of various types of investigation. Would paleontological goals be furthered more by funding further fieldwork, or should our efforts be focused on developing new midrange theories? ("experimental taphonomy" for instance; see Briggs 1995). At least one important factor when considering these questions is the likelihood of meeting the direct aims of the investigation in question. This is the notion of success I am interested in here. As we will see, my claim will be somewhat ambiguous. I will first argue that the optimism I have established gives us reason, in some general sense, to put more emphasis on historical science insofar as it is more likely to pay epistemic dividends than we may have initially thought. However, I will then sound a note of caution about how we might go about funding historical science, which I will expand upon later in the chapter.

What are the direct aims of an investigation? To some extent, direct aims are the stated goals of research: what might be said in a grant proposal or published in a subsequent paper. In fieldwork, such output would include new traces, and perhaps more knowledge generated by those traces.

Obdurodon tharalkooschild's molar was located via fieldwork in the Riverslea region. She was only later identified as a new species. Often the identification of a new species involves digging through already discovered remains, as opposed to digging in the outback. Such was the case for Platyzilla. So some studies involve the sorting, analysis, and reanalysis of existing finds. More theoretical studies can have wider output. Consider work on Cope's rule. Here, scientists are interested in uncovering patterns and putting forward models of causal mechanisms to explain them. Large-scale, highly idealized, and sometimes quite speculative epistemic goods are the result. Moreover, developing midrange theory provides knowledge of how traces form, while (as we have seen) simulations and the examination of analogues can help us extend our epistemic reach into the past. We can compare these goods with indirect goods—that is, unintended, indirect outputs of the study. I will discuss two kinds: those that have upshots for historical investigation and those that have upshots in other scientific arenas. An example of the former would be using *O. tharalkooschild* to help us examine Cope's rule. An example of the latter would be using paleoclimatological data to help us understand how Earth's climate might behave in light of anthropogenic climate change.

The last paragraph may have generated a nagging worry about this talk of "direct" and "indirect" epistemic goods or aims. If so, good. Even if historical scientists have a specific target in mind, distinguishing between direct and indirect outputs of their investigations is tricky. As we saw in chapter 6, historical scientists are methodological omnivores. They squeeze all the empirical juice they can from whatever sources are available. So, although a particular study or paper will typically have an explicit central aim, many further connections and interdependencies will be sought. Consider Pian et al.'s (2013) study of *O. tharalkooschild*, for example. They did not constrain themselves to establishing a new species. They also speculated about her lifeways, suggesting for example that she ate vertebrates. Moreover, Pian et al. also highlight a possible connection with patterns in Australian marsupials—specifically, the increase in individual size during the Cenozoic. Which of these are direct, and which are indirect epistemic goods? Moreover, how do epistemic goods that aid in the generation of new knowledge about the past, such as inference tools and midrange theory, fit in? The distinction between direct and indirect epistemic goods in historical science, then, is messy.

I will get back to the messiness in a second, but for now, let's just go along with the idea that scientific investigations have more or less direct outputs. What I am calling the "chance of success" is effectively how likely it is that direct epistemic goods will be gained. The optimism I have established has obvious upshots for this question and even offers some guidance for more localized cases. My main takeaway message has been that the historical sciences have more epistemic power than it might first appear. Many scientists and philosophers adopt overly trace-centric conceptions of method, that is, they tend to overemphasize evidence provided by the downstream consequences of historical targets. I offer an expanded view of what historical scientists can do. In the last chapter, I identified a group of optimistic theses and shall draw on these to clarify how optimism should lead us to expect the generation of direct epistemic goods (we will get to indirect goods soon).

Recall optimism$_{GD}$. This was a restricted general optimism about our capacity to find out about the past. Except when certain restrictions hold, we should be willing to bet that historical scientists will uncover most epistemic goods pertaining to the past (or perhaps most "important" goods). Insofar as optimism$_{GD}$ has been established, this grounds reason for expecting bountiful dividends in direct goods. Naturally, we may be more optimistic about other disciplines, but suffice to say, showing that the historical sciences have more epistemic grunt than we originally thought increases our expected chance of success, and thus in part how much epistemic resource should be diverted to it.

Moreover, recall Optimism$_C$ from the last chapter. Committing to Optimism$_C$ involved being willing to bet that, for some set of past epistemic goods falling within a particular epistemic context, we will gain those goods. I argued that chapter 5 identified a set of epistemic contexts in which we should be especially optimistic of success: those that look good by the ripple model of evidence—that is, the set of past events, entities, and processes that we expect to be highly dispersed, but not all that gappy or faint. I also distinguished between "unlucky" and "unluckier" circumstances, the latter being unlucky circumstances where the chances of utilizing or constructing surrogates, or exploiting dependencies between past entities, are low. These kinds of distinctions at least give us an idea (albeit in an extremely abstract way) of how we take chance of success into account when solving epistemic resource allocation problems within historical disciplines. I would presume that the properties of the past event—its gappiness and so forth—would be ideally balanced against the cost of the investigation, and the likely interestingness of the potential goods.

However, and here is where we return to the messiness, there is something deeply problematic about using chance of success, particularly when focused on direct epistemic goods, as a way of prioritizing research projects in the historical sciences. To see why, consider two points from earlier chapters. First, historical scientists are methodological omnivores. Second, historical investigations are scaffolded.

As we saw in chapter 6, omnivorous investigations are characterized by a pluralistic, opportunistic methodology. There is no one "method" to historical science. Rather, historical scientists exploit, generate, and weave together as many routes to the past as possible. The upshot of this is that direct success is often not where a study's value lies. Consider Zhang et al.'s study of the fossilized feathers belonging to *Sinosauropteryx*, a tiny theropod. They infer the organism's coloring from features of its fossilized feathers. They are able to do this because Vinther et al. (2008) realized that melanosomes—pigment producing cells—are sometimes preserved in fossils and that (by comparison with living feathers) such cells have distinctive shapes that track the color they produce. This means that by matching the shapes of the fossil melanosomes to extant ones, we can infer the color of the long dead animal's feathers. Here, surely the most exciting aspect of the study is not the color of the extinct bird or dinosaur, but rather the new technique. A whole wealth of knowledge is opened up by our capacity to recover the color of feathers. The new technique is seized upon, developed, and applied widely. So, because historical science is voraciously omnivorous, the value of a study's direct results are often swamped by the indirect benefits.

Certainly knowing the color of an extinct bird's feathers is very cool, but in and of itself this doesn't appear to hold much in the way of scientific interest. However, remember that because investigation is scaffolded, it can be difficult to see what new evidential and investigative upshots might come from some new piece of knowledge. For instance, an animal's coloration can form the basis of new empirically grounded speculation. One could use color to hypothesize about an animal's environment, and such hypotheses could be tested in light of contemporary analogues. Color could also provide inroads to other organismic traits. For instance, assuming that the color has at least some purpose, presumably something should be able to see it. It could provide a basis, then, for consideration of the visual prowess of such animals. Finally, it could provide the basis of evolutionary and behavioral speculation. As with living animals like the peacock, ostentatious coloration is often a signal of selection by mate choice. The benefits that could result from an investigation, then, are extremely tricky to know beforehand, since scaffolding can open whole new vistas of knowledge.

Indeed, the interrelated nature of historical investigation gives us reason to expect surprising indirect dividends to be common. This makes trouble for some of the factors that are often crucial for research funding. Specifically, funding providers often want relatively predictable, relatively precise, and relatively timely results. But this conflicts with the unpredictable, often ambiguous, and often time-consuming nature of good historical science. Moreover, it conflicts with the empirically grounded speculation that, I have argued, is the key to progress in historical science. Such speculation often involves taking epistemic risks without any guarantee of any particular payoff.

All this means that if we are to take the epistemic claims I have made about historical science seriously, we may want to rethink such criteria for dividing epistemic resources. As we will see in the next section, funding is sometimes justified in rather narrow ways: Scientists must have an explicit hypothesis and an explicit set of outcomes. Such an approach would make sense if the main benefits of a study were its direct goods. But often this is not the case, and it is difficult to ascertain just what the epistemic goods of a study will be. This suggests to me that funding decisions ought to be less specific, and autonomy be given to researchers in terms of how research proceeds. Funding, moreover, should have a broad base. A narrow focus on a few, large, projects, would be disastrous, even if such projects have a high chance of success vis-à-vis direct outputs. Of course, this may be very difficult to achieve in practice—and indeed, there may be other good reasons for preferring relatively narrow research goals—but this discussion at least gives reason to sound a note of caution in that regard.

Overall, then, the optimistic stance I have argued for should lead us to expect historical investigation to pay both direct and indirect dividends. Prima facie, this gives us reason to grant it more weight in deciding where to spend our epistemic resources. Moreover, it also gives reason to emphasize a less constrained and more pluralistic approach to divvying out epistemic resources. Before switching to indirect benefits reaching outside historical science, it is worth pausing to consider an example of the kind of optimistic, exploratory study omnivory recommends.

12.2   Exploratory Research

Getting science funding often requires the identification of clear epistemic dividends, clear problems, and clear criteria for success. In other words, the focus is on expected direct returns. The NSF's Sedimentary Geology and Paleobiology funding, for instance, is explicitly geared toward the assessment of particular questions through the development of specific tools or the generation of specific data. To its credit, the funding body emphasizes the integration of different domains across geology and paleontology—its aim being to "stimulate synergistic activities and teams of multidisciplinary scientists to address critical questions about Earth-Life interactions in Deep Time" (NSF 2017, 3). However, at least as stated (and I am very aware of the distance between how criteria are worded and how they are applied), there are some aspects of the criteria that potentially discourage exploratory research. Before going any further, let me note two cautions. First, I am treating the example shallowly: A lot of thought and theory goes into the design of science funding, and there isn't space here to cover it with the depth it deserves. Second, it is an empirical question whether the criteria I will describe actually discourage exploratory research. Rather, my claim is simply that prima facie they are discouraging.

The criteria for being awarded these grants turns both on intellectual and societal aspects, that is, the capacity to advance knowledge and benefit society. This grants priority to research agendas with specific direct goals. Moreover, the criteria require specific quantifiable metrics to measure success. Indeed, this seems fair enough: We want to maximize the epistemic bang we get for our buck, after all. However, this clashes with the indirect nature of progress in historical science. Given that "transformative" discoveries often come from surprising domains and are unintended, it seems that overly focused funding—and overly restrictive criteria—are a bad idea. I am not suggesting that the NSF's criteria are problematic (that would take significantly more careful study than can be managed in this basic treatment), but it does suggest that we should wonder whether demanding that success be measured in such direct terms is the best way to motivate the *indirect* success that is characteristic of scaffolded, methodologically omnivorous science. This is an empirical question—indeed, it may be that all things considered this approach is the best way to go—but it is worth considering how to encourage the production of indirect goods, too. As to how to actually develop and implement policies that encourage this kind of work, we have drifted so far from my wheelhouse that I had better leave this line of thought here.

I do have an example of the kind of decentralized, determinedly vague vis-à-vis success, and exploratory research that I think my picture of the epistemology of historical science should encourage, however: the (currently ongoing) Flood Mitigation Project undertaken by the Royal Tyrrell Museum in Alberta (see Sanchez & Borkovic 2016).

In June 2013, a highly concentrated and voluminous rainstorm hit Alberta, dropping over 200 mm of water in less than two days. This caused an unprecedented increase in river water, leading many of the province's major rivers to burst their banks with devastating consequences: damage estimates at around 5 billion Canadian dollars; more than 100,000 people displaced; 5 people dead. When I arrived in Calgary almost exactly a year later, the city was still affected by the disaster.

Floodwaters also affected Alberta's riverbanks, and this presented an opportunity for new fossil finds. Alberta is fossil-rich, and so there is good reason to think that all of that freshly revealed strata will reveal a paleontological bounty. The Tyrrell Museum quickly launched a project whose aim was not to dig out a particular cite, but—roughly—to see what was there.

Joe Sanchez and Ben Borkovic, who carried out the bulk of the work of the project, have prospected 175 kilometers of riverbank as of early 2016. Over those two years, they identified 144 new sites of paleontological interest, and collected more than 60 new specimens. Much of the fieldwork involved a kind of fossil triage. It was necessarily to quickly determine which specimens were easily removable, which could be returned to in good time, and which were in danger of degradation due to being exposed to the elements—and indeed, to ascertain their scientific importance.

Consider, for instance, two noteworthy finds. One was an enormous (and potentially unique) ceratopsid skull that required a month to remove. In this circumstance, folks at the Tyrrell were called in to extract the fossil.

Somewhat ironically, sometimes we need a "rapid response" team to extract a fossil, millions of years in the making. Another was a large trackways assemblage involving more than 75 tracks from a range of different critters. Here, a latex cast is used on the spot as opposed to extraction.

In this context, conducting a rough-and-ready field search was preferable to something more directed, careful, and methodical. Such a search lacks specific success criteria, and a particularly well-formed question or hypothesis. It is true that, owing to Alberta's fossil richness, there is good reason to expect rewards in this context—we're not entirely in the dark—but nonetheless this kind of research illustrates the kind of flexibility and freedom that should generate successful research in methodological omnivorous science.

The flood mitigation project is *exploratory* insofar as (1) there is no specific epistemic dividend, (2) success criteria are vague, and (3) it is *flexible*. On this final point, note that flexibility extends to commitments in terms of time and expense. The relatively cheap labor of the two fieldworkers was sometimes supplemented with more resources (boats, helicopters, more expertise, more people on the ground) as and when triage called for it. On the location of epistemically valuable (or fragile) finds, the team at the Tyrrell can jump to action (as they did with the enormous ceratopsid skull). In my view, encouraging more science of this exploratory nature would potentially encourage success in the kinds of contexts I have focused on. In this case, some success was basically guaranteed. Often, exploratory research is likely to come to nothing, which suggests that funding models that focus on a few big projects will be less successful than those that follow a more decentralized, spread-out strategy.

12.3   Indirect Epistemic Benefit

We can distinguish between two kinds of indirect epistemic goods: those that aid in historical research and those that can help us understand our present and future. Let's discuss the latter. Historical research can provide indirect epistemic benefits to nonhistorical science. Although the range of possible benefits is conceivably open-ended, I will simply consider two types. In brief, I highlight the use of knowledge of the past to understand the present and future—the use of the past as analogues—and the development of tools helpful for both directions of inquiry.

Figure 12.1

Images from the Tyrrell's Flood Recovery Project. On top, Joseph Sanchez and Don Henderson prepare a latex mold of theropod trackways preserved on a sandstone slab along the St. Mary River. On the bottom, a large boulder with part of a Hadrosaur skull visible in Castle River, in a net prepared to be airlifted. (Photographs courtesy of the Royal Tyrrell Museum.)

It is no surprise that the past can be a guide to the present and future. Geology's utility in mineral resource extraction industries is founded on this very notion. However, using the past as an analogue sometimes founders on the thought that in the present we face unprecedented challenges. We have seen that Earth has cooled and warmed in the past, and that mass extinctions have occurred. We currently face apparently analogous challenges. However, it is sometimes thought that the capacity of such historical episodes to inform us about contemporary climate change and biodiversity loss is dramatically limited. It is limited because anthropogenic factors play an essential causal role in these events, and we therefore have no analogue in the past. We are dealing with a new, surprising, and extremely complex set of causes—human intervention at a global scale. As such, events from the deep past, which are only shallowly similar to our present predicaments, are potentially of little real use.

This attitude, I think, is a mistake. It is just to say that some of the anthropogenic-driven processes we want to understand approach Tucker's notion of "unique." As we saw in chapter 8, an event is unique insofar as it cannot be unified with other events in order to generate understanding. As we saw in that chapter, the "exquisite corpse" methodology often still pays dividends. As Gavin Schmidt has said in the context of climate science, "There are no true palaeoclimate analogues for the global changes projected for the twenty-first century in the Intergovernmental Panel for Climate Change (IPCC) Fourth Assessment Report (AR4). However, many of the uncertainties highlighted in that report do involve aspects of the climate that have certainly changed in the past" (2009, 79).

Schmidt develops a group of proposals for incorporating knowledge of the deep past into the validation of climate simulations focused on future predictions. In effect, he provides a roadmap for using paleoclimate data as a way of understanding future climate, which involves navigating between several imperfect analogues, highlighting the matching aspects. In effect, Schmidt applies the exquisite corpse method. Naturally, how the components react when combined in each case is both crucial and tricky, but regardless such analogues form lines of evidence. Citing the unprecedented nature of anthropogenic impacts on climate and biodiversity provides no reason to think that the past cannot inform us about the future.

A set of paleontological examples draws on patterns exhibited in biota during planetary warming events. For instance, the boundary between the

Paleocene and the Eocene, approximately 56 million years ago, is associated with a sudden, but relatively brief, warming event—the Paleocene/Eocene Thermal Maxima, or PETM. The PETM lasted around one hundred thousand years and involved global temperature increases of between five and nine degrees centigrade. Rankin et al. (2015) examined decreases in mammalian body size in two sites over the time period, and argued that they exhibited both individual-level selection for small size (driven in part by immigration) and species-level selection for larger sizes. They tie their conclusions to both climate effects on evolution generally, and to the specific circumstances facing us today:

Rapid global climate change can lead to rapid evolutionary responses. On short time scales of ten to hundreds of generations, responses to climate change are likely to be dominated by within-lineage evolution. On longer time scales, our results illustrate that differentiated speciation and extinction can be just as important as within-lineage forces. Indeed, as current climate change continues to outpace within-lineage adaptation, evolutionary responses at short time scales may be governed by species selection in the form of differential extinction. (Rankin et al. 2015, 6)

Examining past episodes like the PETM can aid in understanding the processes operating now—the relationship between group and individual-level selection over different timeframes and climates, the causes of extinction, and so forth. Undoubtedly the PETM and the warming we face today were profoundly different, but the former can still be revelatory of the latter.

As you can imagine, there is an enormous amount more to be said on this issue, but I will content myself with claiming that my position on the use of analogues from chapter 8 should lead us to think that the past has enormous potential to bolster our understanding of the future and present. That is, it will pay indirect epistemic dividends.

Another way the study of history can aid the study of the present and future is in the development of tools relevant to both concerns. Peter Galison (1996) has discussed the notion of a "trading zone," and Alison Wylie (1999) has applied it to archaeology. Galison identifies an unusual kind of scientific unity or integration that emerged around the development of the computer. Many scientists, doing wildly different activities and focused on wildly different things, sometimes require the same basic knowledge and skill-sets to work effectively, and so come together for that purpose. As Galison says, "What they shared was not common laws, and most certainly not a common ontology. They held a new cluster of skills in common, a new mode of producing scientific knowledge that was rich enough to coordinate highly diverse subject matter" (1996, 190).

This is what Galison calls a "trading zone": "an arena in which radically different activities could be *locally*, but not globally, coordinated" (1996, 190). Even if it is the case that anthropogenic influences create qualitative differences between the deep past and our near future, it does not follow from this that historical investigation cannot produce the kinds of epistemic tools that are the currency of trading zones. That is, tools developed for studying the past might still be of service in other contexts (and vice versa, of course—as we saw in the case of LiDAR for archaeology).

Conservation biologists and paleobiologists who are interested in mass extinctions, and paleoclimatologists and climate scientists, are undoubtedly interested in different questions at different scales (albeit about similar subjects). However, given what I have called the "omnivorous" nature of such research, the construction of models, simulations, and inference tools, it would be astounding if trading zones did not develop. That is, there are likely to be both skill-sets and epistemic tools that would be helpful to both investigations. We already saw an example of this: Hyde et al. co-opted an ice sheet model developed to understand Modern and Pleistocene climate events in order to understand the very different conditions of the Neoproterozoic climate.

The past, then, can also be a source of new skills, tools, and inspiration for those interested in the contemporary or future world. Drawing these together is important if we are to maximize our epistemic reach—and not just into the past.

I have made three points bolstering the "why should we do it" value of historical science. First, I have given reason to be optimistic about historical science, and surely how successful we expect an investigation to be will have some weight in which resources we assign to it. Second, I have pointed out that even rather indirect and messy analogues can play an important role in validating future-focused science—historical science, then, can aid in our understanding of our past and our future. Third, I have highlighted the possibility of "trading zones"—the sharing of various skills and tools toward different tasks. I have also argued that science funding should encourage more "exploratory" studies: that is, more diverse, less centralized, flexible studies with vaguer success criteria.

As I was at pains to point out earlier, I do not take this to be anything like a proper treatment of such issues. My aim instead has been to show how the epistemic considerations of this book have more applied upshots. For me, of course, the value of historical science doesn't need any further bolstering. That it reveals worlds of such alien character in such sophistication and depth, despite the enormous epistemic challenges, is value enough.

13 P.S.: A Note on Progress and Realism

Twentieth-century philosophy of science was marked by some Big Debates—on the natures of representation, explanation, confirmation, and so on. One set of Big Questions divided realists from antirealists. Another concerned the nature of progress. I am going to close by discussing how reflection on historical science allows us to reconceptualize these questions.

I have characterized the method and epistemic resources of historical scientists facing unlucky situations. That is, I have explained how historical reconstruction succeeds when the going gets tough: when evidence is scattered, incomplete, biased, and hard to access. There are three aspects of my view that are relevant going forward. First, historical scientists are often *methodological omnivores*. They do not rely on a particular method, approach, technique, or pattern of reasoning. Instead, they opportunistically exploit whatever epistemic crutches are at hand, often co-opting, developing, and constructing tailored tools suitable to the context (consider the adoption and configuration of LiDAR to the needs of Central American archaeology). Second, historical investigation is *scaffolded*. Evidential relevance depends upon the existence of particular hypotheses, tools, and midrange theories. Some of these scaffolds can be subsequently discharged (the suggestion that sauropod gigantism was an adaptation to low-nutrient Jurassic foliage) while others are retained, becoming the furniture of further research (such as the increasingly precise estimates of sauropod tonnage). Third, progress is often driven by *empirically grounded speculation*. That is, a hypothesis is not simply judged on its plausibility but also on the new empirical avenues and tests that it opens up, how good a scaffold it is (obliquity explanations of Neoproterozoan glaciation are false—and pretty implausible from the outset—but enabled a deeper understanding of the relationship between orbital dynamics and global environment).

These features add up to a picture of scientific progress that is not linear, and where success is tricky, often obscure, and pluralistic.[1] In the next two sections, my aim is to draw some general philosophical lessons from this, beginning with progress and moving on to realism.

Note that I will be treating this view about historical reconstruction in unlucky circumstances as if that is what all science is like. I am officially agnostic about my view's scope—whether it turns out to be true of all of science, just historical science, or apply patchily, is a question that can only be answered by going and looking. Because these features are in part responses to the epistemic situation such historical scientists face, I anticipate that similarities can be found where situations are similar (Currie & Turner 2016)—but that is not my main concern here. So, instead of the careful project that would be required to work out where and to what extent my view applies, join me in a little fantasy. What would the philosophy of science look like if our major examples were, say, recent attempts to explain sauropod gigantism, or Snowball Earth, or reconstructions of ancient Mayan civilization, as opposed to the Copernican revolution, the discovery of oxygen, the Michelson-Morley experiment, and so on? How might things be different if our standard examples of evidence were rock, bone, and ruin, not readouts from mass spectrometers or paths through cloud chambers? I am not sure how much the path of twentieth-century philosophy of science was due to our choice of examples, but there is at least some reason to think that a counterfactual history focused on historical science in unlucky circumstances would be a very different beast. That is, if we were to "rewind" philosophy of science's "tape," resetting our starting place to archaeology, paleontology, or geology, both debates about progress and realism would have turned out quite differently.

To be less dramatic, insofar as my sketch of historical science is characteristic of science, let's see how it contrasts with the Big Views of the twentieth century.

13.1   It's Not a March but a Mess, but It's Progress

That science progresses is almost a datum—something to be explained. And indeed, it is hard to consider the sciences, be they physics, medicine, molecular biology, and engineering or archaeology, paleontology, and geology, without being struck by the success of such enterprises. It is not

surprising then, that twentieth-century philosophy of science is often characterized in terms of central figures presenting Big Views of science built
around accounts of its progress, or at least change. An account of progress
provides a model of scientific change and an account of the rules of the
game. For some, the rules and the changes generate truth; for others, the
changes are not about truth per se. Regardless, these Big Views exhibit a
set of common features. And because these features are common across
pictures where science in fact progresses and those where science merely
changes, I will happily conflate the two notions in this section. A further
happy conflation is between normative and descriptive theories. Whether
you are telling me how science works or how it should work is irrelevant
in this context.

I am going to contrast the picture of progress that I have drawn from the
historical sciences with those common features. Of necessity, the discussion
is going to be coarse, and I apologize in advance for running roughshod
over the subtleties of each view.

First, the locus of progress for many philosophers were what I will call
Theories. Theories are large-scale structures that present the world as being a
certain way. According to Rudolf Carnap, for instance, mature sciences construct formalized languages that express the empirical and analytic content
of scientific theories (Carnap 1937). Imre Lakatos distinguished between a
scientific domain's "hardcore" of claims, methods, and so on, and the belt
of "auxiliary hypotheses" that insulate the core from falsification (Lakatos
1978). The hardcore of his "research programs" played the role of Theory.
Thomas Kuhn replaced Theories with paradigms, roughly a set of shared
problem-solving procedures (and problems) maintained by institutional
learning procedures. To a greater or lesser extent, Theory-based pictures
see scientific progress as a process of aligning observations with Theory, or
altering Theory so it better handles observations.

Second (with the exception of Kuhn), Big Views construe knowledge—
the gains of progress—narrowly. That is, science produces truth, or some
analogue of truth. These analogues can be distinguished by (1) the epistemic good in particular and (2) what the good pertains to. For instance,
structural realists and constructive empiricists both believe that science
aims for truth (the same epistemic good) but come apart insofar as the
former takes scientific truth to range over structural features, and the latter
takes it to range over observations. Narrow, monist construal of scientific

aims and outputs comes hand in hand with the Theory-centric account of science. After all, Theories (and hypotheses) bear some representational relationship with the world, or with observations. They cohere with, or represent veridically, or capture. Scientific knowledge is the relationship between Theory and world, or perhaps observations: that is, truth or some analogue.

Third, scientific progress occurs in a kind of linear evolution (see Hull 1988). A scientific discipline fits our traditional conception of a biological lineage: It more or less gradually evolves (except in Kuhn's punctuated-equilibria revolutions) as its internal features shift, typically in a process of improvement. Here, and fitting with the previous two features, "improvements" are understood in terms of generating knowledge by fitting observations to Theories. On Popper's view, we accumulate falsified hypotheses while corroborating the victors. On Kuhn's view we align results to our paradigm while gradually accumulating anomalies (which will ultimately trigger a revolution). And so on.

If our exemplar sciences are to be historical sciences in unlucky circumstances, then all three aspects of the Big Views are false.

To begin, it is unclear where to find Theories in the historical sciences. There are *theories*, of course: paleobiology was founded on the idea that paleontology had something to say about evolutionary theory—that macroevolutionary forces somehow outstrip microevolutionary theory,[2] or that the fossil record reveals patterns of stasis followed by rapid change, not gradual accumulation; geology has plate tectonics and the rock cycle; archaeologists sometimes appeal to general models of cultural change or evolution. Moreover, all of these sciences are built upon midrange theories. But none of these are Theories in the relevant sense: They do not provide a kind of overarching structure that produces a clear research program. It's true that, for instance, theories of evolution, or plate tectonics, underlie a lot of work in the historical sciences. But these roles are typically in the background. There is no recognizable "normal science," and although we could characterize some aspects (such as plate tectonics) as part of the Lakatosian "hard core" of historical geology, this becomes quite forced in light of methodological omnivory. The historical sciences are better characterized as a patchwork of theories, hypotheses, midrange theories, Theory from other sciences, and so forth. The epistemic structure of these sciences is closer to Peter Galison's cases of partial integration (mentioned in the preceding chapter). There is knowledge that practitioners share in historical science—every paleontologist is familiar with stratification, for instance—but labor is significantly divided in these disciplines, such that a common paradigm is not only difficult to discern but even to look for one misses the point. It is in its disunity that historical science is strong.

And indeed, given the discussion in chapter 7, this is just what we should expect. Theories are good—when they're good—when they have broad, general "lawlike" regularities to capture. They're simply not the sort of thing that is appropriate or useful when you're dealing with fragile regularities. Moreover—to draw from chapter 6—the Kuhnian notion of a paradigm, the idea that there is a set of problems, and techniques and procedures for dealing with those problems, breaks down in the face of methodological omnivory. The success of historical scientists is not located in their learning a particular method—they are not obligates—but rather in their capacity to co-opt, construct, and adapt methods to suit their epistemic situation.

Next, my understanding of science's epistemic products is explicitly pluralist. Scientists do not generate a single kind of epistemic good—they are after a wide range of different products. This is becoming a common idea in philosophy of science (see Potochnik, forthcoming), but it is exaggerated by the opportunism exhibited by historical reconstruction. The adaptability required lends itself to the generation of a variety of goods. On the discovery of a new species, paleontologists do not simply add it to our store of taxa, but they also consider whether its existence can tell us about contemporaneous flora and fauna, use it to better understand the era's ecology, and examine it for clues about the relevant macroevolutionary processes. They will also use the find as an opportunity to develop mid-range theory—exploring whether new techniques could be developed to grant new knowledge of this and other traces.

And moreover, this adaptability leads to a nonlinear picture of progress. On the Big Views, the evolution of a science looks like how we often picture the evolution of large eukaryotes like ourselves. Lineages follow a treelike structure wherein evolution is a tinkerer, leading to a stately unfolding with occasional branching. The tendency of historical scientists to co-opt and adapt tools from other domains, the localized nature of these tools, and the exploitation of interdependencies of various sorts, undermines this simple picture. The scientific version of lateral gene transfer occurs. Instead of the sciences possessing more or less separate histories, marching in a line—the vertical transfer of "traits"—with occasional cross-chatter (perhaps in the form of "interfield theories"[3] or "reductions") we see a pattern more marked by the horizontal transfer of "traits." The history and evolution of science is not treelike in these cases.

None of this, I think, is particularly original—I take it that the Big Views that dominated twentieth-century philosophy of science have become by and large outdated (yet still tend to form the basis of undergraduate teaching in the philosophy of science, for some reason), but perhaps their eclipse is a worthy thing to remind ourselves of. Science is not always—or even typically—about Theories generating truth in a linear fashion. It is often—perhaps usually—pluralistic, indirect, and deviant. That is, it's not a march but a mess, but it's progress.

13.2   Against Realism and Antirealism

As with any long-running philosophical debate, positions regarding realism and antirealism have developed a subtlety making them resistant to bumper sticker slogans. Nevertheless, we can roughly take the realist to think that (1) the scientist aims at truth and (2) she sometimes more or less gets there. That is, realism consists in a claim about the epistemic good scientists are after—truth—and a claim pertaining to their success in generating that good. Stathis Psillos (2005), for instance, takes a comprehensive realism to include a metaphysical thesis (the world is mind-independent and structured by natural kinds), a semantic thesis (theories are intended to be literally true, that is, they refer to real things) and an epistemic thesis (some theories—the mature ones—are approximately true). These are logically independent. We might think science aims for, but doesn't achieve, truth. We might think science doesn't aim for truth, but sometimes gets it anyway. We might think that science aims for something other than truth—there is some other epistemic good on the table—and then argue about whether we should be realists about *that* or not. I am going to argue that the picture of science I have developed undermines a common set of arguments regarding realism and antirealism. The arguments I am targeting—let's call them *historical* arguments—draw on science's history to support some stripes of realism or antirealism. There are three themes running through this book that, I think, make trouble for these arguments. It is important to note that historical argument is not the only way to support realism or antirealism. I will also briefly discuss debates about realism that converge on my discussion of optimism. These rely on what I will call "situated" arguments pertaining to realism or antirealism, arguments that rely on particular epistemic contexts.

So, we begin by articulating the three themes that put pressure on historical arguments for realism.

The first theme is the *context sensitivity* of good method. What works and what doesn't depends crucially on your epistemic situation. If I am lost in a city, whether I should ask for directions or consult a map (or whatever) depends crucially on some of my features—whether I speak the local language, whether I am any good with maps, and so forth, and on features of the case itself—the city's design, how good the local maps are, and the like. The same goes for science. The paleontologist arguing that *Obdurodon tharalkooschild* is a new species of platypus is in a very different situation to one claiming that Cope's rule is a driven trend, or one trying to determine the thermoregulative system of sauropods. As such, if I am going to make a claim about what works, what doesn't, and what is good about scientific investigation, I had better make sure that what I say both identifies and is sensitive to the relevant contexts.

An important aspect of these contexts is the epistemic good being produced, and this is the second theme. As we have seen, I am a pluralist about epistemic goods.[4] It's just a mistake to characterize science's target in monist terms. Whether a scientific investigation generates metaphysically robust Truth, or some less ambitious kind of truth, or to recover observational phenomena, generate understanding, or whatever, depends upon the details of their investigation. The same goes for scientists' intentions.[5]

The third theme is the *indirectness* of scientific success. It would be a mistake to judge the success of a scientific investigation by looking only at its direct outputs. This indirectness has two flavors. Developments of LiDAR technology were originally for work in meteorology (measuring cloud density) and exogeography (topographic mapping of the moon). These were co-opted by archaeologists to identify Mayan ruins. First, then, scientific investigation often has wide-ranging and unexpected upshots. The theory that Neoproterozoic glaciation was caused by a wobble in Earth's obliquity turned out to be false, but it led to a set of models that increased our general understanding of the relationship between obliquity and global temperatures. Second, then, getting things wrong *directly*—backing a false theory or hypothesis, for instance—should not be considered a failure when it scaffolds the development of new knowledge. (Moreover, figuring out that something doesn't work is often as important as figuring out that something does.)

Have I directly argued for any of these general claims? Not exactly, but neither have I merely assumed them. I think context sensitivity, pluralism about epistemic goods, and indirect success have been motivated and demonstrated throughout this book. Discussion of how historical scientists—methodological omnivores—adapt themselves to local context, it seems to me, demonstrates the necessity of understanding that local context: There is no other way of grasping the different approaches taken by the paleontologist reconstructing a critter from a single tooth, and the paleontologist weaving together a bewildering array of evidence to explain sauropod size. Moreover, we have seen the indirectness of progress: Things often don't turn out how historical scientists would like, but we've seen how knowledge generation marches on regardless—and in hard-to-predict directions. And this in itself underwrites pluralism about epistemic goods: given the sheer variety of aims, approaches, and methods in historical science, it would be odd indeed if these were all in service of the same kind of output.

So, to the arguments for realism and antirealism.

We can discern two broad ways of arguing for realism or antirealism. The first I will call *situated* arguments. These identify a set of epistemic virtues or vices and argue that science exhibits that set and from there derives a claim about what our stance should be toward science. Derek Turner (2007), for instance, appeals to two features of historical reconstruction—the inability to manipulate the past, and the degradation of evidence—and then, using experimental science as his yardstick, argues that the kinds of arguments that encourage realism in the latter fail for the former—as I will suggest, insofar as my optimistic picture of historical method and success improves these prospects, it constitutes a response to Turner's arguments. Kyle Stanford (2006, 2013) argues that science exhibits a kind of conservatism that renders it unlikely to explore alternative theories. Peer-review approaches to publication and grants, for instance, are systems that discourage false positives—they are conservative systems, more likely to not publish something legitimate than they are to publish something illegitimate. This means that radical ideas—major alternative theories, for instance—are unlikely to be legitimized, encouraged, or funded. A third example is Ian Hacking's appeal to the role unobservable entities play in the design and construction of physical interventions (1982). By his lights, some experimental practices in physics simply wouldn't make sense unless we take some of the properties of the entities in question to exist.

Such arguments are "situated" due to their context sensitivity. Turner's argument applies to those sciences that lack certain types of epistemic resources. Stanford's arguments constitute a critique of how science is currently organized—this could potentially be fixed, and they would no longer apply. A situated argument for realism or antirealism identifies an epistemic context and then provides reasons to think that, in this context, science will succeed (or fail) to generate the epistemic goods. We can contrast these with "historical" arguments.

A historical argument seeks to establish or undermine realism by appeal to patterns in the history of science. The classic arguments for or against realism take these forms. The "no miracles" argument (Putnam 1980) appeals to scientific success. Given how successful science has been, it would be extremely unlikely, a miracle even, if it turned out that science didn't generate truth.[6] In contrast, the pessimistic induction (sometimes called a "metainduction"[7]) points to the common *failures* of scientific theories over time (Laudan 1981). Most theories have turned out to be false, so why should we think our current theories are any better? Historical arguments about realism have taken various forms—structural realists, for instance, often argue for their position on the basis that it is the structural (rather than concrete) features of theories that are maintained over time (Ladyman 1998; Worrall 1989). I am not interested in comparing the for-or-against of science's history. Rather, I want to argue that the nature of science invalidates this general class of arguments. Of course, my objections turn on the historical sciences being in some respects representative of the sciences generally—one might object that other sciences (perhaps condescendingly referred to as "mature" sciences) have neither context sensitive warrant nor indirect success. If so, I look forward to this being demonstrated.

Historical arguments rely on projecting from a history of success or failure to future or current success or failure. But if good method—that which generates success or failure—depends crucially on contextual matters, then these inductions are unjustified. Some experimental studies have been enormously successful—lab-based work on model organisms aiming at probing the molecular properties of developmental systems, for instance. But such success occurs under particular conditions. First, the relevant technologies are available and understood. Second, the system itself is relatively tame, it is amenable to the controlled environment of the lab and genetic switches are great for intervention. Third, the aims of these studies are largely unambitious. Although this work is often cashed out in terms of understanding the roots of disease, behavior, psychology, and so on, in actual practice molecular genetics is often about understanding the causal relations between relatively simple mechanisms in highly artificial, isolated circumstances.[8] Compare this to, say, paleontological studies of Cope's rule. Here, we are interested in identifying and explaining large-scale patterns across life's shape. Databases are created from the less biased trace sets (ammonites, for instance) and bought to bear on ambitious questions. Now, it doesn't seem to me that you could use an instance of the first kind of study to inform a philosophical view regarding the second kind, or that in combination they might tell you something general about science. They seek to generate different kinds of goods, and they do so in different epistemic situations.

The point is that historical arguments rely on inferring across science, abstract from context. But if the context is where the action is, then these inferences will often be uninformative. My argument relies on a basic point about ampliative induction. The trustworthiness of an induction turns in part on (1) your sample size and (2) the expected heterogeneity of the class you are inferring across. Highly heterogeneous classes require large sample sizes. Science's context-sensitivity and pluralism about epistemic goods should lead us to expect science to be heterogeneous, and this provides reason to be suspicious of such inductions.

Further, making an induction of the historical type involves identifying successes or failures—creating a "tally" if you want[9]—and inferring on that basis. But this relies on our being able to identify successes and failures: We have to know them when we see them. And I'm not sure we do. Moreover, successes and failures must be on a similar scale for a tally to be constructed. Pluralism about epistemic goods suggests that this might not be the case. Here is a simple historical example (see Linton 2004; Saliba 2009). Late medieval Islamic mathematicians and astronomers were working with a wonky theory of the solar system, one that put the earth in the center and got everything else to whirl around it in complex, beautiful patterns. They kept this wonkiness going thanks to the high-powered, sophisticated mathematics they developed. It was this very mathematics that Copernicus relied on in his development of a less wonky (but still not *quite* right) sun-centered model. Did the earlier scholars fail, then? In a sense, yes, but in another and perhaps more important sense, of course they did not. They generated epistemic goods that were of enormous value. Is this episode a victory for realists or antirealists? Well, certainly the basic theory was false—Earth ain't the center of the solar system—but it provided scaffolding that enabled the development of the mathematics that allowed Copernicus to develop his less false theory. The often indirect success and progress of science makes identifying success and failure a much more fraught task than it might appear. And pluralism about epistemic goods can undermine comparisons.

Success and failure in science are interwoven and indeterminate. As such, attempting to pry them apart in order to argue for realism or antirealism is artificial and, I think, unjustified.

So, the indirectness of scientific success, the plurality of its goods, and the context-sensitivity of its method undermine attempting to support either realism or antirealism using the kind of general historical arguments exemplified by the no-miracles argument and the pessimistic induction. Such arguments treat science as if it were one thing. It isn't. Does this mean that the history of science cannot inform us philosophically then? A school of historians and philosophers of science—contextualists—seem to think so.[10] I think they go too far.

What makes for good scientific practice is context-sensitive, but contexts are not unique. The same goes for success—just because it is an elusive target does not mean that we cannot learn from its generation. But in doing so we need to be sensitive to the local details. That is, if we are going to draw conclusions across science, we need to identify the relevant contexts; we need to be somewhat situated. However, this need not be complete—there can still be rich, informative similarities between scientific episodes. Isaac Newton's optical experiments and the work of molecular geneticists are in extraordinarily different contexts. Newton was working with minimal assistance and relatively simple equipment (holes in boards, glass prisms), while molecular biologists work in decked out, populous laboratories with powerful technologies. The kinds of systems they are interested in differ, and the social context differs (Newton didn't quite have the same worries about getting grants or surviving peer reviews). However, aspects of their epistemic situations are continuous. For instance, both target systems that have some properties in common—they are easily controlled and isolatable,[11] and there are good strategies that can be adopted in those circumstances. They involve establishing both causal properties and probing the natures of their targets via finely tuned, careful, repeated interventions. Even if other contexts—reconstructing *O. tharalkooschild*, for instance—diverge from this situation, it doesn't follow that those that do overlap cannot inform each other or, for that matter, that their divergences cannot prove revelatory as well.

This is, in effect, the application of chapter 8's exquisite corpse method to philosophy. Just as it is possible to reconstruct an apparently unique critter like *T. atrox* by appealing to a set of imperfect—but relevant—analogues, so too might we seek patterns in science's history (and, by my approach, in live science as well) and construct models of science based on those cases. If they are to promote realism or antirealism, however, our arguments will need to be context-sensitive.

This brings me to positions like Turner's, which rest on situated arguments. What is the relationship between my arguments for optimism and his arguments for antirealism? First, it is worth noting how exactly Turner's arguments are situated. In effect, he argues that defenses of realism about unobservables are more successful in experimental contexts than historical contexts. His antirealism is of the Finean variety: Committing to the truth of our scientific theories about the past—where "truth" is meant in some appropriately robust, non-deflationary way—outruns our available empirical evidence, that is, we have insufficient evidence to tell whether our theories are true or not (where, presumably, we have sufficient evidence for theories regarding experimentally accessible unobservable entities such as electrons). There are three things to note here.

First, insofar as I have argued that historical scientists have more resources at their command than Turner allows, I put pressure on his claim that we lack sufficient evidence to think historical hypotheses are approximately true. However, there is a disconnect here. Turner takes the resources available to experimental science as his comparative baseline. I have not argued that historical scientists have the *same* resources as experimental scientists. Indeed, although they use simulations and surrogates in experiment-like ways, I don't think these qualify as bona fide experiments. It seems that comparing historical and experimental science is similar to comparing proverbial fruit. As such, it is unclear how to determine the extent to which this undermines Turner's position.

However, and this brings me to my second point, I doubt that questions of realism or antirealism are worth asking at the grain at which Turner asks them. Once we agree that scientists are in the business of generating a plurality of epistemic goods, using a variety of resources and tools, in a variety of contexts, it is not obvious what can be gained by comparing "experimental" and "historical" science in terms of the truth of their theories. For one thing, the focus on truth looks extremely narrow. For another, surely establishing the truth or otherwise of particular hypotheses, or the existence of particular classes of entities, is a more fruitful line of thought.

Having said this, I don't think these considerations strip situated arguments for realism or antirealism of their value—however, as in historical science, their value is often indirect—and this is my third point. I have, in effect, treated Turner's situated argument for antirealism as a situated argument for pessimism. On the one hand, this is unfair, since he never intended to support such a position; on the other hand, the epistemic points he makes certainly support pessimistic attitudes. This is to say that rejecting general—or coarse-grained—discussions of realism need not lead us to lose both baby and bathwater. Indeed, these discussions might provide rich resources and epistemic machinery, once properly contextualized. And indeed, none of this properly speaking signals the end of the realism debate: rather, it should become more local, less ambitious, and more context-sensitive. For indeed, even as a pluralist about epistemic goods, I still want to know when I should take science as reaching truth and when I should believe in scientific entities.

Do I, then, consider myself a realist about the historical sciences? Well, do historical scientists aim for truth? Sometimes—but they provide much more than a bunch of approximately true claims. They also promote understanding of the processes that shape our world, explanations of past histories, lessons for our own present and future. Do historical scientists get truth? Sometimes, certainly. Finally, are the messy, indirect, opportunistic patterns we see in historical reconstruction progress, or just change? In my view, it constitutes progress, which is sufficient realism for me.

Notes

Chapter 1

1. Tooth anatomy is an extraordinarily detailed and useful but jargon-rich body of knowledge. For instance, the evidence I describe here is put by Pian et al. as follows: the "mimicry of the trigonid by the talonid in terms of lophid width and height" (1256). I have simplified matters for the purposes of explanation here.

2. See Currie (2016c) for discussion of species delineation in paleobiology.

3. This is made on the basis of a "homologous" or "phylogenetic" inference: Where two lineages are related, it is sometimes safe to trace their traits along those lines of ancestry. See Levy and Currie (2015) for explicit discussion.

4. See, for instance, Turner 2009b.

5. For discussion of the role of convergence in historical reconstruction, see Currie 2013, 2012; Conway-Morris 2003; McGhee 2011; Pearce 2012; Powell 2012.

6. In Currie (2015b) I discuss a similar study by Wroe et al. (2005) where bite force calculations are used to reconstruct *Thylacoleo carnifex*, the marsupial lion of the Australian Miocene.

7. A welcome shift in this discussion has occurred in light of Kyle Stanford's arguments for antirealism (2006, 2015). He focuses on the attitudes of scientists and the social structure of science regarding antirealism. Roughly, scientists are unlikely to explore alternatives to their theories and thus are unlikely to exhaust the space of possibilities. Such an argument is not based on an induction from past scientific success or failure. In the postscript, I will call these "situated" arguments and discuss how they relate to pessimism and optimism. An earlier (realist) approach with similar virtues is Ian Hacking's (1982).

8. See Leonelli (2016) for a discussion of a very similar concept, drawing inspiration from Dewey.

9. See Currie (2015a) for a more detailed discussion of the role of case studies in philosophical method.

10. Hempel, of course, distinguished between explanations that used probabilistic and deterministic laws, but these are species of the same genus.

Chapter 2

1. However, see Ganse et al. (2011, 112–113) for an opposing view. Their reconstruction predicts a lower energetic requirement for neck-raising. This illustrates the inherent difficulties of such reconstructions.

2. A character displays positive allometry if the ratio between increases in general body size and the size of that character are not isometric. As my body-size increases, my neck length increases at a greater rate.

3. There is certainly a lot of debate about the exact role of the tail in sexual selection—whether it is the elaborateness or the size of the tail, whether it is an honest signal of viability, and so on. But that the tail is sexually selected is not questioned.

4. Coprolites can tell us what sauropods were eating, but are much less revealing of the operation of their digestive systems.

5. There are layers of complexity to the basic definitions "endothermic" (keeping the body at a steady temperature through internal regulation) and "ectothermic" (using outside heat sources), which I will largely ignore for the purpose of clarity.

6. Derek Turner argued in 2007 that Snowball Earth scenarios and Williams-style explanations are underdetermined. I am inclined to think otherwise. I will leave reexamining the cases and considering to what extent Derek's and my differing interpretations are driven by the conclusions we wish to draw as an exercise for the reader.

Chapter 3

1. See Woodward (2003), Lewis (1973), and Salmon (1984).

2. See Wylie (2017) and Chapman and Wylie (2016, chapter 3) for an excellent discussion of refinement in archaeology, which she divides into three classes: first, "secondary retrieval," using new technical tools to reinterpret old data; second, "recontextualization," finding new links between existing data; third, the use of experimental simulations to go beyond data. On my view, these second and third types of refinement start to grade into the genuine generation of evidence—see, in particular, my discussion of "coherence" in chapter 6 and simulations in chapters 9 and 10.

3. Clearly, this account is influenced by recent work on the nature of causation and explanation that has attended to scientific intervention, most obviously Waters (2007) and Woodward (2001, 2010).

4. Such counterfactuals can be thought of probabilistically as follows. Considering the set of worlds where the nonactual antecedent is true, there is a dependency relationship just in case the subset of those worlds with true consequents are more probable than those with false consequents.

5. This is a version of what David Lewis has called "the principal principle" (1981).

Chapter 4

1. Of course, some philosophers and physicists have questioned whether causation is necessarily temporally asymmetric (see Faye 2010). However, these rely on examples far away from the mundane relationships of cause and effect that concern us here.

2. Or, at least, make highly likely: Terms like necessity are tricky when we are dealing with a posteriori theories, but I trust you get my drift.

3. See Elga (2001) for a nice critique of this move.

4. Thanks to Daniel Nolan for guidance here.

5. Kirsten Walsh points out that thermodynamic theory can allow me to infer some past states from present states. For instance, Lord Kelvin (and many others) attempted to infer the age of Earth from its current average temperature, and there was nothing wrong in principle with this. This shows that thermodynamics can be an important midrange theory in some contexts. It does not establish global, systematic overdetermination, however.

6. As an aside, I think Cleland's argument is too quick. Even if the past is overdetermined by the present in both thermodynamic and wave-theoretic systems, it does not follow immediately from this that (1) all macro-level phenomena supervene on thermodynamic or wave-theoretical systems, nor that (2) a system that supervenes on an overdetermined system must itself be overdetermined. Both of these claims would need to be made to claim that the past is overdetermined by the present *on the macroscale*.

7. Another way involves our choice of hypotheses. For instance, it may be that for any particular hypothesis that scientists have in fact entertained there will be at least one hypothesis that is equally well supported by the available evidence. This may be a way of connecting discussion of unconsidered alternative hypotheses (Stanford 2006) with underdetermination.

8. For instance, I suppose it is logically possible to transport ourselves to the part of the universe where Earth's light from the time has reached, construct a supertelescope, and observe the individual.

9. I have a soft spot for Newton's *Experimentum Crucis*. Prior to the experiment, whether light was composed of homogeneous or heterogeneous rays was underdetermined. Newton isolated a single ray of light, and showed that *which* part of the light was isolated made a difference to the light's subsequent behavior (specifically its "refrangability," or disposition to refract). This experimental effect resolved the underdetermination by showing that sunlight was heterogeneous: composed of rays of different refrangibility (see Walsh 2012; Currie & Levy, under review).

10. Indeed, Kim Sterelny and I have suggested that the practice of using parallel taxonomies in paleobiology functions to avoid the unproductive speculation of connecting trackways to the fossil record (Currie & Sterelny, forthcoming).

11. For a similar point about species delineation in paleobiology, see Currie (2016c).

12. I suspect this case could be a counterexample to Kosso's claim that midrange theory is *required* for evidential relevance. Schweitzer et al.'s explanation of how the soft tissue fossilized is highly speculative. However, even if we had no idea how the soft tissue was preserved, surely this would not stop us from utilizing the remains when theorizing about *T. rex*'s squishy bits.

13. Thanks to Kathryn Reese-Taylor for information on LiDAR in archaeology.

14. As I will discuss further in chapter 11, in more recent work Turner is sensitive to this worry. He distinguishes between bets based on our current techniques and knowledge and those without such restrictions (2016).

Chapter 5

1. This, however, would involve costs in other places: How the organism's remains are affected over time can tell us about various shifts in climate, environment, and mineral composition. You can't have it all, apparently.

2. What exactly this turns on is an extremely difficult and subtle question; see Sober (1999).

3. Thanks to Colin Klein for pushing this point.

4. See Turner (2013) for criticism.

Chapter 6

1. Maureen O'Malley (2016) has recently made a further objection to Cleland's picture of historical method, which focuses on uncovering new traces. In the context of

molecular phylogenetics, she points out that progress is not so much made by finding new data (the problem, if anything, is the sheer amount of data!) but rather by variations in how various models present such data and the methods used to understand it. To an extent, progress in midrange theory and our general understanding of evolutionary processes is doing the work, not uncovering new traces.

2. Elliott Sober (1984), in response to Bas van Fraassen's (1982) criticisms of the "screening off" view, has developed a weaker notion of common causes that nonetheless provides some a priori reason to prefer single to multiple causes. However, this account is open to the same criticisms as I supply against any other a priori account: The ceteris paribus clause never holds in actual practice, and all the work is done by midrange theory. Local a posteriori justification is where the action is. Moreover, Sober and van Fraassen's discussions are explicitly geared toward questions of realism about unobservables, making their concerns somewhat tangential to mine.

3. Other than how we go about delineating characters and selecting outgroups—but that's a whole other story ...

4. Having said this, I suspect a relatively general defense of common causes might be given for events of high dispersal. These tend to have both heterogeneous and rich trace sets, and so more traces should expected to be unified. If we have independent reason to expect high dispersal vis-à-vis some event in the past, then the principle of common cause might gain traction.

5. Note that with another collaborator (and in a different context) Forber makes a similar point (see Forber & Epstein 2013).

Chapter 7

1. Narrative explanations account for some particular event using a "narrative" form: They situate the explananda in terms of some particular causal history. I am avoiding discussion of the nature of historical explanation here, but see Currie 2014a for discussion.

2. For discussion of such restricted regularities see the following: Cooper (2003) and Lange (2004) for ecology; Cartwright (1993, 1999) for physics and economics; Waters (2007) and Woodward (2010) for biology; Currie (2014, 2015b) for paleobiology; and Potochnik (manuscript) and Woodward (2001, 2003) for more general discussion.

3. Excepting of course the Pacific Northwest tree octopus ...

4. This kind of conclusion is not necessarily avoided by adopting a weaker notion of kinds. Boyd's (1991) homeostatic property clusters, for instance, characterize a "kind" (such as a species) as a group of co-occurring properties that are maintained

by some common mechanism. Some mechanisms seem to encourage parochialism. For instance, on Griffith's view of species (1999), species are homeostatic clusters, where the properties are maintained by (something like) common ancestry. The reliance on a particular history, on the face of it, ties our explanation to the history of that species.

5. I should point out that this is not identical to the banjo that was stolen in chapter 6—Paul is no instrument thief! Rather, he has very kindly guarded my original banjo's post-burglary replacement.

6. "Fragility" and the related concept of "robustness" are complex terms with layers of meaning. I think that, in this context, an intuitive sense will do (see Calcott 2010).

7. Surely, if this were so, then Cope's rule would only hold when temperatures were cooling.

8. Daniel Nolan (2013) argues that counterfactuals matter for human history as well.

9. I am basing these objections on discussions I had with the Latin American Archaeology reading group at Calgary. Any mistakes, of course, are mine.

10. I considered capturing these notions in reference to "homology" thinking versus "analogy" thinking (Ereshefsky 2012), where "homologous" thinking tracks trace-based, historical explanation. I think there is something to this, but decided the biological terms were too contentious (see Brigandt & Griffiths 2007; Currie 2014b).

11. Of course, there is an important sense in which *all* events are causally connected, insofar as it appears that the universe began with a single event, and so all events are connected via common cause. This is not particularly helpful to geologists, however.

12. Turner is in the business of defending his position that arguments against realism that work for small unobservables, such as electrons, are less effective for past unobservables, such as dinosaurs. One advantage that (some) unobservables in the past hold but tiny unobservables lack is that we can observe things like them. Although I cannot observe Snowball Earth, I can observe glaciers; in lieu of extinct dinosaurs, birds and reptiles are available. This option is (at least prima facie) unavailable for protons.

13. Thanks to Liz Irvine for the example.

Chapter 8

1. See Callearts et al. 1997, for a thorough, if dated, overview.

2. Slutsky (2011) argues that although history matters, it matters for everyone, and so there is nothing distinctive here. This is fine for my project, since I am not in the demarcation game. Rather, I focus on what we can know of the past.

3. Alison McConwell and I have recently argued against "source-independent" notions of contingency like Desjardins's. Our position only applies to arguments claiming that some theories are inapplicable to particular domains (i.e., whether evolution by natural selection is sufficient for explaining macroevolutionary patterns), and so it is not directly relevant here.

4. Stuart Glennan discusses something like this in his (2011). There, he discusses how mechanistic accounts of explanation may be extended to include narratives. I argue in my (2014a) that, in effect, Glennan misses the importance of path dependence in narratives.

5. "Right as diverse pathes leden the folk the righte wey to Rome," in Geoffrey Chaucer, *Treatise on the Astrolabe* (Prologue, ll. 39–40).

6. The taxonomic affiliation of *Sparassodonta* is somewhat contentious; it is not clear whether they are true marsupials or a sister-clade (Rougier et al. 1998).

7. Wroe et al. (2013), for instance, were unable to put their study into "full phylogenetic contexts. This is because there are no living close relatives known for *Thylacosmilus atrox*" (5).

8. Taken largely from van Valkenburg & Jenkins 2002.

Chapter 9

1. Note that Cleland is contrasting two different, legitimate ways of doing science. That is, she presents "paradigm," or ideal, archetypes of the method in question. Historical scientists' sometimes using experimental methods does not in itself undermine her distinction.

2. Wendy Parker (2009) suggests that the physicality of simulations—that they produce novel physical information about the physical system on which the simulation runs—provides in-principle grounds for the confirmatory nature of the simulations (or, at least she is read in these terms by Dardashti, Thebault, & Winsberg, [forthcoming]). I see my arguments here as cohering with but logically independent from this point.

3. This probably isn't the whole story, however. Zachos and Sprinkle were unable to generate some of the more irregular morphologies. This may be a limitation of the model itself. It generates round or spherical structures, so it might just be too constrained and simple to manage more outlandish geometries.

4. Recall our discussion of confirmation holism in chapter 6. Given that any test relies on various auxiliary hypotheses—themselves empirically fragile—for evidential relevance, the idea that any one observation alone provides the epistemic goods to unequivocally establish or undermine a proposition is dubious.

5. John Norton (manuscript) for instance, argues that scientific evidence is often irreducibly vague and ambiguous. In these circumstances, strict formal treatments should not be taken literally.

6. See Sepkoski (2016) for discussion of the MBL model in evidential terms.

7. We can contrast theoretically grounded models from "speculative" (or phenomenal) models that rely on largely imaginative and a priori theoretical underpinnings, such as Lokta-Volterra models from ecology and agent-based models of, say flocking behavior or traffic flow (see Winsberg 2003).

8. I take Oreskes, Shrader-Frechette, and Blitza's (1994) skepticism to fall along the same lines.

9. Whether we take this as an analogue or a sample depends on whether we read the target as Earth's climate/ice pack system, or the snowball events themselves. I don't think much turns on this.

Chapter 10

1. Since Hyde et al.'s work, the possibility of the Neoproterozoic's climate being largely methane-fueled has somewhat undermined the relevance of this.

2. In a recent paper Dardashti, Thebault, and Winsberg (forthcoming) have argued for what they call "analog simulations." They take these to be cases where two systems may be successfully modeled, and the models of those systems are related via an approximate isomorphism. Their main reason for separating such cases from standard analogous reasoning is that "the strength or quality of the inferences one can draw by analog simulation is much greater than is that of those which can be drawn via analogical reasoning." This departs from my distinction, which relies instead on the constructed and controlled nature of the simulation studies I am interested in (they note in passing the differences and similarities between their phenomena and simulation studies). They also provide an argument in favor of the confirmatory power of analogue simulations, which appeals to what they call "model external" and "empirically grounded" arguments. I suspect this view is quite similar to mine, but the enormous difference in scientific contexts (they are interested in physics) makes it difficult to tell.

3. For more discussion on the importance of materiality in modeling, see Harre (2003), Morgan (2003, 2005), Parke (2014), and Parker (2009).

4. See, for instance, Godfrey-Smith (2009), Levy (2012), and Weisberg (2013).

5. See, for instance, Woodward (2001) and Strevens (2008) for the role of difference-making in explanation, Lipton and van Fraassen (1980) for contrastive explanation, and Jackson and Pettit (1992) and Sterelny (1996) for discussions of scientific autonomy driven by explanatory adequacy.

6. It is worth briefly comparing models in historical science and in meteorology or climate science. They differ because of how evidential feedback operates. Often both target inference rather than explanation. However, where the meteorologist can adopt an instrumentalist stance on their model's dynamical fidelity (whether it is structurally similar to its target), historical scientists often cannot. This is because meteorologists can rely on continual feedback from the world, and can tweak their model as they go. Historical scientists do not have this privilege. And so, where model development in meteorology is often data-driven, historical scientists rely upon causal hypothesizing. In order to successfully infer from object to target, they need to get the dynamical fidelity approximately right.

7. Van Loon cites two pieces of evidence, first that packs of velociraptor probably preyed on larger pterosaurs (as evidenced by fossil remains) and second, that there are marks on some sauropod remains showing that such animals ate them. Sauropods are a tad bigger than pterosaurs, and I'm guessing the raptor-sauropod interaction was postmortem on the latter's part.

8. Although whether parasitism lowers maximum population number, or instead decides which individuals compose the maximal population, is a contentious issue in ecology (see Cooper, 2003, 27–95).

9. Kim Sterelny's work (2003, 2012) applies notions of evolutionary scaffolds to the evolution of human cognition.

10. Indeed, Megan Delahanty has pointed out to me that such piecemeal progress is very common in experimental biological sciences. Uncovering molecular and other micro-level mechanisms also requires that we establish relatively coarse hypotheses from which to scaffold further investigation. One example could be establishing the chromosomal theory of genetics prior to actually identifying DNA as the bearer of heredity.

11. See, for example, Batterman (2002), Cartwright (1994), Strevens (2008), Weisberg (2007), and Woodward (2000).

12. Kirsten Walsh (under review) provides a fascinating example of reasoning wherein theoretical scaffolds are intended to be later removed.

13. I am not quite sure whether evidence could decrease from a scaffold, although perhaps, as Megan Delahanty has suggested, we might discover that the whole line of investigation has been folly.

Chapter 11

1. This is perhaps a little quick, since betting decisions turn not only on our credence in the relevant proposition but also on the potential risk of getting things wrong and the reward of getting things right.

2. Kirsten Walsh points out that, given Napoleon's importance in later life, there is a better chance of questions about breakfasting habits being remembered and recorded. For human history at least, later importance can sometimes boost the dispersal, and thus the retrievability, of apparently inconsequential events.

3. My reservations about "importance" are twofold. First, I'm not sure that there isn't a kind of circularity here. If "importance" tracks "retrievability," then am I not just saying that we should optimistic about the retrievable facts? Second, I worry about the role of our interests here. By virtue of what is some fact "important"? If this simply relies on what we, as human agents, find important, then this seems like the wrong kind of distinction to do the epistemic work I am after.

4. See Currie and Sterelny (forthcoming) for further discussion of the role of storytelling in historical reconstruction.

5. Derek Turner has rightly sounded a note of caution in this regard by pointing to the public dissemination of such hypotheses (http://www.extinctblog.org/extinct/2016/1/7/waynes-razor-at-the-smithsonian). Given that many speculative hypotheses are often rather dramatic, they tend to be picked up in the media. This means that the general public may find it difficult to discriminate between well-supported hypothesis—justified on the basis of their empirical support—and speculative hypothesis, which are justified by their empirical grounding.

6. Naturally, this has some affinities with Popper's views on hypotheses, which "stick their necks out," the "bold ideas, unjustified anticipations, and speculative thought" (Popper 1959, 280) that he emphasizes. However, the details are quite different: Popper's main concern (although I am not sure he would approve of the term) is confirmation. For him, interesting results are those which either support (i.e., do not falsify) a bold, speculative hypothesis, or falsify a conservative, apparently well-supposed hypothesis. I'm interested in a much wider set of epistemic goods: bold hypotheses might be falsified, yet in so doing, lead to indirect epistemic gains.

Chapter 12

1. For much more sophisticated reflection on science and its role in society, see Jasnoff (2011), Kitcher (2003), and Longino (1990). For rich discussion focusing on how we might structure funding and other incentive structures, see Avin (2015), Kitcher (2003), Muldoon (2013), and Strevens (2003).

Chapter 13

1. Recently, Heather Douglas (2014) has argued that (at least Kuhn's) Big Views of progress have been marred by a problematic distinction between pure and applied science. I think Douglas is right and see the position sketched here as largely complementary.

2. See McConwell and Currie (2016) and the papers in Sepkoski and Ruse (2009).

3. See Darden and Maull (1977).

4. Hasok Chang (2004) expresses and embraces both an unapologetic pluralism and a nonlinear account of progress.

5. Thanks to Derek Turner for helping me see how pluralism about epistemic goods challenges traditional conceptions of realism.

6. It is worth noting that not all presentations of the "no miracles" argument are historical. Some establish realism about particular claims, or in certain contexts. If so, they are situated, and the points I make here are limited in their application.

7. Kirsten Walsh explains to me that if science is a sequence of inductions, then Laudan's argument is an induction over those arguments—and hence is a metainduction. Fair enough.

8. For discussion of the practice of lab-based molecular genetics, see Waters (2000) and Weber (2005).

9. Larry Laudan perhaps took this most seriously—attempting to create a kind of database of episodes in scientific history against which philosophical theories could be tested (see Laudan et al. 1986).

10. See my discussion of "Natural Histories" of science and the references therein (Currie 2015b).

11. See Currie and Walsh (under review) for a more careful discussion of this kind of point.

References

Akersten, W. (1985). Canine function in Smilodon (Mammalia, Felidae, Machairodontinae). *Los Angeles County Museum Contributions in Science, 356*, 1–22.

Alexander, B. K., Beyerstein, B. L., Hadaway, P. F., & Coambs, R. B. (1981). The effects of early and later colony housing on oral ingestion of morphine in rats. *Pharmacology, Biochemistry, and Behavior, 15*, 571–576.

Alroy, J. (1998). Cope's rule and the dynamics of body mass evolution in North American fossil mammals. *Science, 280*(5364), 731–734.

Avin, S. (2015). Funding science by lottery. In U. Maki, I. Votsis, S. Ruphy, & G. Schurz (Eds.), *Recent Developments in the Philosophy of Science: EPSA13 Helsinki* (pp. 111–126). Heidelberg: Springer International Publishing.

Ayala, F. J. (2009). Molecular evolution vis- à- vis paleontology. In D. Sepkoski & M. Ruse (Eds.), *The paleobiological revolution* (pp. 176–198). Chicago: University of Chicago Press.

Bailey, M. (1984). Astronomy: Nemesis for nemesis? *Nature, 311*, 602–603.

Bartha, P. (2013). Analogy and analogical reasoning. In E. N. Zalta (Ed.), *The Stanford encyclopedia of philosophy* (Fall 2013 ed.). Retrieved from http://plato.stanford.edu/archives/fall2013/entries/reasoning-analogy/

Bartha, P. (2010). *By parallel reasoning: The construction and evaluation of analogical arguments*. Oxford, UK: Oxford University Press.

Batterman, R. (2002). Asymptotics and the role of minimal models. *British Journal for the Philosophy of Science, 8*, 21–38.

Beatty, J. (2006). Replaying life's tape. *Journal of Philosophy, 103*(7), 336–362.

Beatty, J., & Desjardins, E. C. (2009). Natural selection and history. *Biology & Philosophy, 24*(2), 231–246.

Ben-Menahem, Y. (1997). Historical contingency. *Ratio, 10*(2), 99–107.

Benton, M. J. (2009). The fossil record: Biological or geological signal? In D. Sepkoski & M. Ruse (Eds.), *The paleobiological revolution* (pp. 43–59). Chicago: University of Chicago Press.

Berner, R. A., VandenBrooks, J. M., & Ward, P. D. (2007). Oxygen and Evolution *Science, 316*(5824), 557–558.

Binford, L. (1977). General introduction. In L. Binford (Ed.), *for theory building in archaeology* (pp. 1–16). New York: Academic Press.

Boyd, R. (1991). Realism, anti-foundationalism and the enthusiasm for natural kinds. *Philosophical Studies, 61*, 127–148.

Brigandt, I., & Griffiths, P. E. (2007). The importance of homology for biology and philosophy. *Biology & Philosophy, 22*(5), 633–641.

Briggs, D. E. (1995). Experimental taphonomy. *Palaios, 10*, 539–550.

Brotchie, A., & Gooding, M. (1991). *Surrealist games*. London: Redstone Press.

Brusatte, S. L. (2011). Calculating the tempo of morphological evolution: Rates of discrete character change in a phylogenetic context. In Ashraf M. T. Elewa (Ed.), *Computational Paleontology* (pp. 53–74). Heidelberg: Springer International Publishing.

Burness, G., Diamond, J., & Flannery, T. (2001). Dinosaurs, dragons, and dwarfs: The evolution of maximal body size. *Proceedings of the National Academy of Sciences of the United States of America, 98*(25), 14518–14523.

Calcott, B. (2011). Wimsatt and the robustness family: Review of Wimsatt's re-engineering philosophy for limited beings. *Biology & Philosophy, 26*(2), 281–293.

Callearts, P., Halder, G., & Gehring, W. J. (1997). Pax-6 in development and evolution. *Annual Review of Neuroscience, 20*, 483–532.

Callender, C. (1997). What is "The problem of the direction of time"? *Philosophy of Science, 64*(Supplement), S223–S234.

Caporael, L. R., Griesemer, J. R., & Wimsatt, W. C. (Eds.). (2014). *Developing scaffolds in evolution, culture, and cognition*. Cambridge, MA: MIT Press.

Carnap, R. (1937). *Logical syntax of language* (Vol. 4). Psychology Press.

Carpenter, K. (2006). Biggest of the big: A critical re–evaluation of the mega–sauropod Amphicoelias fragillimus Cope, 1878. *New Mexico Museum of Natural History and Science Bulletin, 36*, 131–137.

Carr, M. (2012). The fluvial history of Mars. *Phil. Trans. R. Soc. A, 370*(1966), 2193–2215.

Cartwright, N. (1999). *The dappled world: A study of the boundaries of science.* Cambridge: Cambridge University Press.

Cartwright, N. (1994). *Nature's capacities and their measurement.* Oxford, UK: Oxford University Press.

Cartwright, N. (1983). *How the laws of physics lie.* Oxford, UK: Clarendon Press.

Chang, H. (2004). *Inventing temperature: Measurement and scientific progress.* Oxford, UK: Oxford University Press.

Chapman, R., & Wylie, A. (2016). *Evidential reasoning in archaeology.* London: Bloomsbury Publishing.

Chase, A. F., Chase, D. Z., Awe, J. J., Weishampel, J. F., Iannone, G., Moyes, H., et al. (2014). Ancient Maya regional settlement and inter-site analysis: The 2013 west-central Belize LiDAR survey. *Remote Sensing, 6*(9), 8671–8695.

Chaucer, G. (2002). *A treatise on the astrolabe* (Vol. 6). Norman, OK: University of Oklahoma Press.

Christian, A., & Dzemski, G. (2011). Neck posture in sauropods. In N. Klein, K. Remes, C. T. Gee, & P. M. Sander (Eds.), *Biology of the sauropod dinosaurs: Understanding the life of giants* (pp. 251–262). Bloomington, IN: Indiana University Press.

Christian, A. (2010). Some sauropods raised their necks: Evidence for high browsing in Euhelopus zdanskyi. *Biology Letters, 6,* 823–825.

Christiansen, P. (2011). A dynamic model for the evolution of sabrecat predatory bite mechanics. *Zoological Journal of the Linnean Society, 162,* 220–242.

Clauss, M. (2011) Sauropod biology and the evolution of gigantism: What do we know? Neck posture in sauropods. In N. Klein, K. Remes, C. T. Gee, & P. M. Sander (Eds.), *Biology of the sauropod dinosaurs: Understanding the life of giants* (pp. 4–11). Bloomington, IN: Indiana University Press.

Cleland, C. E. (2013). Common cause explanation and the search for a smoking gun. In V. Baker (Ed.), *125th anniversary volume of the Geological Society of America: Rethinking the fabric of geology, Special Paper 502* (pp. 1–9).

Cleland, C. E. (2011). Prediction and explanation in historical natural science. *British Journal for the Philosophy of Science, 62,* 551–582.

Cleland, C. E. (2002). Methodological and epistemic differences between historical science and experimental science. *Philosophy of Science, 69*(3), 447–451.

Cleland, C. E. (2001). Historical science, experimental science, and the scientific method. *Geology, 29*(11), 987–990.

Colyvan, M., & Steele, K. (2011). Environmental ethics and decision theory: Fellow travellers or bitter enemies? In B. Brown, K. Laplante, & K. Peacock (Eds.), *Philosophy of ecology : Handbook of the philosophy of science.* Amsterdam, Holland: North Holland/ Elsevier.

Conway Morris, S. (2003). *Life's solution: Inevitable humans in a lonely universe.* Cambridge, UK: Cambridge University Press.

Cooper, G. (2003). *The science of the struggle for existence: On the foundations of ecology.* Cambridge, UK: Cambridge University Press.

Cope, E. (1887). *The origin of the fittest.* New York: Appleton.

Craver, C. F. (2007). *Explaining the brain: Mechanisms and the mosaic unity of neuroscience.* Oxford, UK: Oxford University Press/Clarendon Press.

Currie, A. (forthcoming). The argument from surprise. *Canadian Journal of Philosophy.*

Currie, A. (2016a). Hot-blooded gluttons: Dependency, coherence & method in the historical sciences. *British Journal for the Philosophy of Science.* Online First.

Currie, A. (2016b). Ethnographic analogy, the comparative method, and archaeological special pleading. *Studies in History and Philosophy of Science, 55,* 84–94.

Currie, A. (2016c). The mystery of the Triceratops' mother: How to be a realist about the species category. *Erkenntnis, 81,* 795–816.

Currie, A. (2015a). Philosophy of science and the curse of the case study. In C. Daly (Ed.), *The Palgrave Handbook of Philosophical Methods* (pp. 553–572). London: Palgrave Macmillan.

Currie, A. (2015b). Marsupial lions and methodological omnivory: Function, success and reconstruction in paleobiology. *Biology & Philosophy, 30,* 187–209.

Currie, A. M. (2014a). Narratives, mechanisms and progress in historical science. *Synthese, 191*(6), 1163–1183.

Currie, A. (2014b). Venomous dinosaurs and rear-fanged snakes: Homology and homoplasy characterized. *Erkenntnis, 79*(3), 701–727.

Currie, A. (2013). Convergence as evidence. *British Journal for the Philosophy of Science, 64*(4), 763–786.

Currie, A. (2012). Convergence, contingency & morphospace. *Biology & Philosophy, 27*(4), 583–593.

Currie, A., & Levy, A. (under review). Why experiments matter.

Currie, A., & Sterelny, K. (forthcoming). In defence of story-telling. *Studies in History and Philosophy of Science.*

Currie, A., & Turner, D. (2016). Introduction: Scientific knowledge of the deep past. *Studies in History and Philosophy of Science, 55,* 43–46.

Dardashti, R., Thebault, K. P., & Winsberg, E. (2016). Confirmation via analogue simulation: What dumb holes could tell us about gravity. *British Journal for the Philosophy of Science*. Online first. https://academic.oup.com/bjps/.

Darden, L., & Maull, N. (1977). Interfield theories. *Philosophy of Science, 43*, 44–64.

Dawkins, R. (1983). Universal Darwinism. In D. S. Bendall (Ed.), *Evolution from molecules to man* (pp. 403–425). Cambridge: Cambridge University Press.

Dennett, D. C. (1995). *Darwin's dangerous idea*. New York: Simon & Schuster.

Desjardins, E. (2011). Historicity and experimental evolution. *Biology & Philosophy, 26*(3), 339–364.

Diamond, J., & Robinson, J. A. (Eds.). (2010). *Natural experiments of history*. Harvard University Press.

Donnadieu, Y., Godderis, Y., Ramstein, G., Nedelec, A., & Meert, J. (2004). A "snowball Earth" climate triggered by continental break-up through changes in runoff. *Nature, 428*, 303–306.

Douglas, H. (2014). Pure science and the problem of progress. *Studies in History and Philosophy of Science, 46*, 55–63.

Dowe, P. (1992). Process causality and asymmetry. *Erkenntnis, 37*, 179–196.

Dretske, F. (1981). *Knowledge and the flow of information*. Cambridge, MA: MIT Press.

Duhem, P. (1954). *The aim and structure of physical theory*. Princeton, NJ: Princeton University Press.

Elga, A. (2001). Statistical mechanics and the asymmetry of counterfactual dependence. *Philosophy of Science, 68*, S313–S324.

Elliott-Graves, A. (2016). The problem of prediction in invasion biology. *Biology & Philosophy, 31*(3), 373–393.

Eng, C. M., Ward, S. R., Vinyard, C. J., & Taylor, A. B. (2009). The morphology of the masticatory apparatus facilitates muscle force production at wide jaw gapes in tree-gouging common marmosets (Callithrix jacchus). *The Journal of Experimental Biology, 212*, 4040–4055.

Ereshefsky, M. (2014). Species, historicity, and path dependency. *Philosophy of Science, 81*(5), 714–726.

Ereshefsky, M. (2012). Homology thinking. *Biology & Philosophy, 27*(3), 381–400.

Evans, D. A. D. (2000). Stratigraphic, geochronological, and paleomagnetic constraints upon the Neoproterozoic climatic paradox. *American Journal of Science, 300*, 347–433.

Farlow, J. O., Coroian, I. D., & Foster, J. R. (2010). Giants on the landscape: Modelling the abundance of megaherbivorous dinosaurs of the Morrison Formation (Late Jurassic, western USA). *Historical Biology, 22*, 403–429.

Farlow, J. O. (1990). Dinosaur energetics and thermal biology. In D. B. Weishampel, P. Dodson, & H. Osmolska (Eds.), *The dinosauria* (pp. 43–55). Berkeley, CA: University of California Press.

Faye, J. (2010). Backward causation. In E. N. Zalta (Ed.), *The Stanford encyclopedia of philosophy* (Spring 2010 ed.). Retrieved from http://plato.stanford.edu/archives/spr2010/entries/causation-backwards.

Fernandez-Diaz, J. C., Carter, W. E., Shrestha, R. L., & Glennie, C. L. (2014). Now you see it ... now you don't: Understanding airborne mapping LiDAR collection and data product generation for archaeological research in Mesoamerica. *Remote Sensing, 6*(10), 9951–10001.

Fetzer, J. H. (1981). Probability and explanation. *Synthese, 48*(3), 371–408.

Forber, P. (2009). Spandrels and a pervasive problem of evidence. *Biology & Philosophy, 24*(2), 247–266.

Forber, P., & Epstein, B. (2013). The perils of tweaking: How to use macrodata to set parameters in complex simulations. *Synthese, 190*, 203–218.

Forber, P., & Griffith, E. (2011). Historical reconstruction: Gaining epistemic access to the deep past. *Philosophy and Theory in Biology, 3*, 1–19.

Fortelious, M., & Kappelman, J. (1993). The largest land mammal ever imagined. *Zoological Journal of the Linnean Society, 108*(1), 85–101.

Frigg, R., and Reiss, J. (2009). The philosophy of simulation: Hot new issues or same old stew. *Synthese, 169*, 593–613.

Galison, P. (1996). Computer simulations and the trading zone. In P. Galison & D. J. Stump (Eds.), *The disunity of science: Boundaries, contexts, and power* (pp. 118–157). Palo Alto, CA: Stanford University Press.

Ganse, B., Stahn, A., Stoinski, S., Suthau, T., & Gunga, H.-C. (2011). Body mass estimation, thermoregulation and cardiovascular physiology of large sauropods. In N. Klein, K. Remes, G. T. Gee, & P. M. Sander (Eds.), *Biology of the sauropod dinosaurs: Understanding the life of giants* (pp. 105–118). Bloomington: Indiana University Press.

Gao, T. P., Shih, C. K., Xu, X., Wang, S., & Ren, D. (2012). Mid-Mesozoic flea-like ectoparasites of feathered or haired vertebrates. *Current Biology, 22*(8), 732–735.

Gee, H. (2000). *Deep time: Cladistics, the revolution in evolution.* London: Fourth Estate.

Gillooly, J. F., Allen, A. P., & Charnov, E. L. (2006). Dinosaur fossils predict body temperatures. *PLoS Biology, 4*, 1467–1469.

Glennan, S. (2010). Ephemeral mechanisms and historical explanation. *Erkenntnis, 72*(2), 251–266.

Godfrey-Smith, P. (2008). Recurrent, transient underdetermination and the glass half-full. *Philosophical Studies, 137*, 141–148.

Godfrey-Smith, P. (2006). The strategy of model-based science. *Biology & Philosophy, 21*(5), 725–740.

Goodman, N. (1955). *Fact, fiction, and forecast.* Indianapolis: Bobbs-Merrill.

Gould, S. J. (1996). *Full house: The spread of excellence from Plato to Darwin.* New York: Harmony Books.

Gould, S. J. (1989). *The Burgess Shale and the nature of history.* New York: Norton.

Gould, S. J., & Eldredge, N. (1977). Punctuated equilibria: The tempo and mode of evolution reconsidered. *Paleobiology, 3*(2), 115–151.

Gould, S. J., Raup, D. M., Sepkoski, J., Schopf, T., & Simberloff, D. (1977). The shape of evolution: A comparison of real and random clades. *Paleobiology*, (3), 23–40.

Griebeler, E. M., & Werner, J. (2011). The life-cycle of sauropod dinosaurs. In N. Klein, K. Remes, G. T. Gee, & P. M. Sander (Eds.), *Biology of the sauropod dinosaurs: Understanding the life of giants* (pp. 263–275). Bloomington: Indiana University Press.

Griffiths, P. E. (2007). Evo-devo meets the mind: Towards a developmental evolutionary psychology. In R. B. Sansom & N. Robert (Eds.), *Integrating evolution and development: From theory to practice* (pp. 195–226). Cambridge, MA: MIT Press.

Griffiths, P. E. (2006). Function, homology, and character individuation. *Philosophy of Science, 73*, 1–25.

Griffiths, P. E. (1999). Squaring the circle: Natural kinds with historical essences. In R. A. Wilson (Ed.), *Species: New interdisciplinary essays* (pp. 208–228). Cambridge, MA: MIT Press.

Griffiths, P. E. (1996). The historical turn in the study of adaptation. *British Journal for the Philosophy of Science, 47*, 511–532.

Griffiths, P. E. (1994). Cladistic classification and functional explanation. *Philosophy of Science, 61*, 206–227.

Gunga, H. Suthau, T. Bellmann, Anke. Stoinski, S. Friedrich, A. Trippel, T. Kirsch, K. Hellwich, O. (2008). A new body mass estimation of *Brachiosaurs bancai* Janensch, 1919 mounted and exhibited at the Museum of Natural History (Berlin, Germany). *Fossil, 11*(1), 33–38.

Hacking, I. (1982). Experimentation and scientific realism. *Philosophical Topics*, *13*(1), 71–87.

Hacking, I. (1965). *Logic of statistical inference*. Cambridge, UK; New York: Cambridge University Press.

Hall, B. K. (2003). Descent with modification: The unity underlying homology and homoplasy as seen through an analysis of development and evolution. *Biological Reviews of the Cambridge Philosophical Society*, *78*(3), 409–433.

Harré, R. (2003). The materiality of instruments in a metaphysics for experiments. In H. Radder (Ed.), *The philosophy of scientific experimentation* (pp. 19–38). Pittsburgh, PA: University of Pittsburgh Press.

Hawkes, C. (1954). Archeological theory and method: Some suggestions from the old world. *American Anthropologist*, *56*, 155–168.

Hayter, H. (1994). Hunter-gatherers and the ethnographic analogy: Theoretical perspectives. *Totem: The University of Western Ontario Journal of Anthropology*, *1*(1).

Hesse, M. (1966). *Models and analogies in science*. Notre Dame, IN: University of Notre Dame Press.

Hoffman, P. F. (1999). Snowball Earth theory still stands. *Nature*, *400*, 708.

Hoffman, P. F., & Schrag, D. P. (2002). The snowball Earth hypothesis: Testing the limits of global change. *Terra Nova*, *14*(3), 129–155.

Hone, D. W., & Benton, M. J. (2005). The evolution of large size: How does Cope's rule work? *Trends Ecol. Evol. (Amst.)*, *20*(1), 4–6.

Hone, D. W. E., Keesey, T. M., Pisani, D., & Purvis, A. (2005). Macroevolutionary trends in the Dinosauria: Cope's rule. *Journal of Evolutionary Biology*, *18*, 587–595.

Horwich, P. (1987). *Asymmetries in time*. Cambridge, MA: MIT Press.

Hull, D. L. (1988). *Science as a process: An evolutionary account of the social and conceptual development of science*. Chicago: University of Chicago Press.

Hull, D. L. (1976). Are species really individuals? *Systematic Zoology*, *25*, 174–191.

Hummel, J., & Clauss, M. (2011). Sauropod feeding and digestive physiology. In N. Klein, K. Remes, G. T. Gee, & P. M. Sander (Eds.), *Biology of the sauropod dinosaurs: Understanding the life of giants* (pp. 11–33). Bloomington: Indiana University Press.

Hummel, J., Gee, C. T., Südekum, K.-H., Sander, P. M., Nogge, G., & Clauss, M. (2008). In vitro digestibility of fern and gymnosperm foliage: Implications for sauropod feeding ecology and diet selection. *Proceedings. Biological Sciences*, *275*, 1015–1021.

Hunt, G., & Roy, K. (2006). Climate change, body size evolution, and Cope's rule in deap-sea ostracods. *Proceedings of the National Academy of Sciences of the United States of America, 103*(5), 1347–1352.

Huss, J. (2009). The shape of evolution: The MBL model and clade shape. In D. Sepkoski & M. Ruse (Eds.), *The paleobiological revolution*. Chicago: University of Chicago Press.

Hutson, S. R. (2015). Adapting LiDAR data for regional variation in the tropics: A case study from the northern Maya lowlands. *Journal of Archaeological Science: Reports, 4*, 252–263.

Hyde, W. T., Crowley, T. J., et al. (2000). Neoproterozoic /'snowball Earth'/simulations with a coupled climate/ice-sheet model. *Nature, 405*(6785), 425–429.

Jackson, F., & Pettit, P. (1992). In defense of explanatory ecumenism. *Economics and Philosophy, 8*(1), 1–21.

Jasanoff, S. (2011). *Designs on nature: Science and democracy in Europe and the United States*. Princeton, NJ: Princeton University Press.

Jeffares, B. (2010). Guessing the future of the past. *Biology & Philosophy, 25*(1), 125–142.

Jeffares, B. (2008). Testing times: Regularities in the historical sciences. *Studies in History and Philosophy of Science Part C, 39*(4), 469–475.

Jerolmack, D. (2013). Pebbles on Mars. *Science, 340*(6136), 1055–1056.

Kim, J. (2000). *Mind in a physical world: An essay on the mind-body problem and mental causation*. Cambridge, MA: MIT Press.

Kirschvink, J. L. (1992). Late Proterozoic low-latitude glaciation: The snowball Earth. In J. W. Schopf & C. Klein (Eds.), *The proterozoic biosphere* (pp. 51–52). Cambridge, UK: Cambridge University Press.

Kitcher, P. (2003). *Science, truth, and democracy*. Oxford, UK: Oxford University Press.

Kleinhans, M. G., Buskes, C., & de Regt, H. (2010). Philosophy of earth science. In F. Allhoff, (Eds.), *Philosophies of the sciences*. Oxford, UK: Wiley-Blackwell.

Kleinhans, M. G., Buskes, C., & de Regt, H. (2005). Terra incognita: Explanation and reduction in earth science. *International Studies in the Philosophy of Science, 19*, 289–317.

Kosso, P. (2001). *Knowing the past: Philosophical issues of history and archaeology*. San Jose, CA: Humanity Books.

Kukla, A. (1996). Does every theory have empirically equivalent rivals? *Erkenntnis, 44*, 137–166.

Lacovara, K. J., Lamanna, M. C., Ibiricu, L. M., Poole, J. C., Schroeter, E. R., Ullmann, P. V., et al. (2014). A gigantic, exceptionally complete Titanosaurian sauropod dinosaur from southern Patagonia, Argentina. *Scientific Reports, 4*(7). DOI: 10.1038/srep06196.

Ladyman, J. (1998). What is structural realism? *Studies in History and Philosophy of Science, 29*(3), 409–424.

Lakatos, I. (1978). *The methodology of scientific research programmes: Philosophical papers* (Vol. 1). Cambridge, UK; New York: Cambridge University Press.

Lange, M. (2005). Ecological laws: What would they be and why would they matter? *Oikos, 110*(2), 394–403.

Laudan, L. (1990). Demystifying underdetermination. In C. W. Savage (Ed.), *Scientific Theories* (pp. 267–297). Minneapolis: University of Minnesota Press.

Laudan, L. (1981). A confutation of convergent realism. *Philosophy of Science, 48*(1), 19–49.

Laudan, L., Donovan, A., Laudan, R., Barker, P., Brown, H., Leplin, J., et al. (1986). Scientific change: Philosophical models and historical research. *Synthese, 69*(2), 141–223.

Laudan, L., & Leplin, J. (1991). Empirical equivalence and underdetermination. *The Journal of Philosophy, 88*(9), 449–472.

Lehane, M. J. (2005). *The biology of blood-sucking in insects.* Cambridge, UK; New York: Cambridge University Press.

Leonelli, S. (2016). *Data-centric biology: A philosophical study.* Chicago: University of Chicago Press.

Levy, A., & Currie, A. (2015). Model organisms are not (theoretical) models. *British Journal for the Philosophy of Science, 66*(2), 327–348.

Levy, A. (2012). Models, fictions, and realism: Two packages. *Philosophy of Science, 79*(5), 738–748.

Lewis, D. (2000). Causation as influence. *Journal of Philosophy, 97*(4), 182–197.

Lewis, D. (1986). Causal explanation. In D. Lewis (Ed.), *Philosophical papers: Vol. II* (pp. 214–240). Oxford, UK: Oxford University Press.

Lewis, D. (1981). A subjectivist's guide to objective chance. In W. L. Harper, R. Stalnaker, and G. Pearce (Eds.), *Ifs* (pp. 267–297). Dordrecht: Springer International Publishing.

Lewis, D. (1979). Counterfactual dependence and time's arrow. *Noûs, 13*(4): 455–476.

Lewis, D. (1973). Causation. *Journal of Philosophy, 70,* 556–567.

Lewontin, R. C. (2002). Directions in evolutionary biology. *Annual Review of Genetics, 36*(1), 1–18.

Lewontin, R. C. (1998). The evolution of cognition: Questions we will never answer. In D. Scarborough & S. Sternberg (Eds.), *An invitation to cognitive science, vol 4: Methods, models, and conceptual issues*. Cambridge, MA: MIT Press.

Lewontin, R. (1967). The principle of historicity in evolution. In P. Moorhead (Ed.), *Mathematical challenges to the neo-Darwinian interpretation of evolution*. Philadelphia: Wistar Institute.

Linton, C. M. (2004). *From Eudoxus to Einstein—A history of mathematical astronomy*. Cambridge, UK: Cambridge University Press.

Lipton, P. (1990). Contrastive explanation. In D. Knowles (Ed.), *Explanation and its limits* (pp. 247–266). Cambridge, UK: Cambridge University Press.

Long, J., Archer, N., Flannery, T., & Hand, S. (2003). *Prehistorical mammals of Australia and New Guinea*. Sydney: UNSW Press.

Longino, H. E. (1990). *Science as social knowledge: Values and objectivity in scientific inquiry*. Princeton, NJ: Princeton University Press.

Lyon, A. (2011). Deterministic probability: Neither chance nor credence. *Synthese, 182*(3), 413–432.

MacArthur, R., & Wilson, E. O. (1967). *The theory of island biogeography*. Princeton, NJ: Princeton University Press.

Matthewson, J., & Weisberg, M. (2009). The structure of tradeoffs in model building. *Synthese, 170*(1), 169–190.

Mazariegos, O. C., Tiesler, V., Gomez, O., & Price, T. D. (2015). Myth, ritual and human sacrifice in early classic Mesoamerica: Interpreting a cremated double burial from Tikal, Guatemala. *Cambridge Archaeological Journal, 25*(01), 187–210.

Martin, J., Martin–Rolland, V., & Frey, E. (1998). Not cranes or masts, but beams: The biomechanics of sauropod necks. *Oryctos, 1*, 113–120.

Martin, L. D., Babiarz, J. P., Naples, V. L., & Hearst, J. (2000). Three ways to be a saber-toothed cat. *Naturwissenschaften, 87*, 41–44.

McConwell, A. K., & Currie, A. (2016). Gouldian arguments and the sources of contingency. [Online First.]. *Biology & Philosophy, December*, 1–19.

McDermott, M. (1995). Redundant causation. *British Journal for the Philosophy of Science, 46*(4), 523–544.

McGhee, G. R. (2011). *Convergent evolution: Limited forms most beautiful*. Cambridge, MA.: MIT Press.

McIntyre, L. (1997). Gould on laws in biological science. *Biology & Philosophy, 12*(3), 357–367.

McNab, B. K. (2009). Resources and energetics determined dinosaur maximal size. *Proceedings of the National Academy of Sciences of the United States of America, 106,* 12184–12188.

McShea, D. W. (1994). Mechanisms of large-scale evolutionary trends. *Evolution; International Journal of Organic Evolution, 48,* 1747–1763.

Meketa, I. (2012). *The false dichotomy between experiment and observation: The case of comparative cognition.* Unpublished manuscript.

Midgley, J. J., Midgley, G., & Bond, W. J. (2002). Why were dinosaurs so large? A food quality hypothesis. *Evolutionary Ecology Research, 4,* 1093–1095.

Mill, J. S. (1904). On nature (1874). In *Collected works* (vol. 10), 400.

Mitchell, S. D. (2003). *Biological complexity and integrative pluralism.* Cambridge, UK; New York: Cambridge University Press.

Mitchell, S. D. (2002). Integrative pluralism. *Biology & Philosophy, 17*(1), 55–70.

Mitchell, S. (1997). Pragmatic laws. *Philosophy of Science, 64*(4), 479.

Morgan, M. (2005). Experiments versus models: New phenomena, inference and surprise. *Journal of Economic Methodology, 12*(2), 317–329.

Morgan, M. (2003). Experiments without material intervention: Model experiments, virtual experiments and virtually experiments. In H. Radder (Ed.), *The philosophy of scientific experimentation* (pp. 216–235). Pittsburgh, PA: University of Pittsburgh Press.

Morrison, M., & Morgan, M. (Eds.). (1999). *Models as mediators.* Cambridge, UK; New York: Cambridge University Press.

Muldoon, R. (2013). Diversity and the division of cognitive labor. *Philosophy Compass, 8*(2), 117–125.

Nolan, D. (2013). Why historians (and everyone else) should care about counterfactuals. *Philosophical Studies, 163*(2), 317–335.

Norton, J. D. (draft). The material theory of induction.

Norton, J. D. (2003). A material theory of induction. *Philosophy of Science, 70*(4), 647–670.

Okasha, S. (2011). Experiment, observation and the confirmation of laws. *Analysis, 71,* 222–232.

O'Malley, M. A. (2016). Histories of molecules: Reconciling the past. *Studies in History and Philosophy of Science, 55,* 69–83.

Oreskes, N., Shrader-Frechette, K., & Belitz, K. (1994). Verification, validation and confirmation of numerical models in the earth sciences. *Science, 263,* 641–642.

Parke, E. (2014). Experiments, simulations, and epistemic privilege. *Philosophy of Science, 81*(4), 516–536.

Parker, W. S. (2011). When climate models agree: The significance of robust model predictions. *Philosophy of Science, 78*(4), 579–600.

Parker, W. (2009). Confirmation and adequacy-for-purpose in climate modelling. *Aristotelian Society Supplementary, 83*(1), 233–249.

Parker, W. (2008). Franklin, Holmes, and the epistemology of computer simulation. *International Studies in the Philosophy of Science, 22*(2), 165–183.

Paul, G. S. (1994). Is Garden Park home to the world's largest known land animal? *Garden Park Paleontology Society, 4,* 5.

Pearce, T. (2012). Convergence and parallelism in evolution: A Neo-Gouldian account. *British Journal for the Philosophy of Science, 63,* 429–448.

Peltier, W. R., Liu, Y., & Crowley, J. W. (2007). Snowball Earth prevention by dissolved organic carbon remineralization. *Nature, 450,* 813–818.

Pian, R., Archer, M., & Hand, S. J. (2013). A new, giant platypus, *Obdurodon tharalkooschild,* SP. Nov. (Monotremata, Ornithorhynchidae) from the Riversleigh world heritage area, Australia. *Journal of Vertebrate Paleontology, 33,* 1255–1259.

Poinar, G. O. (2012). Palaeontology: The 165-million-year itch. *Current Biology, 22*(8), R278–R280.

Popper, K. (1959). *The logic of scientific discovery.* Abingdon, UK: Routledge.

Potochnik, A. (forthcoming). *Idealization and the aims of science.* Chicago: Chicago University Press.

Potochnik, A. (2015). The diverse aims of science. *Studies in History and Philosophy of Science, 53,* 71–80.

Powell, R. (2012). Convergent evolution and the limits of natural selection. *European Journal for Philosophy of Science, 2*(3), 355–373.

Psillos, S. (2005). *Scientific realism: How science tracks truth.* Abingdon, UK: Routledge.

Putnam, H. (1982). Three kinds of scientific realism. *The Philosophical Quarterly, 32*(128), 195–200.

Putnam, H. (1980). *Mind. language, and reality.* Cambridge, UK: Cambridge University Press.

Quine, W. (1951). Two dogmas of empiricism. *Philosophical Review, 60,* 20–43.

Railton, P. (1978). A deductive-nomological model of probabilistic explanation. *Philosophy of Science, 45*(2), 206–226.

Rankin, B. D., Fox, J. W., Barrón-Ortiz, C. R., Chew, A. E., Holroyd, P. A., Ludtke, J. A., et al. (2015). The extended price equation quantifies species selection on mammalian body size across the Palaeocene/Eocene thermal maximum. In *Proceedings of the Royal Society B 282*(1812).

Rauhut, O., Fechner, R., Remes, K., & Reis, K. (2011). How to get big in the Mesozoic: The evolution of the sauropodomorph body plan. In N. Klein, K. Remes, G. T. Gee, & P. M. Sander (Eds.), *Biology of the sauropod dinosaurs: Understanding the life of giants* (pp. 119–149). Bloomington: Indiana University Press.

Raup, D. M. (1993). *Extinction: Bad genes or bad luck?* Oxford, UK: Oxford University Press.

Raup, D. M., & Sepkoski, J. J. (1984). Periodicity of extinctions in the geologic past. *Proceedings of the National Academy of Sciences of the United States of America, 81*(3), 801–805.

Reichenbach, H. (1956). *The direction of time*. Berkeley, CA: University of Los Angeles Press.

Rosales, A. 2014. *The narrative structure of scientific theorizing* (Unpublished doctoral dissertation). University of British Columbia, Canada.

Rosenberg, A. (2006). *Darwinian reductionism: Or, how to stop worrying and love molecular biology*. Chicago: University of Chicago Press.

Rosenberg, A. (2000). *Darwinism in philosophy, social science, and policy*. Cambridge, UK: Cambridge University Press.

Rougier, G., Wible, J., & Novacek, M. (1998). First implications of Deltatheridium specimens for early marsupial history. *Nature, 396*, 459–463.

Runnegar, B. (2000). Palaeoclimate: Loophole for snowball earth. *Nature, 405*, 403–404.

Ruxton, G., & Wilkinson, D. (2011). The energetics of low browsing in sauropods. *Biology Letters, 7*(5), 779–781.

Saliba, G. (2009). Islamic reception of Greek astronomy. In D. Valls-Gabaud & A. Boksenberg (Eds.), The Role of Astronomy in Society and Culture. Proceedings IAU Symposium No. 260, pp. 149–165.

Salmon, W. (1984). *Scientific explanation and the causal structure of the world*. Princeton, NJ: Princeton University Press.

Salmon, W. (1975) Theoretical explanation. In S. Korner (Ed.), *Explanation* (pp. 118–145). Oxford, UK: Basil Blackwell.

Sanchez, J., & Borkovic, B. (2016). The palaeontology flood mitigation project: The impact of flooding on fossils in southern Alberta rivers. Alberta Palaeontological Society Twentieth Anniversary Symposium, abstracts.

Sander, P. M., & Clauss, M. (2008). Sauropod gigantism. *Science, 322*, 200–201.

Sander, P. M., Christian, A., Clauss, M., Fechner, R., Gee, C. T., Griebeler, E.-M., et al. (2011). Biology of the sauropod dinosaurs: The evolution of gigantism. *Biological Reviews of the Cambridge Philosophical Society, 86*(1), 117–155.

Sanders, F., Manley, K., & Carpenter, K. (2001). Gastroliths from the lower Cretaceous sauropod Cedarosaurus weiskopfae. In D. H. Tanke & K. Carpenter (Eds.), *Mesozoic vertebrate life: New research inspired by the paleontology of Philip J. Currie* (pp. 166–180). Bloomington: Indiana University Press.

Schmidt, G. (2010). Enhancing the relevance of palaeoclimate model/data comparisons for assessments of future climate change. *Journal of Quaternary Science, 25*(1), 79–87.

Schwarz, D., Ikejiri, T., Breithaupt, B. H., Sander, P. M., & Klein, N. (2007). A nearly complete skeleton of an early juvenile diplodocid (Dinosauria: Sauropoda) from the lower Morrison Formation of north central Wyoming and its implications for early ontogeny and pneumaticity in sauropods. *Historical Biology, 19*, 225–253.

Schweitzer, M., Wittmeyer, J., Horner, J., & Toporski, J. (2005). Soft-tissue vessels and cellular preservation in Tyrannosaurus rex. *Science, 25*, 1952–1955.

Sellers, W. I., Margetts, L., Coria, R. A., & Manning, P. L. (2013). March of the titans: The locomotor capabilities of sauropod dinosaurs. *PLoS One, 8*(10), e78733.

Senter, P. (2007). Necks for sex: Sexual selection as an explanation for sauropod dinosaur neck elongation. *Journal of Zoology, 271*(1), 45–53.

Sepkoski, D. (2016). "Replaying life's tape": Simulations, metaphors, and historicity in Stephen Jay Gould's view of life. *Studies in History and Philosophy of Science Part C Studies in History and Philosophy of Biological and Biomedical Sciences.* DOI: 10.1016/j. shpsc.2015.12.009.

Seymour, R. S. (2009). Raising the sauropod neck: It costs more to get less. *Biology Letters, 5*, 317–319.

Sklar, L. (1977). What might be right about the causal theory of time. *Synthese, 35*(2), 155–171.

Slutsky, D. (2012). Confusion and dependence in uses of history. *Synthese, 184*(3), 261–286.

Smart, J. J. C. (2008). The tenseless theory of time. In T. Sider, J. Hawthorne, & D. W. Zimmerman (Eds.), *Contemporary debates in metaphysics* (pp. 211–225). Hoboken, NJ: Wiley-Blackwell Publishing.

Smart, J. J. C. (1963). *Philosophy and scientific realism*. Abingdon, UK: Routledge.

Smart, J. J. C. (1959). Can biology be an exact science? *Synthese, 11*(4), 359–368.

Smith, A. B., Zamora, S., & Álvaro, J. J. (2013). The oldest echinoderm faunas from Gondwana how that echinoderm body plan diversification was rapid. *Nature Communications, 4*, 1385.

Sober, E., & Steel, M. (2014). Time and knowability in evolutionary processes. *Philosophy of Science, 81*(4), 558–579.

Sober, E. (2009). Absence of evidence and evidence of absence: Evidential transitivity in connection with fossils, fishing, fine-tuning, and firing squads. *Philosophical Studies, 143*(1), 63–90.

Sober, E. (1990). Explanation in biology: Let's razor Ockham's razor. *Royal Institute of Philosophy*, (Supplement, 27), 73–93.

Sober, E. (1988). *Reconstructing the past: Parsimony, evolution, and inference*. Cambridge, MA: MIT Press.

Sober, E. (1984). Common cause explanation. *Philosophy of Science, 56*, 275–287.

Sober, E. (1981). The principle of parsimony. *British Journal for the Philosophy of Science, 314*(5801), 145–156.

Sodergren, E., Weinstock, G. M., Davidson, E. H., Cameron, R. A., Gibbs, R. A., Angerer, R. C., et al. (2006). The genome of the sea urchin Strongylocentrotus purpuratus. *Science, 32*(2), 941–952.

Sohl, L. E., Christie-Blick, N., & Kent, D. V. (1999). Paleomagnetic polarity reversals in Marinoan (ca 600 Ma) glacial deposits of Australia: Implications for the duration of low-latitude glaciation in Neoproterozoic time. *Geological Society of America Bulletin, 111*, 1120–1139.

Soinski, S., Suthau, T., & Gunga, H. (2011). Reconstructing body volume and surface area of dinosaurs using laser scanning and photogrammetry. In N. Klein, K. Remes, G. T. Gee, & P. M. Sander (Eds.), *Biology of the sauropod dinosaurs: Understanding the life of giants* (pp. 94–104). Bloomington: Indiana University Press.

Stanford, P. K. (2015). Unconceived alternatives and conservatism in science: The impact of professionalization, peer-review, and Big Science. *Synthese.* DOI: 10.1007/s11229-015-0856-4.

Stanford, P. K. (2009). Underdetermination of scientific theory. In E. N. Zalta (Ed.), *The Stanford encyclopedia of philosophy*. https://plato.stanford.edu/archives/fall2009/entries/scientific-underdetermination.

Stanford, P. K. (2006). *Exceeding our grasp: Science, history, and the problem of unconceived alternatives*. Oxford, UK: Oxford University Press.

Steele, K., & Werndl, C. (2013). Climate models, calibration, and confirmation. *British Journal for the Philosophy of Science, 64*(3), 609–635.

Sterelny, K. (2012). *The evolved apprentice*. Cambridge, MA: MIT Press.

Sterelny, K. (2003). *Thought in a hostile world: The evolution of human cognition*. Oxford, UK: Wiley-Blackwell.

Sterelny, K. (1996). Explanatory pluralism in evolutionary biology. *Biology & Philosophy, 11*(2), 193–214.

Strevens, M. (2008). *Depth: An account of scientific explanation*. Cambridge, MA: Harvard University Press.

Strevens, M. (2003). The role of the priority rule in science. *Journal of Philosophy, 100*(2), 55–79.

Tarasov, L., & Peltier, W. R. (1997). Terminating the 100 kyr ice age cycle. *Journal of Geophysical Research, 102*(D18), 21665–21693.

Tucker, A. (2011). Historical science, over- and underdetermined: A study of Darwin's inference of origins. *British Journal for the Philosophy of Science, 62*(4), 805–829.

Tucker, A. (2004). *Our knowledge of the past: A philosophy of historiography*. Cambridge, UK: Cambridge University Press.

Tucker, A. (1998). Unique events: The underdetermination of explanation. *Erkenntnis, 48*, 59–80.

Turner, D. (2016). Another look at the color of dinosaurs. *Journal of Studies in History & Philosophy of Science Part A, 55*, 60–68.

Turner, D. (2013). Historical geology: Methodology and metaphysics. In V. Baker (Ed.), *125th anniversary volume of the Geological Society of America: Rethinking the fabric of geology, special paper 502* (pp. 11–18).

Turner, D. (2011a). Gould's replay revisited. *Biology & Philosophy, 26*(1), 65–79.

Turner, D. (2011b). *Paleontology: A philosophical introduction*. Cambridge, UK: Cambridge University Press.

Turner, D. (2009a). How much can we know about the causes of evolutionary trends? *Biology & Philosophy, 24*(3), 341–357.

Turner, D. (2009b). Beyond detective work: Empirical testing in paleontology. In D. Sepkoski & M. Ruse (Eds.), *The paleobiological revolution: Essays on the growth of modern paleontology* (pp. 201–214). Chicago: University of Chicago Press.

Turner, D. (2007). *Making prehistory: Historical science and the scientific realism debate*. Cambridge: Cambridge University Press.

Turner, D. (2005). Local underdetermination in historical science. *Philosophy of Science, 72*(1), 209–230.

Turner, D. (2004). The past vs. the tiny: Historical science and the abductive arguments for realism. *Studies in History and Philosophy of Science, 35*(1), 1–17.

Tutken, T. (2011). The diet of sauropod dinosaurs: Implications of carbon isotope analysis on teeth, bones, and plants In N. Klein, K. Remes, G. T. Gee, & P. M. Sander (Eds.), *Biology of the sauropod dinosaurs: Understanding the life of giants* (pp. 57–82). Bloomington: Indiana University Press.

Underwood, A. J. (1990). Experiments in ecology and management: Their logics, functions and interpretations. *Australian Journal of Ecology, 15*, 365–389.

Van Fraassen, B. C. (1982). The charybdis of realism: The epistemological implications of Bell's inequality. *Synthese, 52*, 35–38.

Van Fraassen, B. C. (1980). *The scientific image.* Oxford, UK: Oxford University Press.

Van Loon, A. (2012). Were sauropod dinosaurs responsible for the warm Mesozoic climate? *Journal of Palaeogeography, 1*(2), 138–148.

VanPool, C. S. (2009). The signs of the sacred: Identifying shamans using archaeological evidence. *Journal of Anthropological Archaeology, 28*(2), 177–190.

Van Valkenburgh, B., & Jenkins, I. (2002). Evolutionary patterns in the history of Permo-Triassic and Cenozoic synapsid predators. *Paleontological Society Papers, 8*, 267–288.

Vinther, J., Briggs, D. E. G., Prum, R. O., & Saranathan, V. (2008). The colour of fossil feathers. *Biology Letters, 4*, 522–525.

Wagner, A. (2013). *Robustness and evolvability in living systems.* Princeton, NJ: Princeton University Press.

Walsh, K., & Currie, A. (2016). Caricatures, myths & white lies. *Metaphilosophy, 46*(3), 414–435.

Walsh, K. (in press). Newton's corpuscular scaffolding.

Walsh, K. (2012). Did Newton feign the corpuscular hypothesis? In J. Maclaurin (Ed.), *Rationis defensor* (pp. 97–110). Netherlands: Springer International Publishing.

Ward, P. D. (1990). The Cretaceous/Tertiary extinctions in the marine realm: A 1990 perspective. In V. L. Sharpton & P. D. Ward (Eds.), *Global Catastrophes in Earth History; An Interdisciplinary Conference on Impacts, Volcanism, and Mass Mortality: Geological Society of America Special Paper 247* (pp. 425–432).

Ward, P. D. (1983). The extinction of the ammonites. *Scientific American, 249*, 136–147, 1083–1136.

Waters, C. K. (2007). Causes that make a difference. *Journal of Philosophy, CIV*(11), 551–579.

Waters, C. K. (2000). Molecules made biological. *Revue Internationale de Philosophie, 54*, 214.

Weber, M. (2004). *Philosophy of experimental biology.* Cambridge, UK; New York: Cambridge University Press.

Wedel, M. J. (2009). Evidence for bird-like air sacs in saurischian dinosaurs. *Journal of Experimental Zoology, 311A*, 611–628.

Weisberg, M. (2013). *Simulation and similarity: Using models to understand the world.* Oxford, UK: Oxford University Press.

Weisberg, M. (2007). Who is a modeler? *British Journal for the Philosophy of Science, 58*(2), 207–233.

Weisberg, M. (2006). Robustness analysis. *Philosophy of Science, 73*(5), 730–742.

Wilkinson, D., Nisbet, E., & Ruxton, G. (2012). Could methane produced by sauropod dinosaurs have helped drive Mesozoic climate warmth? *Current Biology, 22*(9), R292–R293.

Williams, R., Grotzinger, J., Dietrich, W., Gupta, S., Sumner, D., Weins, R. Mangold, N., Malin, M., Edgett, K., Maurice, S., Forni, O., Gasnault, O., Ollilia, A., Newsom, H., Dromart, G., Palucis, M., Yingst, R., Anderson, R., Herkenhoff, K., Mouelic, S., Goetz, W., Madsen, M., Koefoed, A., Jensen, J., Bridges, J., Schwenzer, S., Lewis, K., Stack, K., Rubin, D., Kah, L., Bell, J., Farmer, J., Sullivan, R., Van Beek, T., Blaney, D., Pariser, O., Deen, R., & MSL Science Team. (2013). Martian fluvial conglomerates at Gale Crater. *Science, 340*(6136), 1068–1072.

Williams, G. E. (2000). Geological constraints on the Precambrian history of Earth's rotation and the moon's orbit. *Reviews of Geophysics, 38*(1), 37–59.

Williams, G. E. (1975). Late Precambrian glacial climate and the earth's obliquity. *Geological Magazine, 112*, 441–544.

Wills, M. A. (2007). Fossil ghost ranges are most common in some of the oldest and some of the youngest strata. *Proceedings of the Royal Society of London. Series B, Biological Sciences, 274*(1624), 2421–2427.

Wimsatt, W. C. (2007). *Re-engineering philosophy for limited beings: Piecewise approximations to reality.* Cambridge, MA: Harvard University Press.

Wimsatt, W. C. (1987). False models as means to truer theories. In M. Nitecki (Ed.), *Neutral models in biology* (pp. 23–55). Oxford, UK: Oxford University Press.

Wings, O., & Sander, P. M. (2007). No gastric mill in sauropod dinosaurs: New evidence from analysis of gastrolith mass and function in ostriches. *Proceedings. Biological Sciences, 274*, 635–640.

Winsberg, E. (2010). *Science in the age of computer simulation*. Chicago: University of Chicago Press.

Winsberg, E. (2009). A tale of two methods. *Synthese, 169*(3), 575–592.

Winsberg, E. (2003). Simulated experiments: Methodology for a virtual world. *Philosophy of Science, 70*, 105–125.

Winsberg, E. (1999). The hierarchy of models in simulation. In L. Magnani, N. J. Nersessian, & P. Thagard (Eds.), *Model-based reasoning in scientific discovery* (pp. 255–269). Dordrecht, the Netherlands: Kluwer Academic/Plenum Publishers.

Woodward, J. (2010). Causation in biology: Stability, specificity, and the choice of levels of explanation. *Biology & Philosophy, 25*(3), 287–318.

Woodward, J. (2003). Experimentation, causal inference, and instrumental realism. In H. Radder (Ed.), *The philosophy of scientific experimentation* (pp. 87–118). Pittsburgh: University of Pittsburgh Press.

Woodward, J. (2003). *Making things happen: A theory of causal explanation*. Oxford, UK: Oxford University Press.

Woodward, J. (2000). Explanation and invariance in the special sciences. *British Journal for the Philosophy of Science, 51*(2), 197–254.

Worrall, J. (1989). Structural realism: The best of both worlds? *Dialectica, 43*(1–2), 99–124.

Wouters, A. (1995). Viability explanation. *Biology & Philosophy, 10*(4), 435–457.

Wroe, S., Chamoli, U., Parr, W. C., Clausen, P., Ridgely, R., & Witmer, L. (2013). Comparative biomechanical modeling of metatherian and placental saber-tooths: A different kind of bite for an extreme pouched predator. *PLoS One, 8*(6), e66888.

Wroe, S., McHenry, C., & Thomason, J. (2005). Bite club: Comparative bite force in big biting mammals and the prediction of predatory behaviour in fossil taxa. *Proceedings. Biological Sciences, 272*, 619–625.

Wylie, A. (2016). How archeological evidence "bites back": Strategies for putting old data to work in new ways. *Science, Technology & Human Values, 42*(2), 203–225.

Wylie, A. (2011). Critical distance: Stabilising evidential claims in archaeology. In P. Dawid, W. Twining, & M. Vasilaki (Eds.), *Evidence, inference and enquiry* (pp. 371-394). Oxford, UK: Oxford University Press/British Academy.

Wylie, A. (2010). Archaeological facts in transit: The "eminent mounds" of central North America. In P. Howlett and M. S. Morgan (Eds.), *How well do "facts" travel?: The dissemination of reliable knowledge* (pp. 301–322). Cambridge, UK: Cambridge University Press.

Wylie, A. (1999). Rethinking unity as a "working hypothesis" for philosophy of science: How archaeologists exploit the disunities of science. *Perspectives on Science, 7*(3), 293–317.

Wylie, A. (1985). The reaction against analogy. In M. Schiffer (Ed.), *Advances in archaeological method and theory*, 63–111. New York: Academic Press.

Wylie, C. D. (2014). "The artist's piece is already in the stone": Constructing creativity in paleontology laboratories. *Social Studies of Science, 45*(1), 31–55. DOI: 10.1177/0306312714549794.

Zachos, L., & Sprinkle, J. (2011). Computational model of growth and development in paleozoic echinoids. In T. Elewa (Ed.), *Computational Paleontology* (pp. 75–94). Heidelberg: Springer International Publishing.

Zachos, L. (2009). A new computational growth model for sea urchin skeletons. *Journal of Theoretical Biology, 259*, 646–657.

Zhang, F., et al. (2010). Fossilized melanosomes and the colour of Cretaceous dinosaurs and birds. *Nature, 463*, 1075–1078.

Index